The EM Algorithm and Extensions

WILEY SERIES IN PROBABILITY AND STATISTICS

Established by WALTER A. SHEWHART and SAMUEL S. WILKS

A complete list of the titles in this series appears at the end of this volume.

The EM Algorithm and Extensions

GEOFFREY J. McLACHLAN
University of Queensland, Brisbane

THRIYAMBAKAM KRISHNAN
Indian Statistical Institute

A Wiley-Interscience Publication
JOHN WILEY & SONS, INC.
New York • Chichester • Brisbane • Toronto • Singapore • Weinheim

This text is printed on acid-free paper.

Library of Congress Cataloging in Publication Data:

McLachlan, Geoffrey J., 1946–
The EM algorithm and extensions / G.J. McLachlan and T. Krishnan.
p. cm. — (Wiley series in probability and statistics.
Applied probability and statistics)
Includes bibliographical references and index.
ISBN 0-471-12358-7 (alk. paper)
1. Expectation-maximization algorithms. 2. Estimation theory.
3. Missing observations (Statistics) I. Krishnan, T.
(Thriyambakam), 1938– . II. Title. III. Series.
QA276.8.M394 1996
519.5'44—DC20 96-38417
CIP

Printed in the United States of America
10 9 8 7

To
Beryl, Jonathan, and Robbie

Contents

Preface

This book deals with the Expectation-Maximization algorithm, popularly known as the EM algorithm. This is a general-purpose algorithm for maximum likelihood estimation in a wide variety of situations best described as *incomplete-data* problems. The name EM algorithm was given by Dempster, Laird, and Rubin in a celebrated paper read before the Royal Statistical Society in 1976 and published in its journal in 1977. In this paper, a general formulation of the EM algorithm was presented, its basic properties established, and many examples and applications of it provided. The idea behind the EM algorithm is intuitive and natural and so algorithms like it were formulated and applied in a variety of problems even before this paper. However, it was in this seminal paper that the ideas in the earlier papers were synthesized, a general formulation and a theory developed, and a host of traditional and non-traditional applications indicated. Since then, the EM algorithm has become a standard piece in the statistician's repertoire. The incomplete-data situations where the EM algorithm has been successfully applied include not only evidently incomplete-data situations, where there are missing data, truncated distributions, censored or grouped observations, but also a whole variety of situations where the incompleteness of the data is not natural or evident. Thus, in some situations, it requires a certain amount of ingenuity on the part of the statistician to formulate the incompleteness in a suitable manner to facilitate the application of the EM algorithm in a computationally profitable manner. Following the paper of Dempster, Laird, and Rubin (1977), a spate of applications of the algorithm has appeared in the literature.

The EM algorithm is not without its limitations, many of which came to light in attempting to apply it in certain complex incomplete-data problems and some even in innocuously simple incomplete-data problems. However, a number of modifications and extensions of the algorithm has been developed to overcome some of these limitations. Thus there is a whole battery of EM-related algorithms and more are still being developed. The current developments are, however, in the direction of iterative simulation techniques or Markov Chain Monte Carlo methods, many of which can be looked upon as simulation-based versions of various EM-type algorithms.

Incomplete-data problems arise in all statistical contexts. Hence in these problems where maximum likelihood estimates usually have to be computed iteratively, there is the scope and need for an EM algorithm to tackle them. Further, even if there are no missing data or other forms of data incompleteness, it is often profitable to express the given problem as an incomplete-data problem within an EM framework. For example, in some multiparameter problems like in random effects models, where an averaging over some parameters is to be carried out, an incomplete-data approach via the EM algorithm and its variants has been found useful. No wonder then that the EM algorithm has become an ubiquitous statistical tool, is a part of the entire spectrum of statistical methods, and has found applications in almost all fields where statistical techniques have been applied. The EM algorithm and its variants have been applied in such fields as medical imaging, dairy science, correcting census undercount, and AIDS epidemiology, to mention a few. Articles containing applications of the EM algorithm and even some with some methodological content have appeared in a variety of journals on statistical theory and methodology, statistical computing, and statistical applications in engineering, biology, medicine, social sciences, etc. Meng and Pedlow (1992) list a bibliography of more than 1000 items and now there are at least 1700 publications related to the EM algorithm.

It is surprising that despite the obvious importance of the technique and its ramifications, no book on the subject has so far appeared. Indeed, many modern books dealing with some aspect of statistical estimation have at least some EM algorithm content. The books by Little and Rubin (1987), Tanner (1991, 1993), and Schafer (1996) have substantial EM algorithm content. But still, there seems to be a need for a full-fledged book on the subject. In our experience of lecturing to audiences of professional statisticians and to users of statistics, it appears that there is a definite need for a unified and complete treatment of the theory and methodology of the EM algorithm and its extensions, and their applications. The purpose of our writing this book is to fulfill this need. The various extensions of the EM algorithm due to Rubin, Meng, C. Liu, and others that have appeared in the last few years, have made this need even greater. Many extensions of the EM algorithm in the direction of iterative simulation have also appeared in recent years. Inclusion of these techniques in this book may have resulted in a more even-handed and comprehensive treatment of the EM algorithm and its extensions. However, we decided against it, since this methodology is still evolving and rapid developments in this area may make this material soon obsolete. So we have restricted this book to the EM algorithm and its variants and have only just touched upon the iterative simulation versions of it.

The book is aimed at theoreticians and practitioners of statistics and its objective is to introduce to them the principles and methodology of the EM algorithm and its tremendous potential for applications. The main parts of the book describing the formulation of the EM algorithm, detailing its methodology, discussing aspects of its implementation, and illustrating its application in many simple statistical contexts, should be comprehensible to graduates with Statistics as their major subject. Throughout the book, the theory and methodology are illustrated with a number of examples. Where relevant, analytical examples are followed up with numerical examples. There are about thirty examples in the book. Some parts of the book, especially examples like factor analysis and variance components analysis, will need basic knowledge of these techniques to comprehend the full impact of the use of the EM algorithm. But our treatment of these examples is self-contained, although brief. However, there examples can be skipped without losing continuity.

Chapter 1 begins with a brief discussion of maximum likelihood (ML) estimation and standard approaches to the calculation of the maximum likelihood estimate (MLE) when it does not exist as a closed form solution of the likelihood equation. This is followed by a few examples of incomplete-data problems for which an heuristic derivation of the EM algorithm is given. The EM algorithm is then formulated and its basic terminology and notation established. The case of the regular exponential family (for the complete-data problem) for which the EM algorithm results in a particularly elegant solution, is specially treated. Throughout the treatment, the Bayesian perspective is also included by showing how the EM algorithm and its variants can be adapted to compute maximum *a posteriori* (MAP) estimates. The use of the EM algorithm and its variants in maximum penalized likelihood estimation (MPLE), a technique by which the MLE is smoothed, is also included.

Chapter 1 also includes a summary of the properties of the EM algorithm. Towards the end of this chapter, a comprehensive discussion of the history of the algorithm is presented, with a listing of the earlier ideas and examples upon which the general formulation is based. The chapter closes with a summary of the developments in the methodology since the Dempster et al. (1977) paper and with an indication of the range of applications of the algorithm.

In Chapter 2, a variety of examples of the EM algorithm is presented, following the general formulation in Chapter 1. These examples include missing values (in the conventional sense) in various experimental designs, the multinomial distribution with complex cell structure as used in genetics, the multivariate t-distribution for the provision of a robust estimate of a location parameter, Poisson regression models in a computerized image reconstruction process such as SPECT/PET, and the fitting of normal mixture models to grouped and truncated data as in the modeling of the volume of red blood cells.

In Chapter 3, the basic theory of the EM algorithm is systematically presented, and the monotonicity of the algorithm, convergence, and rates of convergence properties are established. The Generalized EM (GEM) algorithm and its properties are also presented. The principles of Missing Information and Self-Consistency are discussed. In this chapter, attention is inevitably given to mathematical details. However, mathematical details and theoretical points are explained and illustrated with the help of earlier and new examples. Readers not interested in the more esoteric aspects of the EM algorithm may only study the examples in this chapter or omit the chapter altogether without losing continuity.

In Chapter 4, two issues which have led to some criticism of the EM algorithm are addressed. The first concerns the provision of standard errors, or the full covariance matrix in multiparameter situations, of the MLE obtained via the EM algorithm. One initial criticism of the EM algorithm was that it does not automatically provide an estimate of the covariance matrix of the MLE, as do some other approaches such as Newton-type methods. Hence we consider a number of methods for assessing the covariance matrix of the MLE $\hat{\boldsymbol{\Psi}}$ of the parameter vector $\boldsymbol{\Psi}$, obtained via the EM algorithm. Most of these methods are based on the observed information matrix. A coverage is given of methods such as the Supplemented EM algorithm that allow the observed information matrix to be calculated within the EM framework. The other common criticism that has been leveled at the EM algorithm is that its convergence can be quite slow. We therefore consider some methods that have been proposed for accelerating the convergence of the EM algorithm. They include methods based on Aitken's acceleration procedure and the generalized conjugate gradient approach, and hybrid methods that switch from the EM algorithm after a few iterations to some Newton-type method. We consider also the use of the EM gradient algorithm as a basis of a quasi-Newton approach to accelerate convergence of the EM algorithm. This algorithm approximates the M-step of the EM algorithm by one Newton-Raphson step when the solution does not exist in closed form.

In Chapter 5, further modifications and extensions to the EM algorithm are discussed. The focus is on the Expectation-Conditional Maximum (ECM) algorithm and its extensions, including the Expectation-Conditional Maximum Either (ECME) and Alternating ECM (AECM) algorithms. The ECM algorithm is a natural extension of the EM algorithm in situations where the maximization process on the M-step is relatively simple when conditional on some function of the parameters under estimation. The ECM algorithm therefore replaces the M-step of the EM algorithm by a number of computationally simpler conditional maximization (CM) steps. These extensions of the EM algorithm typically produce an appreciable reduction in total computer time. More importantly, they preserve the appealing convergence properties of the EM algorithm, such as its monotone convergence.

In Chapter 6, very brief overviews are presented of iterative simulation techniques such as the Monte Carlo E-step, Stochastic EM algorithm, Data Augmentation, and the Gibbs sampler and their connections with the various versions of the EM algorithm. Then, a few methods such as Simulated Annealing, which are considered competitors to the EM algorithm, are described and a few examples comparing the performance of the EM algorithm with these competing methods are presented. The book is concluded with a brief account of the applications of the EM algorithm in such topical and interesting areas as Hidden Markov Models, AIDS epidemiology, and Neural Networks.

One of the authors (GJM) would like to acknowledge gratefully financial support from the Australian Research Council. Work on the book by one of us (TK) was facilitated by two visits to the University of Queensland, Brisbane, and a visit to Curtin University of Technology, Perth. One of the visits to Brisbane was under the Ethel Raybould Fellowship scheme. Thanks are due to these two Universities for their hospitality. Thanks are also due to the authors and owners of copyrighted material for permission to reproduce tables and figures. The authors also wish to thank Ramen Kar and Amiya Das of the Indian Statistical Institute, Calcutta, and Pauline Wilson of the University of Queensland, Brisbane, for their help with LATEX and word processing, and Rudy Blazek of Michigan State University and David Peel and Matthew York of the University of Queensland for assistance with the preparation of some of the figures.

Brisbane, Australia GEOFFREY J. MCLACHLAN
Calcutta, India THRIYAMBAKAM KRISHNAN

CHAPTER 1

General Introduction

1.1 INTRODUCTION

The Expectation–Maximization (EM) algorithm is a broadly applicable approach to the iterative computation of maximum likelihood (ML) estimates, useful in a variety of incomplete-data problems, where algorithms such as the Newton–Raphson method may turn out to be more complicated. On each iteration of the EM algorithm, there are two steps—called the *expectation step* or the E-step and the *maximization step* or the M-step. Because of this, the algorithm is called the EM algorithm. This name was given by Dempster, Laird, and Rubin (1977) in their fundamental paper. (We hereafter refer to this paper as the DLR paper or simply DLR.) The situations where the EM algorithm is profitably applied can be described as *incomplete-data problems*, where ML estimation is made difficult by the absence of some part of data in a more familiar and simpler data structure. The EM algorithm is closely related to the *ad hoc* approach to estimation with missing data, where the parameters are estimated after filling in initial values for the missing data. The latter are then updated by their predicted values using these initial parameter estimates. The parameters are then reestimated, and so on, proceeding iteratively until convergence.

This idea behind the EM algorithm being intuitive and natural, algorithms like the EM have been formulated and applied in a variety of problems even before the DLR paper. But it was in the seminal DLR paper that the ideas were synthesized, a general formulation of the EM algorithm established, its properties investigated, and a host of traditional and nontraditional applications indicated.

The situations where the EM algorithm can be applied include not only evidently incomplete-data situations, where there are missing data, truncated distributions, censored or grouped observations, but also a whole variety of situations where the incompleteness of the data is not all that natural or evident. These include statistical models such as random effects, mixtures, convolutions, log linear models, and latent class and latent variable structures. Hitherto intractable ML estimation problems for these situations have been solved or complicated ML estimation procedures have been simplified using the EM algorithm. The EM algorithm has thus found applications in almost all statistical contexts and in almost all fields where statistical techniques have been applied—medical imaging, dairy science, correcting census undercount, and AIDS epidemiology, to mention a few.

Data sets with missing values, censored and grouped observations, and models with truncated distributions, etc., which result in complicated likelihood functions cannot be avoided in practical situations. The development of the EM algorithm and related methodology together with the availability of inexpensive and rapid computing power have made analysis of such data sets much more tractable than they were earlier. The EM algorithm has already become a standard tool in the statistical repertoire.

The basic idea of the EM algorithm is to associate with the given *incomplete-data problem*, a *complete-data problem* for which ML estimation is computationally more tractable; for instance, the complete-data problem chosen may yield a closed-form solution to the maximum likelihood estimate (MLE) or may be amenable to MLE computation with a standard computer package. The methodology of the EM algorithm then consists in reformulating the problem in terms of this more easily solved complete-data problem, establishing a relationship between the likelihoods of these two problems, and exploiting the simpler MLE computation of the complete-data problem in the M-step of the iterative computing algorithm.

Although a problem at first sight may not appear to be an incomplete-data one, there may be much to be gained computationwise by artificially formulating it as such to facilitate ML estimation. This is because the EM algorithm exploits the reduced complexity of ML estimation given the complete data. For many statistical forms, the complete-data likelihood has a nice form. The E-step consists in manufacturing data for the complete-data problem, using the observed data set of the incomplete-data problem and the current value of the parameters, so that the simpler M-step computation can be applied to this "completed" data set. More precisely, it is the log likelihood of the complete-data problem that is "manufactured" in the E-step. As it is based partly on unobservable data, it is replaced by its conditional expectation given the observed data, where this E-step is effected using the current fit for the unknown parameters. Starting from suitable initial parameter values, the E- and M-steps are repeated until convergence. Of course, the complete-data problem is to be suitably chosen from the point of view of simplicity of the complete-data MLEs; it may even be a hypothetical problem from the point of view of practical implementation. For instance, a complete-data problem defined in the context of factor analysis has data on unobservable latent variables.

Although the EM algorithm has been successfully applied in a variety of contexts, it can in certain situations be painfully slow to converge. This has resulted in the development of modified versions of the algorithm as well as many simulation-based methods and other extensions of it. This area is still developing. An initial criticism of the EM algorithm was that it did not produce estimates of the covariance matrix of the MLEs. However, developments subsequent to the DLR paper have provided methods for such estimation, which can be integrated into the EM computational scheme.

In the next section, we shall provide a very brief description of ML estimation. However, in this book, we do not discuss the question of why use MLEs; excellent treatises are available wherein the attractive properties of MLEs are established (for instance, Rao, 1973; Cox and Hinkley, 1974; Lehmann, 1983; Stuart and Ord, 1994). In Section 1.3, we briefly define the standard techniques before the advent of the EM algorithm for the computation of MLEs, which are the Newton–Raphson method and its variants such as Fisher's scoring method and quasi-Newton methods. For a more detailed account of these and other numerical methods for statistical computation, the reader is referred to books that discuss these methods; for example, Everitt (1987) and Thisted (1988) discuss ML estimation and Dennis and Schnabel (1983) and Ratschek and Rokne (1988) discuss optimization methods in general.

1.2 MAXIMUM LIKELIHOOD ESTIMATION

Although we shall focus on the application of the EM algorithm for computing MLEs in a frequentist framework, it can be equally applied to find the mode of the posterior distribution in a Bayesian framework.

We let $\boldsymbol{Y}$ be a p-dimensional random vector with probability density function (p.d.f.) $g(\boldsymbol{y};\, \boldsymbol{\Psi})$ on $\mathbb{R}^p$, where $\boldsymbol{\Psi} = (\Psi_1, \ldots, \Psi_d)^T$ is the vector containing the unknown parameters in the postulated form for the p.d.f. of $\boldsymbol{Y}$. Here (and also elsewhere in this book) the superscript T denotes the transpose of a vector or a matrix. The parameter space is denoted by $\boldsymbol{\Omega}$. Although we are taking $\boldsymbol{Y}$ to be a continuous random vector, we can still view $g(\boldsymbol{y};\, \boldsymbol{\Psi})$ as a p.d.f. in the case where $\boldsymbol{Y}$ is discrete by the adoption of counting measure.

For example, if $\boldsymbol{w}_1, \ldots, \boldsymbol{w}_n$ denotes an observed random sample of size n on some random vector $\boldsymbol{W}$ with p.d.f. $f(\boldsymbol{w};\, \boldsymbol{\Psi})$, then

$$\boldsymbol{y} = (\boldsymbol{w}_1^T, \ldots, \boldsymbol{w}_n^T)^T$$

and

$$g(\boldsymbol{y};\, \boldsymbol{\Psi}) = \prod_{j=1}^{n} f(\boldsymbol{w}_j;\, \boldsymbol{\Psi}).$$

The vector $\boldsymbol{\Psi}$ is to be estimated by maximum likelihood. The likelihood function for $\boldsymbol{\Psi}$ formed from the observed data $\boldsymbol{y}$ is given by

$$L(\boldsymbol{\Psi}) = g(\boldsymbol{y};\, \boldsymbol{\Psi}).$$

An estimate $\hat{\boldsymbol{\Psi}}$ of $\boldsymbol{\Psi}$ can be obtained as a solution of the likelihood equation

$$\partial L(\boldsymbol{\Psi}/\partial \boldsymbol{\Psi}) = \mathbf{0},$$

or equivalently,

$$\partial \log L(\boldsymbol{\Psi})/\partial \boldsymbol{\Psi} = \mathbf{0}. \tag{1.1}$$

Briefly, the aim of ML estimation (Lehmann, 1983) is to determine an estimate for each n ($\hat{\boldsymbol{\Psi}}$ in the present context), so that it defines a sequence of roots of the likelihood equation that is consistent and asymptotically efficient. Such a sequence is known to exist under suitable regularity conditions (Cramér, 1946). With probability tending to one, these roots correspond to local maxima in the interior of the parameter space. For estimation models in general, the likelihood usually has a global maximum in the interior of the parameter space. Then typically a sequence of roots of the likelihood equation with the desired asymptotic properties is provided by taking $\hat{\boldsymbol{\Psi}}$ for each n to be the root that globally maximizes the likelihood; that is, $\hat{\boldsymbol{\Psi}}$ is the MLE. We shall henceforth refer to $\hat{\boldsymbol{\Psi}}$ as the MLE, even though it may not globally maximize the likelihood. Indeed, in some of the examples on mixture models to be presented, the likelihood is unbounded. However, for these models there may still exist under the usual regularity conditions a sequence of roots of the likelihood equation with the properties of consistency, efficiency, and asymptotic normality; see McLachlan and Basford (1988, Chapter 1). We let

$$\boldsymbol{I}(\boldsymbol{\Psi}; \boldsymbol{y}) = -\partial^2 \log L(\boldsymbol{\Psi})/\partial \boldsymbol{\Psi} \, \partial \boldsymbol{\Psi}^T$$

be the matrix of the negative of the second-order partial derivatives of the log likelihood function with respect to the elements of $\boldsymbol{\Psi}$. Under regularity conditions, the expected (Fisher) information matrix $\mathcal{I}(\boldsymbol{\Psi})$ is given by

$$\begin{aligned}\mathcal{I}(\boldsymbol{\Psi}) &= E_{\boldsymbol{\Psi}}\{\boldsymbol{S}(\boldsymbol{Y}; \boldsymbol{\Psi})\boldsymbol{S}^T(\boldsymbol{Y}; \boldsymbol{\Psi})\} \\ &= -E_{\boldsymbol{\Psi}}\{\boldsymbol{I}(\boldsymbol{\Psi}; \boldsymbol{Y})\},\end{aligned}$$

where

$$\boldsymbol{S}(\boldsymbol{y}; \boldsymbol{\Psi}) = \partial \log L(\boldsymbol{\Psi})/\partial \boldsymbol{\Psi}$$

is the gradient vector of the log likelihood function; that is, the score statistic. Here and elsewhere in this book, the operator $E_{\boldsymbol{\Psi}}$ denotes expectation using the parameter vector $\boldsymbol{\Psi}$.

The asymptotic covariance matrix of the MLE $\hat{\boldsymbol{\Psi}}$ is equal to the inverse of the expected information matrix $\mathcal{I}(\boldsymbol{\Psi})$, which can be approximated by $\mathcal{I}(\hat{\boldsymbol{\Psi}})$; that is, the standard error of $\hat{\Psi}_i = (\hat{\boldsymbol{\Psi}})_i$ is given by

$$SE(\hat{\Psi}_i) \approx (\mathcal{I}^{-1}(\hat{\boldsymbol{\Psi}}))_{ii}^{1/2} \quad (i = 1, \ldots, d), \tag{1.2}$$

where the standard notation $(\boldsymbol{A})_{ij}$ is used for the element in the ith row and jth column of a matrix $\boldsymbol{A}$.

It is common in practice to estimate the inverse of the covariance matrix of a maximum solution by the observed information matrix $\boldsymbol{I}(\hat{\boldsymbol{\Psi}}; \boldsymbol{y})$, rather than the expected information matrix $\mathcal{I}(\boldsymbol{\Psi})$ evaluated at $\boldsymbol{\Psi} = \hat{\boldsymbol{\Psi}}$. This approach gives the approximation

$$SE(\hat{\Psi}_i) \approx (\boldsymbol{I}^{-1}(\hat{\boldsymbol{\Psi}}; \boldsymbol{y}))_{ii}^{1/2} \quad (i = 1, \ldots, d). \tag{1.3}$$

Efron and Hinkley (1978) have provided a frequentist justification of (1.3) over (1.2) in the case of one-parameter ($d = 1$) families. Also, the observed information matrix is usually more convenient to use than the expected information matrix, as it does not require an expectation to be taken.

Often in practice the log likelihood function cannot be maximized analytically. In such cases, it may be possible to compute iteratively the MLE of $\boldsymbol{\Psi}$ by using a Newton–Raphson maximization procedure or some variant, provided the total number d of parameters in the model is not too large. Another alternative is to apply the EM algorithm. Before we proceed with the presentation of the EM algorithm, we briefly define in the next section the Newton–Raphson method and some variants for the computation of the MLE.

1.3 NEWTON-TYPE METHODS

1.3.1 Introduction

Since the properties of the EM algorithm are to be contrasted with those of Newton-type methods, which are the main alternatives for the computation of MLEs, we now give a brief review of the Newton–Raphson method and some variants.

Like many other methods for computing MLEs, the EM algorithm is a method for finding zeros of a function. In numerical analysis there are various techniques for finding zeros of a specified function, including the Newton–Raphson (NR) method, quasi-Newton methods, and modified Newton methods. In a statistical framework, the modified Newton methods include the scoring algorithm of Fisher and its modified version using the empirical information matrix in place of the expected information matrix.

1.3.2 Newton–Raphson Method

The Newton–Raphson method for solving the likelihood equation

$$\boldsymbol{S}(\boldsymbol{y}; \boldsymbol{\Psi}) = \boldsymbol{0}, \tag{1.4}$$

approximates the gradient vector $\boldsymbol{S}(\boldsymbol{y}; \boldsymbol{\Psi})$ of the log likelihood function $\log L(\boldsymbol{\Psi})$ by a linear Taylor series expansion about the current fit $\boldsymbol{\Psi}^{(k)}$ for $\boldsymbol{\Psi}$. This gives

$$\boldsymbol{S}(\boldsymbol{y}; \boldsymbol{\Psi}) \approx \boldsymbol{S}(\boldsymbol{y}; \boldsymbol{\Psi}^{(k)}) - \boldsymbol{I}(\boldsymbol{\Psi}^{(k)}; \boldsymbol{y})(\boldsymbol{\Psi} - \boldsymbol{\Psi}^{(k)}). \tag{1.5}$$

A new fit $\boldsymbol{\Psi}^{(k+1)}$ is obtained by taking it to be a zero of the right-hand side of (1.5). Hence

$$\boldsymbol{\Psi}^{(k+1)} = \boldsymbol{\Psi}^{(k)} + \boldsymbol{I}^{-1}(\boldsymbol{\Psi}^{(k)}; \boldsymbol{y})\boldsymbol{S}(\boldsymbol{y}; \boldsymbol{\Psi}^{(k)}). \tag{1.6}$$

If the log likelihood function is concave and unimodal, then the sequence of iterates $\{\boldsymbol{\Psi}^{(k)}\}$ converges to the MLE of $\boldsymbol{\Psi}$, in one step if the log likelihood function is a quadratic function of $\boldsymbol{\Psi}$. When the log likelihood function is not concave, the Newton–Raphson method is not guaranteed to converge from an arbitrary starting value. Under reasonable assumptions on $L(\boldsymbol{\Psi})$ and a sufficiently accurate starting value, the sequence of iterates $\boldsymbol{\Psi}^{(k)}$ produced by the Newton-Raphson method enjoys local quadratic convergence to a solution $\boldsymbol{\Psi}^*$ of (1.4). That is, given a norm $||\cdot||$ on $\boldsymbol{\Omega}$, there is a constant h such that if $\boldsymbol{\Psi}^{(0)}$ is sufficiently close to $\boldsymbol{\Psi}^*$, then

$$||\boldsymbol{\Psi}^{(k+1)} - \boldsymbol{\Psi}^*|| \leq h||\boldsymbol{\Psi}^{(k)} - \boldsymbol{\Psi}^*||^2$$

holds for $k = 0, 1, 2, \ldots$. Quadratic convergence is ultimately very fast, and it is regarded as the major strength of the Newton–Raphson method. But there can be potentially severe problems with this method in applications. Firstly, it requires at each iteration, the computation of the $d \times d$ information matrix $\boldsymbol{I}(\boldsymbol{\Psi}^{(k)}; \boldsymbol{y})$ (that is, the negative of the Hessian matrix) and the solution of a system of d linear equations. In general, this is achieved at a cost of $O(d^3)$ arithmetic operations. Thus the computation required for an iteration of the Newton–Raphson method is likely to become expensive very rapidly as d becomes large. One must allow for the storage of the Hessian or some set of factors of it. Furthermore, the Newton–Raphson method in its basic form (1.6) requires for some problems an impractically accurate initial value for $\boldsymbol{\Psi}$ for the sequence of iterates $\{\boldsymbol{\Psi}^{(k)}\}$ to converge to a solution of (1.4). It has the tendency to head toward saddle points and local minima as often as toward local maxima. In some problems, however, Böhning and Lindsay (1988) show how the Newton–Raphson method can be modified to be monotonic.

Since the Newton–Raphson method requires the evaluation of $\boldsymbol{I}(\boldsymbol{\Psi}^{(k)}; \boldsymbol{y})$ on each iteration k, it immediately provides an estimate of the covariance matrix of its limiting value $\boldsymbol{\Psi}^*$ (assuming it is the MLE), through the inverse of the observed information matrix $\boldsymbol{I}^{-1}(\boldsymbol{\Psi}^*; \boldsymbol{y})$. Also, if the starting value is a $\sqrt{n}$-consistent estimator of $\boldsymbol{\Psi}$, then the one-step iterate $\boldsymbol{\Psi}^{(1)}$ is an asymptotically efficient estimator of $\boldsymbol{\Psi}$.

1.3.3 Quasi-Newton Methods

A broad class of methods are so-called quasi-Newton methods, for which the solution of (1.4) takes the form

$$\boldsymbol{\Psi}^{(k+1)} = \boldsymbol{\Psi}^{k)} - \boldsymbol{A}^{-1}\boldsymbol{S}(\boldsymbol{y}; \boldsymbol{\Psi}^{(k)}), \tag{1.7}$$

where $\boldsymbol{A}$ is used as an approximation to the Hessian matrix. This approximation can be maintained by doing a *secant update* of $\boldsymbol{A}$ at each iteration. These updates are typically effected by rank-one or rank-two changes in $\boldsymbol{A}$. Methods of this class have the advantage over the Newton–Raphson method of not requiring the evaluation of the Hessian matrix at each iteration and of being implementable in ways that require only $O(d^2)$ arithmetic operations to solve the system of d linear equations corresponding to (1.6) with $\boldsymbol{I}(\boldsymbol{\Psi}^{(k)}; \boldsymbol{y})$ replaced by $-\boldsymbol{A}$. However, the full quadratic convergence of the Newton–Raphson method is lost, as a sequence of iterates $\boldsymbol{\Psi}^{(k)}$ can be shown to exhibit only local superlinear convergence to a solution $\boldsymbol{\Psi}^*$ of (1.4). More precisely, suppose that the initial value $\boldsymbol{\Psi}^{(0)}$ of $\boldsymbol{\Psi}$ is sufficiently near to the solution $\boldsymbol{\Psi}^*$ and that the initial value $\boldsymbol{A}^{(0)}$ of $\boldsymbol{A}$ is sufficiently near to the Hessian matrix at the solution, that is, $-\boldsymbol{I}(\boldsymbol{\Psi}^*; \boldsymbol{y})$. Then, under reasonable assumptions on the likelihood function $L(\boldsymbol{\Psi})$, it can be shown that there exists a sequence h_k that converges to zero and is such that

$$||\boldsymbol{\Psi}^{(k+1)} - \boldsymbol{\Psi}^*|| \leq h_k ||\boldsymbol{\Psi}^{(k)} - \boldsymbol{\Psi}^*||$$

for $k = 0, 1, 2, \ldots$. For further details, the reader is referred to the excellent account on Newton-type methods in Redner and Walker (1984).

It can be seen that quasi-Newton methods avoid the explicit evaluation of the Hessian of the log likelihood function at every iteration, as with the Newton–Raphson method. Also, they circumvent the tendency of the Newton-Raphson method to lead to saddle points and local minima as often as local maxima by forcing the approximate Hessian to be negative definite. However, as pointed out by Lange (1995b), even with these safeguards, they still have some drawbacks in many statistical applications. In particular, they usually approximate the Hessian initially by the identity, which may be a poorly scaled approximation to the problem at hand. Hence the algorithm can wildly overshoot or undershoot the maximum of the log likelihood along the direction of the current step. This has led to some alternative methods, which we now briefly mention.

1.3.4 Modified Newton Methods

Fisher's method of scoring is a member of the class of modified Newton methods, where the observed information matrix $\boldsymbol{I}(\boldsymbol{\Psi}^{(k)}; \boldsymbol{y})$ for the current fit for $\boldsymbol{\Psi}$, is replaced by $\mathcal{I}(\boldsymbol{\Psi}^{(k)})$, the expected information matrix evaluated at the current fit $\boldsymbol{\Psi}^{(k)}$ for $\boldsymbol{\Psi}$.

In practice, it is often too tedious or difficult to evaluate analytically the expectation of $\boldsymbol{I}(\boldsymbol{\Psi}; \boldsymbol{Y})$ to give the expected information matrix $\mathcal{I}(\boldsymbol{\Psi})$. Indeed, in some instances, one may not wish even to perform the prerequisite task of calculating the second-order partial derivatives of the log likelihood. In that case, for independent and identically distributed (i.i.d.)

data, the method of scoring can be employed with the empirical information matrix $\boldsymbol{I}_e(\boldsymbol{\Psi}^{(k)};\, \boldsymbol{y})$ evaluated at the current fit for $\boldsymbol{\Psi}$. The empirical information matrix $\boldsymbol{I}_e(\boldsymbol{\Psi}^{(k)};\, \boldsymbol{y})$ is given by

$$\boldsymbol{I}_e(\boldsymbol{\Psi};\, \boldsymbol{y}) = \sum_{j=1}^{n} \boldsymbol{s}(\boldsymbol{w}_j;\, \boldsymbol{\Psi})\boldsymbol{s}^T(\boldsymbol{w}_j;\, \boldsymbol{\Psi}) \\ - n^{-1}\boldsymbol{S}(\boldsymbol{y};\, \boldsymbol{\Psi})\boldsymbol{S}^T(\boldsymbol{y};\, \boldsymbol{\Psi}), \tag{1.8}$$

where $\boldsymbol{s}(\boldsymbol{w}_j;\, \boldsymbol{\Psi})$ is the score function based on the single observation $\boldsymbol{w}_j$ and

$$\begin{aligned} S(\boldsymbol{y};\, \boldsymbol{\Psi}) &= \partial \log L(\boldsymbol{\Psi})/\partial \boldsymbol{\Psi} \\ &= \sum_{j=1}^{n} \boldsymbol{s}(\boldsymbol{w}_j;\, \boldsymbol{\Psi}) \end{aligned}$$

is the score statistic for the full sample

$$\boldsymbol{y} = (\boldsymbol{w}_1^T, \ldots, \boldsymbol{w}_n^T)^T.$$

On evaluation at $\boldsymbol{\Psi} = \hat{\boldsymbol{\Psi}}, \boldsymbol{I}_e(\hat{\boldsymbol{\Psi}};\, \boldsymbol{y})$ reduces to

$$\boldsymbol{I}_e(\hat{\boldsymbol{\Psi}};\, \boldsymbol{y}) = \sum_{j=1}^{n} \boldsymbol{s}(\boldsymbol{w}_j;\, \hat{\boldsymbol{\Psi}})\boldsymbol{s}^T(\boldsymbol{w}_j;\, \hat{\boldsymbol{\Psi}}), \tag{1.9}$$

since $\boldsymbol{S}(\boldsymbol{y};\, \hat{\boldsymbol{\Psi}}) = \boldsymbol{0}$.

Actually, Meilijson (1989) recommends forming $\boldsymbol{I}_e(\boldsymbol{\Psi}^{(k)};\boldsymbol{y})$ by evaluating (1.8) at $\boldsymbol{\Psi} = \boldsymbol{\Psi}^{(k)}$, since $S(\boldsymbol{y};\, \boldsymbol{\Psi}^{(k)})$ is not zero. The justification of the empirical information matrix as an approximation to either the expected information matrix $\mathcal{I}(\boldsymbol{\Psi})$ or the observed information matrix $\boldsymbol{I}(\hat{\boldsymbol{\Psi}};\, \boldsymbol{y})$ is to be discussed in Section 4.3.

The modified Newton method, which uses on the kth iteration the empirical information matrix $\boldsymbol{I}_e(\boldsymbol{\Psi}^{(k)};\, \boldsymbol{y})$ in place of $\mathcal{I}(\boldsymbol{\Psi}^{(k)})$, or equivalently, the Newton–Raphson method with $\boldsymbol{I}(\boldsymbol{\Psi}^{(k)};\, \boldsymbol{y})$ replaced by $\boldsymbol{I}_e(\boldsymbol{\Psi}^{(k)};\, \boldsymbol{y})$, requires $O(nd^2)$ arithmetic operations to calculate $\boldsymbol{I}_e(\boldsymbol{\Psi}^{(k)};\, \boldsymbol{y})$ and $O(d^3)$ arithmetic operations to solve the system of d equations implicit in (1.4). As pointed out by Redner and Walker (1984), since the $O(nd^2)$ arithmetic operations needed to compute the empirical information matrix $\boldsymbol{I}_e(\boldsymbol{\Psi}^{(k)};\, \boldsymbol{y})$ are likely to be considerably less expensive than the evaluation of $\boldsymbol{I}(\boldsymbol{\Psi}^{(k)};\, \boldsymbol{y})$ (that is, the full Hessian matrix), the cost of computation per iteration of this method should lie between that of a quasi-Newton method employing a low-rank secant update and that of the Newton–Raphson method. Under reasonable assumptions on $L(\boldsymbol{\Psi})$, one can show that, with probability one, if a solution $\boldsymbol{\Psi}^*$ of (1.4) is sufficiently close to $\boldsymbol{\Psi}$ and if n is sufficiently large, then a sequence of iterates $\{\boldsymbol{\Psi}^{(k)}\}$ generated by the method of scoring or its modified version using the empirical information matrix exhibits local linear convergence to $\boldsymbol{\Psi}^*$. That is, given a norm $||\cdot||$ on $\boldsymbol{\Omega}$, there is a constant h, such that

$$||\boldsymbol{\Psi}^{(k+1)} - \boldsymbol{\Psi}^*|| \leq h||\boldsymbol{\Psi}^{(k)} - \boldsymbol{\Psi}^*||$$

for $k = 0, 1, 2, \ldots$, whenever $\boldsymbol{\Psi}^{(0)}$ is sufficiently close to $\boldsymbol{\Psi}^*$.

It is clear that the modified Newton method using the empirical information matrix (1.9) in (1.5) is an analog of the Gauss–Newton method for nonlinear least-squares estimation. With nonlinear least-squares estimation on the basis of some observed univariate random variables, $w_1, \ldots, w_n$, one minimizes

$$\frac{1}{2}\sum_{j=1}^{n}\{w_j - \mu_j(\boldsymbol{\Psi})\}^2 \tag{1.10}$$

with respect to $\boldsymbol{\Psi}$, where

$$\mu_j(\boldsymbol{\Psi}) = E_{\boldsymbol{\Psi}}(W_j).$$

The Gauss–Newton method approximates the Hessian of (1.10) by

$$\sum_{j=1}^{n}\{\partial\mu_j(\boldsymbol{\Psi})/\partial\boldsymbol{\Psi}\}\{\partial\mu_j(\boldsymbol{\Psi})/\partial\boldsymbol{\Psi}\}^T.$$

1.4 INTRODUCTORY EXAMPLES

1.4.1 Introduction

Before proceeding to formulate the EM algorithm in its generality, we give here two simple illustrative examples.

1.4.2 Example 1.1: A Multinomial Example

We consider first the multinomial example that DLR used to introduce the EM algorithm and that has been subsequently used many times in the literature to illustrate various modifications and extensions of this algorithm. It relates to a classic example of ML estimation due to Fisher (1925) and arises in genetic models for gene frequency estimation. Hartley (1958) also gives three multinomial examples of a similar nature in proposing the EM algorithm in special circumstances.

The data relate to a problem of estimation of linkage in genetics discussed by Rao (1973, pp. 368–369). The observed data vector of frequencies

$$\boldsymbol{y} = (y_1, y_2, y_3, y_4)^T$$

is postulated to arise from a multinomial distribution with four cells with cell probabilities

$$\tfrac{1}{2} + \tfrac{1}{4}\Psi,\ \tfrac{1}{4}(1 - \Psi),\ \tfrac{1}{4}(1 - \Psi),\ \text{and}\ \tfrac{1}{4}\Psi \tag{1.11}$$

with $0 \leq \Psi \leq 1$. The parameter Ψ is to be estimated on the basis of $\boldsymbol{y}$. In the multinomial example of DLR, the observed frequencies are

$$\boldsymbol{y} = (125, 18, 20, 34)^T \tag{1.12}$$

from a sample of size $n = 197$. For the Newton–Raphson and scoring methods (but not the EM algorithm), Thisted (1988, Section 4.2.6) subsequently considered the same example, but with

$$\boldsymbol{y} = (1997, 906, 904, 32)^T \tag{1.13}$$

from a sample of size $n = 3839$. We shall consider here the results obtained for both data sets.

The probability function $g(\boldsymbol{y}; \Psi)$ for the observed data $\boldsymbol{y}$ is given by

$$g(\boldsymbol{y}; \Psi) = \frac{n!}{y_1!y_2!y_3!y_4!}(\tfrac{1}{2} + \tfrac{1}{4}\Psi)^{y_1}(\tfrac{1}{4} - \tfrac{1}{4}\Psi)^{y_2}(\tfrac{1}{4} - \tfrac{1}{4}\Psi)^{y_3}(\tfrac{1}{4}\Psi)^{y_4}.$$

The log likelihood function for Ψ is therefore, apart from an additive term not involving Ψ,

$$\log L(\Psi) = y_1 \log(2 + \Psi) + (y_2 + y_3)\log(1 - \Psi) + y_4 \log \Psi. \tag{1.14}$$

On differentiation of (1.14) with respect to Ψ, we have that

$$\partial \log L(\Psi)/\partial\Psi = \frac{y_1}{2 + \Psi} - \frac{y_2 + y_3}{1 - \Psi} + \frac{y_4}{\Psi} \tag{1.15}$$

and

$$\begin{aligned} I(\Psi; \boldsymbol{y}) &= -\partial^2 \log L(\Psi)/\partial\Psi^2 \\ &= \frac{y_1}{(2 + \Psi)^2} + \frac{y_2 + y_3}{(1 - \Psi)^2} + \frac{y_4}{\Psi^2}. \end{aligned} \tag{1.16}$$

The right-hand side of (1.15) can be rewritten as a rational function, the numerator of which is a quadratic in Ψ. One of the roots is negative, and so it is the other root that we seek.

Although the likelihood equation can be solved explicitly to find the MLE $\hat{\Psi}$ of Ψ, we shall use this example to illustrate the computation of the MLE via Newton-type methods and the EM algorithm. In a later section, we shall give an example of a multinomial depending on two unknown parameters where the likelihood equation cannot be solved explicitly.

Considering the Newton-type methods of computation for the data set in

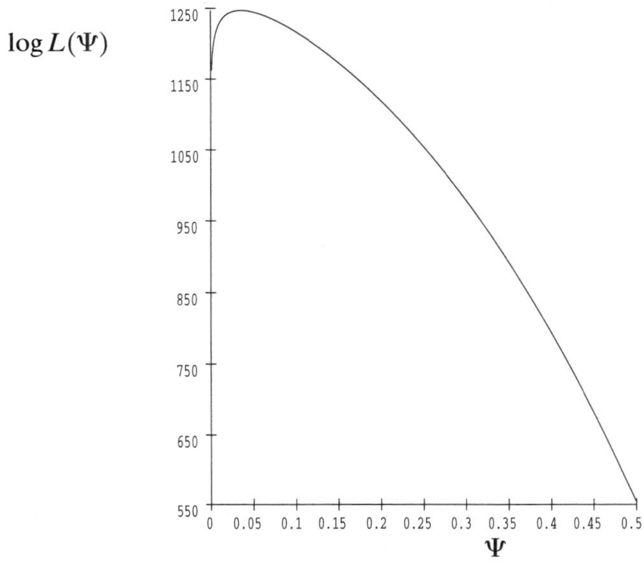

Figure 1.1. Plot of log likelihood function $\log L(\Psi)$ for the multinomial data in Thisted (1988).

Thisted (1988), it can be seen from the plot of $\log L(\Psi)$ as given by (1.14) in Figure 1.1 that a choice of starting value too close to zero or much greater than 0.1 will cause difficulty with these methods. Indeed, if the Newton–Raphson procedure is started from 0.5 (the midpoint of the admissible interval for Ψ), then it converges to the negative root of -0.4668; the method of scoring, however, does converge to the MLE given by $\hat{\Psi} = 0.0357$ (see Thisted, 1988, p. 176).

For this problem, it is not difficult to obtain a reasonable starting value since it can be seen that an unbiased estimator of Ψ is given by

$$\begin{aligned}\tilde{\Psi} &= (y_1 - y_2 - y_3 + y_4)/4 \\ &= 0.0570.\end{aligned}$$

Starting the Newton–Raphson procedure from $\Psi^{(0)} = \tilde{\Psi} = 0.0570$, leads to convergence to the MLE $\hat{\Psi}$, as shown in Table 1.1. For comparative purposes, we have also displayed the iterates for the method of scoring. The latter converges more rapidly at the start, but Newton–Raphson's quadratic convergence takes over in the last few iterations. The method of scoring uses $\mathcal{I}(\Psi^{(k)})$ instead of $I(\Psi^{(k)}; \boldsymbol{y})$ on each iteration k. For this problem, the expected information $\mathcal{I}(\Psi)$ about Ψ is given on taking the expectation of (1.16) by

$$\begin{aligned}\mathcal{I}(\Psi) &= E_{\Psi}\{I(\Psi; \boldsymbol{Y})\} \\ &= \frac{n}{4}\left\{\frac{1}{2+\Psi} + \frac{2}{(1-\Psi)} + \frac{1}{\Psi}\right\}. \end{aligned} \tag{1.17}$$

Table 1.1. Results of Newton–Raphson and Scoring Algorithms for Example 1.1 for Data in Thisted (1988).

	Newton–Raphson		Scoring	
Iteration	$\Psi^{(k)}$	$S(y; \Psi^{(k)})$	$\Psi^{(k)}$	$S(y; \Psi^{(k)})$
0	0.05704611	−387.74068038	0.05704611	−387.74068038
1	0.02562679	376.95646890	0.03698326	− 33.88267279
2	0.03300085	80.19367817	0.03579085	− 2.15720180
3	0.03552250	5.24850707	0.03571717	− 0.13386352
4	0.03571138	0.02527096	0.03571260	− 0.00829335
5	0.03571230	0.00000059	0.03571232	− 0.00051375
6	0.03571230	− 0.00000000	0.03571230	− 0.00003183

Source: Adapted from Thisted (1988).

Suppose now that the first of the original four multinomial cells, which has an associated probability of $\frac{1}{2} + \frac{1}{4}\Psi$, could be split into two subcells having probabilities $\frac{1}{2}$ and $\frac{1}{4}\Psi$, respectively, and let y_{11} and y_{12} be the corresponding split of y_1, where

$$y_1 = y_{11} + y_{12}.$$

Then on modifying the likelihood equation (1.14) according to this split, it is clear that the MLE of Ψ on the basis of this split is simply

$$(y_{12} + y_4)/(y_{12} + y_2 + y_3 + y_4). \tag{1.18}$$

This is because in effect the modified likelihood function for Ψ has the same form as that obtained by considering $y_{12} + y_4$ to be a realization from a binomial distribution with sample size $y_{12} + y_2 + y_3 + y_4$ and probability parameter Ψ.

We now formalize this approach through the application of the EM algorithm. The observed vector of frequencies $\boldsymbol{y}$ is viewed as being incomplete and the complete-data vector is taken to be

$$\boldsymbol{x} = (y_{11}, y_{12}, y_2, y_3, y_4)^T.$$

The cell frequencies in $\boldsymbol{x}$ are assumed to arise from a multinomial distribution having five cells with probabilities

$$\tfrac{1}{2},\ \tfrac{1}{4}\Psi,\ \tfrac{1}{4}(1-\Psi),\ \tfrac{1}{4}(1-\Psi),\ \text{and}\ \tfrac{1}{4}\Psi. \tag{1.19}$$

In this framework, y_{11} and y_{12} are regarded as the unobservable or missing data, since we only get to observe their sum y_1.

If we take the distribution of the complete-data vector $\boldsymbol{X}$ to be multinomial with n draws with respect to now five cells with probabilities specified by (1.19), then it implies that the observable or incomplete-data vector $\boldsymbol{Y}$ has its original multinomial distribution with cell probabilities specified by (1.11). This can be confirmed by verifying that

$$g(\boldsymbol{y};\ \Psi) = \sum g_c(\boldsymbol{x};\ \Psi), \tag{1.20}$$

where

$$g_c(\boldsymbol{x};\ \Psi) = C(\boldsymbol{x})(\tfrac{1}{2})^{y_{11}}(\tfrac{1}{4}\Psi)^{y_{12}}(\tfrac{1}{4} - \tfrac{1}{4}\Psi)^{y_2}(\tfrac{1}{4} - \tfrac{1}{4}\Psi)^{y_3}(\tfrac{1}{4}\Psi)^{y_4} \tag{1.21}$$

and

$$C(\boldsymbol{x}) = \frac{n!}{y_{11}!y_{12}!y_2!y_3!y_4!},$$

and where the summation in (1.20) is over all values of $\boldsymbol{x}$ such that

$$y_{11} + y_{12} = y_1.$$

From (1.21), the complete-data log likelihood is therefore, apart from a term not involving Ψ,

$$\log L_c(\Psi) = (y_{12} + y_4)\log\Psi + (y_2 + y_3)\log(1 - \Psi). \tag{1.22}$$

On equating the derivative of (1.22) with respect to Ψ to zero and solving for Ψ, we find that the complete-data MLE of Ψ is given by (1.18).

Since the frequency y_{12} is unobservable, we are unable to estimate Ψ by (1.18). With the EM algorithm, this obstacle is overcome by the E-step, as it handles the problem of filling in for unobservable data by averaging the complete-data log likelihood over its conditional distribution given the observed data $\boldsymbol{y}$. But in order to calculate this conditional expectation, we have to specify a value for Ψ. Let $\Psi^{(0)}$ be the value specified initially for Ψ. Then on the first iteration of the EM algorithm, the E-step requires the computation of the conditional expectation of $\log L_c(\Psi)$ given $\boldsymbol{y}$, using $\Psi^{(0)}$ for Ψ, which can be written as

$$Q(\Psi;\ \Psi^{(0)}) = E_{\Psi^{(0)}}\{\log L_c(\Psi) \mid \boldsymbol{y}\}.$$

As $\log L_c(\Psi)$ is a linear function of the unobservable data y_{11} and y_{12} for this problem, the E-step is effected simply by replacing y_{11} and y_{12} by their current conditional expectations given the observed data $\boldsymbol{y}$. Considering the random variable Y_{11}, corresponding to y_{11}, it is easy to verify that conditional on $\boldsymbol{y}$, effectively y_1, Y_{11} has a binomial distribution with sample size y_1 and probability parameter

$$\tfrac{1}{2}/(\tfrac{1}{2} + \tfrac{1}{4}\Psi^{(0)}),$$

where $\Psi^{(0)}$ is used in place of the unknown parameter Ψ. Thus the initial conditional expectation of Y_{11} given y_1 is

$$E_{\Psi^{(0)}}(Y_{11} \mid y_1) = y_{11}^{(0)},$$

where

$$y_{11}^{(0)} = \tfrac{1}{2}y_1/(\tfrac{1}{2} + \tfrac{1}{4}\Psi^{(0)}). \tag{1.23}$$

This completes the E-step on the first iteration since

$$\begin{aligned} y_{12}^{(0)} &= y_1 - y_{11}^{(0)} \\ &= \tfrac{1}{4}y_1\Psi^{(0)}/(\tfrac{1}{2} + \tfrac{1}{4}\Psi^{(0)}). \end{aligned} \tag{1.24}$$

The M-step is undertaken on the first iteration by choosing $\Psi^{(1)}$ to be the value of Ψ that maximizes $Q(\Psi; \Psi^{(0)})$ with respect to Ψ. Since this Q-function is given simply by replacing the unobservable frequencies y_{11} and y_{12} with their current conditional expectations $y_{11}^{(0)}$ and $y_{12}^{(0)}$ in the complete-data log likelihood, $\Psi^{(1)}$ is obtained by substituting $y_{12}^{(0)}$ for y_{12} in (1.18) to give

$$\begin{aligned} \Psi^{(1)} &= (y_{12}^{(0)} + y_4)/(y_{12}^{(0)} + y_2 + y_3 + y_4) \\ &= (y_{12}^{(0)} + y_4)/(n - y_{11}^{(0)}). \end{aligned} \tag{1.25}$$

This new fit $\Psi^{(1)}$ for Ψ is then substituted for Ψ into the right-hand sides of (1.23) and (1.24) to produce updated values $y_{11}^{(1)}$ and $y_{12}^{(1)}$ for y_{11} and y_{12} for use in place of $y_{11}^{(0)}$ and $y_{12}^{(0)}$ in the right-hand side of (1.25). This leads to a new fit $\Psi^{(2)}$ for Ψ, and so on. It follows on so alternating the E- and M-steps on the $(k+1)$th iteration of the EM algorithm that

$$\Psi^{(k+1)} = (y_{12}^{(k)} + y_4)/(n - y_{11}^{(k)}) \tag{1.26}$$

where

$$y_{11}^{(k)} = \tfrac{1}{2}y_1/(\tfrac{1}{2} + \tfrac{1}{4}\Psi^{(k)})$$

and

$$y_{12}^{(k)} = y_1 - y_{11}^{(k)}.$$

On putting

$$\Psi^{(k+1)} = \Psi^{(k)} = \Psi^*$$

in (1.26), we can explicitly solve the resulting quadratic equation in Ψ^* to confirm that the sequence of EM iterates $\{\Psi^{(k)}\}$, irrespective of the starting value $\Psi^{(0)}$, converges to the MLE of Ψ obtained by directly solving the (incomplete-data) likelihood equation given by equating (1.15) to zero.

Table 1.2. Results of EM Algorithm for Example 1.1 for Data in DLR.

Iteration	$\Psi^{(k)}$	$\Psi^{(k)} - \hat{\Psi}$	$r^{(k)}$	$\log L(\Psi^{(k)})$
0	0.500000000	0.126821498	—	64.62974
1	0.608247423	0.018574075	0.1465	67.32017
2	0.624321051	0.002500447	0.1346	67.38292
3	0.626488879	0.000332619	0.1330	67.38408
4	0.626777323	0.000044176	0.1328	67.38410
5	0.626815632	0.000005866	0.1328	67.38410
6	0.626820719	0.000000779	0.1328	67.38410
7	0.626821395	0.000000104	0.1328	67.38410
8	0.626821484	0.000000014	—	67.38410

Source: Adapted from Dempster et al. (1977).

Table 1.3. Results of EM Algorithm for Example 1.1 for Data in Thisted (1988).

Iteration	$\Psi^{(k)}$	$\Psi^{(k)} - \hat{\Psi}$	$r^{(k)}$	$\log L(\Psi^{(k)})$
0	0.05704611	0.0213338	—	1242.4180
1	0.04600534	0.0103411	0.48473	1245.8461
2	0.04077975	0.0005067	0.49003	1246.7791
3	0.03820858	0.0002496	0.49261	1247.0227
4	0.03694516	0.0001233	0.49388	1247.0845
5	0.03632196	0.0000610	0.49450	1247.0999
6	0.03605397	0.0000302	0.49481	1247.1047
7	0.03586162	0.0000149	0.49496	1247.1049
8	0.03578622	0.0000074	0.49502	1247.1050
9	0.03574890	0.0000037	0.49503	1247.1050
10	0.03573042	0.0000018	0.49503	1247.1050
11	0.03572127	0.0000009	0.49503	1247.1050
12	0.03571674	0.0000004	0.49503	1247.1050
13	0.03571450	0.0000002	0.49503	1247.1050
14	0.03571339	0.0000001	0.49503	1247.1050
15	0.03571284	0.0000001	0.49503	1247.1050
16	0.03571257	0.0000000	0.49503	1247.1050
17	0.03571243	0.0000000	0.49503	1247.1050
18	0.03571237	0.0000000	0.49503	1247.1050
19	0.03571233	0.0000000	0.49503	1247.1050
20	0.03571232	0.0000000	0.49503	1247.1050
21	0.03571231	0.0000000	0.49503	1247.1050
22	0.03571231	0.0000000	—	1247.1050

In Tables 1.2 and 1.3, we report the results of the EM algorithm applied to the data sets considered by DLR and Thisted (1988), respectively. In Table 1.2, we see that starting from an initial value of $\Psi^{(0)} = 0.5$, the EM algorithm moved for eight iterations. The third column in this table gives the deviation $\Psi^{(k)} - \hat{\Psi}$, and the fourth column gives the ratio of successive deviations

$$r^{(k)} = (\Psi^{(k+1)} - \Psi^{(k)})/(\Psi^{(k)} - \Psi^{(k-1)}).$$

It can be seen that $r^{(k)}$ is essentially constant for $k \geq 3$ consistent with a linear rate of convergence equal to 0.1328. This rate of convergence is to be established in general for the EM algorithm in Section 3.9.

On comparing the results of the EM algorithm in Table 1.3 for the data set in Thisted (1988) with those of the Newton-Raphson and scoring methods in Table 1.1, we see that the EM algorithm takes about 15 more iterations to converge to the MLE.

1.4.3 Example 1.2: Estimation of Mixing Proportions

As to be discussed further in Section 1.8, the publication of the DLR paper greatly stimulated interest in the use of finite mixture distributions to model heterogeneous data. This is because the fitting of mixture models by maximum likelihood is a classic example of a problem that is simplified considerably by the EM's conceptual unification of ML estimation from data that can be viewed as being incomplete.

A wide variety of applications of finite-mixture models are given in McLachlan and Basford (1988) and, more recently, in McLachlan (1997). We consider here an example involving the estimation of the proportions in which the components of the mixture occur, where the component densities are completely specified. In the next chapter, we consider the more difficult case where the component densities are specified up to a number of unknown parameters that have to be estimated along with the mixing proportions. But the former case of completely specified component densities is not unrealistic, as there are situations in practice where there are available separate samples from each of the component distributions of the mixture that enable the component densities to be estimated with adequate precision before the fitting of the mixture model; see McLachlan (1992, Chapter 2).

Suppose that the p.d.f. of a random vector $\boldsymbol{W}$ has a g-component mixture form

$$f(\boldsymbol{w};\, \boldsymbol{\Psi}) = \sum_{i=1}^{g} \pi_i f_i(\boldsymbol{w}), \tag{1.27}$$

where $\boldsymbol{\Psi} = (\pi_1, \ldots, \pi_{g-1})^T$ is the vector containing the unknown parameters, namely the $g-1$ mixing proportions $\pi_1, \ldots, \pi_{g-1}$, since

$$\pi_g = 1 - \sum_{i=1}^{g-1} \pi_i.$$

The component p.d.f.'s $f_i(\boldsymbol{w})$ are completely specified.

This mixture model covers situations where the underlying population is modeled as consisting of g distinct groups $G_1, \ldots, G_g$ in some unknown proportions $\pi_1, \ldots, \pi_g$, and where the conditional p.d.f of $\boldsymbol{W}$ given membership of the ith group G_i is $f_i(\boldsymbol{w})$. For example, in the problem considered by Do and McLachlan (1984), the population of interest consists of rats from g species $G_1, \ldots, G_g$, that are consumed by owls in some unknown proportions $\pi_1, \ldots, \pi_g$. The problem is to estimate the π_i on the basis of the observation vector $\boldsymbol{W}$ containing measurements recorded on a sample of size n of rat skulls taken from owl pellets. The rats constitute part of an owl's diet, and indigestible material is regurgitated as a pellet.

We let

$$\boldsymbol{y} = (\boldsymbol{w}_1^T, \ldots, \boldsymbol{w}_n^T)^T$$

denote the observed random sample obtained from the mixture density (1.27). The log likelihood function for $\boldsymbol{\Psi}$ that can be formed from the observed data $\boldsymbol{y}$ is given by

$$\begin{aligned} \log L(\boldsymbol{\Psi}) &= \sum_{j=1}^{n} \log f(\boldsymbol{w}_j; \boldsymbol{\Psi}) \\ &= \sum_{j=1}^{n} \log \left\{ \sum_{i=1}^{g} \pi_i f_i(\boldsymbol{w}_j) \right\}. \end{aligned} \tag{1.28}$$

On differentiating (1.28) with respect to π_i $(i = 1, \ldots, g-1)$ and equating the result to zero, we obtain

$$\sum_{j=1}^{n} \left\{ \frac{f_i(\boldsymbol{w}_j)}{f(\boldsymbol{w}_j; \hat{\boldsymbol{\Psi}})} - \frac{f_g(\boldsymbol{w}_j)}{f(\boldsymbol{w}_j; \hat{\boldsymbol{\Psi}})} \right\} = 0 \quad (i = 1, \ldots, g-1) \tag{1.29}$$

as the likelihood equation, which clearly does not yield an explicit solution for

$$\hat{\boldsymbol{\Psi}} = (\hat{\pi}_1, \ldots, \hat{\pi}_{g-1})^T.$$

In order to pose this problem as an incomplete-data one, we now introduce as the unobservable or missing data the vector

$$\boldsymbol{z} = (\boldsymbol{z}_1^T, \ldots, \boldsymbol{z}_n^T)^T, \tag{1.30}$$

where $\boldsymbol{z}_j$ is a g-dimensional vector of zero-one indicator variables and where $z_{ij} = (\boldsymbol{z}_j)_i$ is one or zero according as whether $\boldsymbol{w}_j$ arose or did not arise from the ith component of the mixture $(i = 1, \ldots, g;\ j = 1, \ldots, n)$. Of course, in some applications (such as in the rat data one above), the components of the mixture correspond to externally existing groups and so each realization $\boldsymbol{w}_j$ in the observed random sample from the mixture density does have a tangible component membership. But in other applications, component membership of the realizations is just a conceptual device to formulate the problem within the EM framework.

If these z_{ij} were observable, then the MLE of π_i is simply given by

$$\sum_{j=1}^{n} z_{ij}/n \quad (i = 1, \ldots, g), \tag{1.31}$$

which is the proportion of the sample having arisen from the ith component of the mixture.

As in the last example, the EM algorithm handles the addition of the unobservable data to the problem by working with the current conditional expectation of the complete-data log likelihood given the observed data. On defining the complete-data vector $\boldsymbol{x}$ as

$$\boldsymbol{x} = (\boldsymbol{y}^T, \boldsymbol{z}^T)^T, \tag{1.32}$$

the complete-data log likelihood for $\boldsymbol{\Psi}$ has the multinomial form

$$\log L_c(\boldsymbol{\Psi}) = \sum_{i=1}^{g} \sum_{j=1}^{n} z_{ij} \log \pi_i + C, \tag{1.33}$$

where

$$C = \sum_{i=1}^{g} \sum_{j=1}^{n} z_{ij} \log f_i(\boldsymbol{w}_j)$$

does not depend on $\boldsymbol{\Psi}$.

As (1.33) is linear in the unobservable data z_{ij}, the E-step (on the $(k + 1)$th iteration) simply requires the calculation of the current conditional expectation of Z_{ij} given the observed data $\boldsymbol{y}$, where Z_{ij} is the random variable corresponding to z_{ij}. Now

$$\begin{aligned} E_{\boldsymbol{\Psi}^{(k)}}(Z_{ij} \mid \boldsymbol{y}) &= \mathrm{pr}_{\boldsymbol{\Psi}^{(k)}}\{Z_{ij} = 1 \mid \boldsymbol{y}\} \\ &= z_{ij}^{(k)}, \end{aligned} \tag{1.34}$$

where by Bayes' Theorem,

$$z_{ij}^{(k)} = \tau_i(\boldsymbol{w}_j;\ \boldsymbol{\Psi}^{(k)}) \tag{1.35}$$

and

$$\tau_i(\boldsymbol{w}_j;\ \boldsymbol{\Psi}^{(k)}) = \pi_i^{(k)} f_i(\boldsymbol{w}_j)/f(\boldsymbol{w}_j;\ \boldsymbol{\Psi}^{(k)}) \tag{1.36}$$

for $i = 1, \ldots, g;\ j = 1, \ldots, n$. The quantity $\tau_i(\boldsymbol{w}_j;\ \boldsymbol{\Psi}^{(k)})$ is the posterior probability that the jth member of the sample with observed value $\boldsymbol{w}_j$ belongs to the ith component of the mixture.

The M-step on the $(k+1)$th iteration simply requires replacing each z_{ij} by $z_{ij}^{(k)}$ in (1.31) to give

$$\pi_i^{(k+1)} = \sum_{j=1}^{n} z_{ij}^{(k)}/n \tag{1.37}$$

for $i = 1, \ldots, g$. Thus in forming the estimate of π_i on the $(k+1)$th iteration, there is a contribution from each observation $\boldsymbol{w}_j$ equal to its (currently assessed) posterior probability of membership of the ith component of the mixture model. This EM solution therefore has an intuitively appealing interpretation.

The computation of the MLE of π_i by direct maximization of the incomplete-data log likelihood function (1.28) requires solving the likelihood equation (1.29). The latter can be identified with the iterative solution (1.37) provided by the EM algorithm after some manipulation as follows. On multiplying throughout by $\hat{\pi}_i$ in equation (1.29), we have that

$$\sum_{j=1}^{n} \{\tau_i(\boldsymbol{w}_j;\ \hat{\boldsymbol{\Psi}}) - (\hat{\pi}_i/\hat{\pi}_g)\tau_g(\boldsymbol{w}_j;\ \hat{\boldsymbol{\Psi}})\} = 0 \quad (i = 1, \ldots, g-1). \tag{1.38}$$

As (1.38) also holds for $i = g$, we can sum over $i = 1, \ldots, g$ in (1.38) to give

$$\hat{\pi}_g = \sum_{j=1}^{n} \tau_g(\boldsymbol{w}_j;\ \hat{\boldsymbol{\Psi}})/n. \tag{1.39}$$

Substitution now of (1.39) into (1.38) yields

$$\hat{\pi}_i = \sum_{j=1}^{n} \tau_i(\boldsymbol{w}_j;\ \hat{\boldsymbol{\Psi}})/n \tag{1.40}$$

for $i = 1, \ldots, g-1$, which also holds for $i = g$ from (1.39). The resulting equation (1.40) for the MLE $\hat{\pi}_i$ can be identified with the iterative solution (1.36) given by the EM algorithm. The latter solves the likelihood equation by substituting an initial value for $\boldsymbol{\Psi}$ into the right-hand side of (1.40), which yields a new estimate for $\boldsymbol{\Psi}$, which in turn is substituted into the right-hand side of (1.40) to yield a new estimate, and so on until convergence.

Even before the formulation of the EM algorithm by DLR, various researchers have carried out these manipulations in their efforts to solve the likelihood equation for mixture models with specific component densities; see, for example, Hasselblad (1966, 1969), Wolfe (1967, 1970), and Day (1969). As demonstrated above for the estimation of the mixing proportions, the application of the EM algorithm to the mixture problem automatically reveals the iterative scheme to be followed for the computation of the MLE. Furthermore, it ensures that the likelihood values increase monotonically. Prior to DLR, various researchers did note the monotone convergence of the likelihood sequences produced in their particular applications, but were only able to speculate on this monotonicity holding in general.

As the $z_{ij}^{(k)}$ are probabilities, the E-step of the EM algorithm effectively imputes fractional values for the unobservable zero-one indicator variables z_{ij}. In so doing, it avoids the biases associated with *ad hoc* iterative procedures that impute only zero-one values; that is, that insist on outright component membership for each observation at each stage. For example, for each j ($j = 1, \ldots, n$), let

$$\hat{z}_{ij}^{(k)} = 1 \quad \text{if } i = \arg\max_h \tau_h(\boldsymbol{w}_j; \boldsymbol{\Psi}^{(k)}),$$

and be zero otherwise, which is equivalent to assigning the jth observation $\boldsymbol{w}_j$ to the component of the mixture for which it has the highest (currently assessed) posterior probability of belonging. If these zero-one values are imputed for the z_{ij} in updating the estimate of π_i from (1.31), then this in general will produce biased estimates of the mixing proportions (McLachlan and Basford, 1988, Chapter 4).

Numerical Example. As a numerical example, we generated a random sample of $n = 50$ observations $w_1, \ldots, w_n$ from a mixture of two univariate normal densities with means $\mu_1 = 0$ and $\mu_2 = 2$ and common variance $\sigma^2 = 1$ in proportions $\pi_1 = 0.8$ and $\pi_2 = 0.2$. Starting the EM algorithm from $\pi_1^{(0)} = 0.5$, it converged after 27 iterations to the solution $\hat{\pi}_1 = 0.75743$. The EM algorithm was stopped when

$$\mid \pi_1^{(k+1)} - \pi_1^{(k)} \mid < 10^{-5}.$$

It was also started from the moment estimate given by

$$\begin{aligned} \tilde{\pi}_1 &= (\overline{w} - \mu_2)/(\mu_1 - \mu_2) \\ &= 0.86815 \end{aligned}$$

and, using the same stopping criterion, it converged after 30 iterations to $\hat{\pi}_1$.

In Table 1.4, we have listed the value of $\pi_1^{(k)}$ and of $\log L(\pi_1^{(k)})$ for various values of k. It can be seen that it is during the first few iterations that the EM algorithm makes most of its progress in reaching the maximum value of the log likelihood function.

Table 1.4. Results of the EM Algorithm for the Example on Estimation of Mixing Proportions.

Iteration k	$\pi_1^{(k)}$	$\log L(\pi_1^{(k)})$
0	0.50000	−91.87811
1	0.68421	−85.55353
2	0.70304	−85.09035
3	0.71792	−84.81398
4	0.72885	−84.68609
5	0.73665	−84.63291
6	0.74218	−84.60978
7	0.74615	−84.58562
⋮	⋮	⋮
27	0.75743	−84.58562

1.5 FORMULATION OF THE EM ALGORITHM

1.5.1 EM Algorithm

We let $\boldsymbol{Y}$ be the random vector corresponding to the observed data $\boldsymbol{y}$, having p.d.f. postulated as $g(\boldsymbol{y}; \boldsymbol{\Psi})$, where $\boldsymbol{\Psi} = (\Psi_1, \ldots, \Psi_d)^T$ is a vector of unknown parameters with parameter space $\boldsymbol{\Omega}$.

The EM algorithm is a broadly applicable algorithm that provides an iterative procedure for computing MLEs in situations where, but for the absence of some additional data, ML estimation would be straightforward. Hence, in this context, the observed data vector $\boldsymbol{y}$ is viewed as being incomplete and is regarded as an observable function of the so-called complete-data. The notion of 'incomplete data' includes the conventional sense of missing data, but it also applies to situations where the complete data represent what would be available from some hypothetical experiment. In the latter case, the complete data may contain some variables that are never observable in a data sense. Within this framework, we let $\boldsymbol{x}$ denote the vector containing the augmented or so-called complete data, and we let $\boldsymbol{z}$ denote the vector containing the additional data, referred to as the unobservable or missing data.

As will become evident from the many examples of the EM algorithm discussed in this book, even when a problem does not at first appear to be an incomplete-data one, computation of MLE is often greatly facilitated by artificially formulating it to be as such. This is because the EM algorithm exploits the reduced complexity of ML estimation given the complete data. For many statistical problems the complete-data likelihood has a nice form.

We let $g_c(\boldsymbol{x}; \boldsymbol{\Psi})$ denote the p.d.f. of the random vector $\boldsymbol{X}$ corresponding to the complete-data vector $\boldsymbol{x}$. Then the complete-data log likelihood function that could be formed for $\boldsymbol{\Psi}$ if $\boldsymbol{x}$ were fully observable is given by

$$\log L_c(\boldsymbol{\Psi}) = \log g_c(\boldsymbol{x};\ \boldsymbol{\Psi}).$$

Formally, we have two samples spaces $\mathcal{X}$ and $\mathcal{Y}$ and a many-to-one mapping from $\mathcal{X}$ to $\mathcal{Y}$. Instead of observing the complete-data vector $\boldsymbol{x}$ in $\mathcal{X}$, we observe the incomplete-data vector $\boldsymbol{y} = \boldsymbol{y}(\boldsymbol{x})$ in $\mathcal{Y}$. It follows that

$$g(\boldsymbol{y};\ \boldsymbol{\Psi}) = \int_{\mathcal{X}(\boldsymbol{y})} g_c(\boldsymbol{x};\ \boldsymbol{\Psi})d\boldsymbol{x},$$

where $\mathcal{X}(\boldsymbol{y})$ is the subset of $\mathcal{X}$ determined by the equation $\boldsymbol{y} = \boldsymbol{y}(\boldsymbol{x})$.

The EM algorithm approaches the problem of solving the incomplete-data likelihood equation (1.1) indirectly by proceeding iteratively in terms of the complete-data log likelihood function, $\log L_c(\boldsymbol{\Psi})$. As it is unobservable, it is replaced by its conditional expectation given $\boldsymbol{y}$, using the current fit for $\boldsymbol{\Psi}$.

More specifically, let $\boldsymbol{\Psi}^{(0)}$ be some initial value for $\boldsymbol{\Psi}$. Then on the first iteration, the E-step requires the calculation of

$$Q(\boldsymbol{\Psi};\ \boldsymbol{\Psi}^{(0)}) = E_{\boldsymbol{\Psi}^{(0)}}\{\log L_c(\boldsymbol{\Psi}) \mid \boldsymbol{y}\}.$$

The M-step requires the maximization of $Q(\boldsymbol{\Psi},\ \boldsymbol{\Psi}^{(0)})$ with respect to $\boldsymbol{\Psi}$ over the parameter space $\boldsymbol{\Omega}$. That is, we choose $\boldsymbol{\Psi}^{(1)}$ such that

$$Q(\boldsymbol{\Psi}^{(1)};\ \boldsymbol{\Psi}^{(0)}) \geq Q(\boldsymbol{\Psi};\ \boldsymbol{\Psi}^{(0)})$$

for all $\boldsymbol{\Psi} \in \boldsymbol{\Omega}$. The E- and M-steps are then carried out again, but this time with $\boldsymbol{\Psi}^{(0)}$ replaced by the current fit $\boldsymbol{\Psi}^{(1)}$. On the $(k+1)$th iteration, the E- and M-steps are defined as follows:

E-Step. Calculate $Q(\boldsymbol{\Psi};\ \boldsymbol{\Psi}^{(k)})$, where

$$Q(\boldsymbol{\Psi};\ \boldsymbol{\Psi}^{(k)}) = E_{\boldsymbol{\Psi}^{(k)}}\{\log L_c(\boldsymbol{\Psi}) \mid \boldsymbol{y}\}. \tag{1.41}$$

M-Step. Choose $\boldsymbol{\Psi}^{(k+1)}$ to be any value of $\boldsymbol{\Psi} \in \boldsymbol{\Omega}$ that maximizes $Q(\boldsymbol{\Psi};\ \boldsymbol{\Psi}^{(k)})$; that is,

$$Q(\boldsymbol{\Psi}^{(k+1)};\ \boldsymbol{\Psi}^{(k)}) \geq Q(\boldsymbol{\Psi};\ \boldsymbol{\Psi}^{(k)}) \tag{1.42}$$

for all $\boldsymbol{\Psi} \in \boldsymbol{\Omega}$.

The E- and M-steps are alternated repeatedly until the difference

$$L(\boldsymbol{\Psi}^{(k+1)}) - L(\boldsymbol{\Psi}^{(k)})$$

changes by an arbitrarily small amount in the case of convergence of the sequence of likelihood values $\{L(\boldsymbol{\Psi}^{(k)})\}$. DLR show that the (incomplete-data) likelihood function $L(\boldsymbol{\Psi})$ is not decreased after an EM iteration; that is,

$$L(\boldsymbol{\Psi}^{(k+1)}) \geq L(\boldsymbol{\Psi}^{(k)}) \tag{1.43}$$

for $k = 0, 1, 2, \ldots$. Hence convergence must be obtained with a sequence of likelihood values that are bounded above.

Another way of expressing (1.42) is to say that $\boldsymbol{\Psi}^{(k+1)}$ belongs to

$$\mathcal{M}(\boldsymbol{\Psi}^{(k)}) = \arg\max_{\boldsymbol{\Psi}} Q(\boldsymbol{\Psi}; \boldsymbol{\Psi}^{(k)}), \tag{1.44}$$

which is the set of points that maximize $Q(\boldsymbol{\Psi}; \boldsymbol{\Psi}^{(k)})$.

We see from the above that it is not necessary to specify the exact mapping from $\mathcal{X}$ to $\mathcal{Y}$, nor the corresponding representation of the incomplete-data density g in terms of the complete-data density g_c. All that is necessary is the specification of the complete-data vector $\boldsymbol{x}$ and the conditional density of $\boldsymbol{X}$ given the observed data vector $\boldsymbol{y}$. Specification of this conditional density is needed in order to carry out the E-step. As the choice of the complete-data vector $\boldsymbol{x}$ is not unique, it is chosen for computational convenience with respect to carrying out the E- and M-steps. Recently, consideration has been given to the choice of $\boldsymbol{x}$ so as to speed up the convergence of the corresponding EM algorithm; see Section 5.11.

As pointed out by a referee of the DLR paper, the use of the word "algorithm" to describe this procedure can be criticized, 'because it does not specify the sequence of steps actually required to carry out a single E- or M-step'. The EM algorithm is really a generic device.

1.5.2 Example 1.3: Censored Exponentially Distributed Survival Times

We suppose W is a nonnegative random variable having an exponential distribution with mean μ. Thus its probability density function (p.d.f.) is given by

$$f(w; \mu) = \mu^{-1} \exp(-w/\mu)\, I_{(0,\infty)}(w) \quad (\mu > 0), \tag{1.45}$$

where the indicator function $I_{(0,\infty)}(w) = 1$ for $w > 0$ and is zero elsewhere. The distribution function is given by

$$F(w; \mu) = \{1 - \exp(-w/\mu)\} I_{(0,\infty)}(w).$$

In survival or reliability analyses, a study to observe a random sample $W_1, \ldots, W_n$ from (1.45) will generally be terminated in practice before all of these random variables are able to be observed. We let

$$\boldsymbol{y} = (\boldsymbol{y}_1^T, \ldots, \boldsymbol{y}_n^T)^T$$

denote the observed data, where

$$\boldsymbol{y}_j = (c_j, \delta_j)^T$$

and $\delta_j = 0$ or 1 according as the observation W_j is censored or uncensored at c_j $(j = 1, \ldots, n)$. That is, if the observation W_j is uncensored, its realized value w_j is equal to c_j, whereas if it is censored at c_j, then w_j is some value greater than c_j $(j = 1, \ldots, n)$.

In this example, the unknown parameter vector $\boldsymbol{\Psi}$ is a scalar, being equal to μ. We suppose now that the observations have been relabeled so that $W_1, \ldots, W_r$ denote the r uncensored observations and $W_{r+1}, \ldots, W_n$ the $n - r$ censored observations. The log likelihood function for μ formed on the basis of $\boldsymbol{y}$ is given by

$$\log L(\mu) = -r \log \mu - \sum_{j=1}^{n} c_j/\mu. \tag{1.46}$$

In this case, the MLE of μ can be derived explicitly from equating the derivative of (1.46) to zero to give

$$\hat{\mu} = \sum_{j=1}^{n} c_j/r. \tag{1.47}$$

Thus there is no need for the iterative computation of $\hat{\mu}$. But in this simple case, it is instructive to demonstrate how the EM algorithm would work.

The complete-data vector $\boldsymbol{x}$ can be declared to be

$$\begin{aligned} \boldsymbol{x} &= (w_1, \ldots, w_n)^T \\ &= (w_1, \ldots, w_r, \boldsymbol{z}^T)^T, \end{aligned}$$

where

$$\boldsymbol{z} = (w_{r+1}, \ldots, w_n)^T$$

contains the unobservable realizations of the $n - r$ censored random variables. In this example, the so-called unobservable or missing vector $\boldsymbol{z}$ is potentially observable in a data sense, as if the experiment were continued until each item failed, then there would be no censored observations.

The complete-data log likelihood is given by

$$\begin{aligned} \log L_c(\mu) &= \sum_{j=1}^{n} \log g_c(w_j; \mu) \\ &= -n \log \mu - \mu^{-1} \sum_{j=1}^{n} w_j. \end{aligned} \tag{1.48}$$

It can be seen that $L_c(\mu)$ belongs to the regular exponential family. We shall proceed now without making explicit use of this property, but in the next section, we shall show how it can be exploited to simplify the implementation of the EM algorithm.

As $L_c(\mu)$ can be seen to be linear in the unobservable data $w_{r+1}, \ldots, w_n$, the calculation of $Q(\mu; \mu^{(k)})$ on the E-step (on the $(k+1)$ th iteration) simply requires each such w_j to be replaced by its conditional expectation given the observed data $\boldsymbol{y}$, using the current fit $\mu^{(k)}$ for μ. By the lack of memory of the exponential distribution, the conditional distribution of $W_j - c_j$ given that $W_j > c_j$ is still exponential with mean μ. Equivalently, the conditional p.d.f. of W_j given that it is greater than c_j is

$$\mu^{-1} \exp\{-(w_j - c_j)/\mu\} I_{(c_j, \infty)}(w_j) \quad (\mu > 0). \tag{1.49}$$

From (1.49), we have that

$$\begin{aligned} E_{\mu^{(k)}}(W_j \mid \boldsymbol{y}) &= E_{\mu^{(k)}}(W_j \mid W_j > c_j) \\ &= c_j + E_{\mu^{(k)}}(W_j) \\ &= c_j + \mu^{(k)} \end{aligned} \tag{1.50}$$

for $j = r+1, \ldots, n$.

On using (1.50) to take the current conditional expectation of the complete-data log likelihood $\log L_c(\mu)$, we have that

$$\begin{aligned} Q(\mu; \mu^{(k)}) &= -n \log \mu - \mu^{-1} \left\{ \sum_{j=1}^{r} c_j + \sum_{j=r+1}^{n} (c_j + \mu^{(k)}) \right\} \\ &= -n \log \mu - \mu^{-1} \left\{ \sum_{j=1}^{n} c_j + (n - r)\mu^{(k)} \right\}. \end{aligned} \tag{1.51}$$

Concerning the M-step on the $(k+1)$th iteration, it follows from (1.51) that the value of μ that maximizes $Q(\mu; \mu^{(k)})$ is given by the MLE of μ that would be formed from the complete-data, but with each unobservable w_j replaced by its current conditional expectation given by (1.50). Accordingly,

$$\begin{aligned} \mu^{(k+1)} &= \left\{ \sum_{j=1}^{r} c_j + \sum_{j=r+1}^{n} E_{\mu^{(k)}}(W_j \mid \boldsymbol{y}) \right\} \Big/ n \\ &= \left\{ \sum_{j=1}^{r} c_j + \sum_{j=r+1}^{n} (c_j + \mu^{(k)}) \right\} \Big/ n \\ &= \left\{ \sum_{j=1}^{n} c_j + (n - r)\mu^{(k)} \right\} \Big/ n. \end{aligned} \tag{1.52}$$

On putting $\mu^{(k+1)} = \mu^{(k)} = \mu^*$ in (1.52) and solving for μ^*, we have for $r < n$ that $\mu^* = \hat{\mu}$. That is, the EM sequence $\{\mu^{(k)}\}$ has the MLE $\hat{\mu}$ as its unique limit point, as $k \rightarrow \infty$.

In order to demonstrate the rate of convergence of this sequence to $\hat{\mu}$, we can from (1.52) express $\mu^{(k+1)}$ in terms of the MLE $\hat{\mu}$ as

$$\begin{aligned} \mu^{(k+1)} &= \{r\hat{\mu} + (n-r)\mu^{(k)}\}/n \\ &= \hat{\mu} + n^{-1}(n-r)(\mu^{(k)} - \hat{\mu}), \end{aligned}$$

which gives

$$\mu^{(k+1)} - \hat{\mu} = (1 - r/n)(\mu^{(k)} - \hat{\mu}). \tag{1.53}$$

This establishes that $\mu^{(k)}$ converges to $\hat{\mu}$, as $k \rightarrow \infty$, provided $r < n$.

It can be seen for this problem that each EM iteration is linear. We shall see later that in general the rate of convergence of the EM algorithm is essentially linear. The rate of convergence here is $(1 - r/n)$, which is the proportion of censored observations in the observed sample. This proportion can be viewed as the missing information in the sample, as to be made more precise in Section 3.9.

1.5.3 E- and M-Steps for the Regular Exponential Family

The complete-data p.d.f. $g_c(\boldsymbol{x}; \boldsymbol{\Psi})$ is from an exponential family if

$$g_c(\boldsymbol{x}; \boldsymbol{\Psi}) = b(\boldsymbol{x}) \exp\{\boldsymbol{c}^T(\boldsymbol{\Psi})\boldsymbol{t}(\boldsymbol{x})\}/a(\boldsymbol{\Psi}), \tag{1.54}$$

where the sufficient statistic $\boldsymbol{t}(\boldsymbol{x})$ is a $k \times 1$ $(k \geq d)$ vector and $\boldsymbol{c}(\boldsymbol{\Psi})$ is a $k \times 1$ vector function of the $d \times 1$ parameter vector $\boldsymbol{\Psi}$, and $a(\boldsymbol{\Psi})$ and $b(\boldsymbol{x})$ are scalar functions. The parameter space $\boldsymbol{\Omega}$ is a d-dimensional convex set such that (1.54) defines a p.d.f. for all $\boldsymbol{\Psi}$ in $\boldsymbol{\Omega}$; that is,

$$\boldsymbol{\Omega} = \left\{ \boldsymbol{\Psi}: \int_{\mathcal{X}} b(\boldsymbol{x}) \exp\{\boldsymbol{c}^T(\boldsymbol{\Psi})\boldsymbol{t}(\boldsymbol{x})\} d\boldsymbol{x} < \infty \right\}. \tag{1.55}$$

If $k = d$ and the Jacobian of $\boldsymbol{c}(\boldsymbol{\Psi})$ is of full rank, then $g_c(\boldsymbol{x}; \boldsymbol{\Psi})$ is said to be from a regular exponential family. The coefficient $\boldsymbol{c}(\boldsymbol{\Psi})$ of the sufficient statistic $\boldsymbol{t}(\boldsymbol{x})$ in (1.54) is referred to as the natural or canonical parameter (vector). Thus if the complete-data p.d.f. $g_c(\boldsymbol{x}; \boldsymbol{\Psi})$ is from a regular exponential family in canonical form, then

$$g_c(\boldsymbol{x}; \boldsymbol{\Psi}) = b(\boldsymbol{x}) \exp\{\boldsymbol{\Psi}^T \boldsymbol{t}(\boldsymbol{x})\}/a(\boldsymbol{\Psi}). \tag{1.56}$$

The parameter $\boldsymbol{\Psi}$ in (1.56) is unique up to an arbitrary nonsingular $d \times d$ linear transformation, as is the corresponding choice of $\boldsymbol{t}(\boldsymbol{x})$.

The expectation of the sufficient statistic $\boldsymbol{t}(\boldsymbol{X})$ in (1.56) is given by

$$E_{\boldsymbol{\Psi}}\{\boldsymbol{t}(\boldsymbol{X})\} = \partial \log a(\boldsymbol{\Psi})/\partial \boldsymbol{\Psi}. \tag{1.57}$$

Another property of the regular exponential family, which we shall use in a later section, is that the expected information matrix for the natural parameter vector equals the covariance matrix of the sufficient statistic $\boldsymbol{t}(\boldsymbol{X})$. Thus we have for the regular exponential family in the canonical form (1.56) that

$$\text{cov}_{\boldsymbol{\Psi}}\{\boldsymbol{t}(\boldsymbol{X})\} = \mathcal{I}_c(\boldsymbol{\Psi}), \tag{1.58}$$

where since the second derivatives of (1.56) do not depend on the data,

$$\begin{aligned} \mathcal{I}_c(\boldsymbol{\Psi}) &= -\partial^2 \log L_c(\boldsymbol{\Psi})/\partial \boldsymbol{\Psi} \partial \boldsymbol{\Psi}^T \\ &= \partial^2 \log a(\boldsymbol{\Psi})/\partial \boldsymbol{\Psi} \partial \boldsymbol{\Psi}^T. \end{aligned} \tag{1.59}$$

On taking the conditional expectation of $\log L_c(\boldsymbol{\Psi})$ given $\boldsymbol{y}$, we have from (1.56) that $Q(\boldsymbol{\Psi}; \boldsymbol{\Psi}^{(k)})$ is given by, ignoring terms not involving $\boldsymbol{\Psi}$,

$$Q(\boldsymbol{\Psi}; \boldsymbol{\Psi}^{(k)}) = \boldsymbol{\Psi}^T \boldsymbol{t}^{(k)} - \log a(\boldsymbol{\Psi}), \tag{1.60}$$

where

$$\boldsymbol{t}^{(k)} = E_{\boldsymbol{\Psi}^{(k)}}\{\boldsymbol{t}(\boldsymbol{X}) \mid \boldsymbol{y}\}$$

and where $\boldsymbol{\Psi}^{(k)}$ denotes the current fit for $\boldsymbol{\Psi}$.

On differentiation of (1.60) with respect to $\boldsymbol{\Psi}$ and noting (1.57), it follows that the M-step requires $\boldsymbol{\Psi}^{(k+1)}$ to be chosen by solving the equation

$$E_{\boldsymbol{\Psi}}\{\boldsymbol{t}(\boldsymbol{X})\} = \boldsymbol{t}^{(k)}. \tag{1.61}$$

If equation (1.61) can be solved for $\boldsymbol{\Psi}^{(k+1)}$ in $\boldsymbol{\Omega}$, then the solution is unique due to the well-known convexity property of minus the log likelihood of the regular exponential family. In cases where the equation is not solvable, the maximizer $\boldsymbol{\Psi}^{(k+1)}$ of L($\boldsymbol{\Psi}$) lies on the boundary of $\boldsymbol{\Omega}$.

1.5.4 Example 1.4: Censored Exponentially Distributed Survival Times (Example 1.3 Continued)

We return now to Example 1.3. It can be seen that the complete-data distribution has the exponential family form (1.56) with natural parameter μ^{-1} and sufficient statistic

$$t(\boldsymbol{X}) = \sum_{j=1}^{n} W_j.$$

Hence the E-step requires the calculation of

$$\begin{aligned}
t^{(k)} &= E_{\boldsymbol{\Psi}^{(k)}}\{t(\boldsymbol{X}) \mid \boldsymbol{y}\} \\
&= \sum_{j=1}^{r} c_j + \sum_{j=r+1}^{n} (c_j + \mu^{(k)}) \\
&= \sum_{j=1}^{n} c_j + (n-r)\mu^{(k)}
\end{aligned}$$

from (1.50).

The M-step then yields $\mu^{(k+1)}$ as the value of μ that satisfies the equation

$$\begin{aligned}
t^{(k)} &= E_{\mu}\{t(\boldsymbol{X})\} \\
&= n\mu.
\end{aligned}$$

This latter equation can be seen to be equivalent to (1.52), as derived by direct differentiation of the Q-function $Q(\mu;\, \mu^{(k)})$.

1.5.5 Generalized EM Algorithm

Often in practice, the solution to the M-step exists in closed form. In those instances where it does not, it may not be feasible to attempt to find the value of $\boldsymbol{\Psi}$ that globally maximizes the function $Q(\boldsymbol{\Psi};\, \boldsymbol{\Psi}^{(k)})$. For such situations, DLR defined a generalized EM algorithm (GEM algorithm) for which the M-step requires $\boldsymbol{\Psi}^{(k+1)}$ to be chosen such that

$$Q(\boldsymbol{\Psi}^{(k+1)};\, \boldsymbol{\Psi}^{(k)}) \geq Q(\boldsymbol{\Psi}^{(k)};\, \boldsymbol{\Psi}^{(k)}) \tag{1.62}$$

holds. That is, one chooses $\boldsymbol{\Psi}^{(k+1)}$ to increase the Q-function $Q(\boldsymbol{\Psi};\, \boldsymbol{\Psi}^{(k)})$ over its value at $\boldsymbol{\Psi} = \boldsymbol{\Psi}^{(k)}$, rather than to maximize it over all $\boldsymbol{\Psi} \in \boldsymbol{\Omega}$. As to be shown in Section 3.3, the above condition on $\boldsymbol{\Psi}^{(k+1)}$ is sufficient to ensure that

$$L(\boldsymbol{\Psi}^{(k+1)}) \geq L(\boldsymbol{\Psi}^{(k)}).$$

Hence the likelihood $L(\boldsymbol{\Psi})$ is not decreased after a GEM iteration, and so a GEM sequence of likelihood values must converge if bounded above. In Section 3.3, we shall discuss what specifications are needed on the process of increasing the Q-function in order to ensure that the limit of $\{L(\boldsymbol{\Psi}^{(k)})\}$ is a stationary value and that the sequence of GEM iterates $\{\boldsymbol{\Psi}^{(k)}\}$ converges to a stationary point.

1.5.6 GEM Algorithm Based on One Newton–Raphson Step

In those situations where the global maximizer of the Q-function $Q(\boldsymbol{\Psi};\, \boldsymbol{\Psi}^{(k)})$ does not exist in closed form, consideration may be given to using the Newton–Raphson procedure to iteratively compute $\boldsymbol{\Psi}^{(k+1)}$ on the M-step. As remarked above, it is not essential that $\boldsymbol{\Psi}^{(k+1)}$ actually maximizes the Q-function for the likelihood to be increased. We can use a GEM algorithm where $\boldsymbol{\Psi}^{(k+1)}$ need satisfy only (1.62), which is a sufficient condition to guarantee the monotonicity of the sequence of likelihood values $\{L(\boldsymbol{\Psi}^{(k)}\}$. In some instances, the limiting value $\boldsymbol{\Psi}^{(k+1)}$ of the Newton–Raphson method may not be a global maximizer. But if condition (1.62) is confirmed to hold on each M-step, then at least the user knows that $\{\boldsymbol{\Psi}^{(k)}\}$ is a GEM sequence.

Following Wu (1983) and Jørgensen (1984), Rai and Matthews (1993) propose taking $\boldsymbol{\Psi}^{(k+1)}$ to be of the form

$$\boldsymbol{\Psi}^{(k+1)} = \boldsymbol{\Psi}^{(k)} + a^{(k)}\boldsymbol{\delta}^{(k)}, \tag{1.63}$$

where

$$\boldsymbol{\delta}^{(k)} = -[\partial^2 Q(\boldsymbol{\Psi};\, \boldsymbol{\Psi}^{(k)})/\partial\boldsymbol{\Psi}\partial\boldsymbol{\Psi}^T]^{-1}_{\boldsymbol{\Psi}=\boldsymbol{\Psi}^{(k)}}\,[\partial Q(\boldsymbol{\Psi};\, \boldsymbol{\Psi}^{(k)})/\partial\boldsymbol{\Psi}]_{\boldsymbol{\Psi}=\boldsymbol{\Psi}^{(k)}} \tag{1.64}$$

and where $0 < a^{(k)} \leq 1$.

It can be seen that in the case of $a^{(k)} = 1$, (1.63) is the first iterate obtained when using the Newton–Raphson procedure to obtain a root of the equation

$$\partial Q(\boldsymbol{\Psi};\, \boldsymbol{\Psi}^{(k)})/\partial\boldsymbol{\Psi} = \mathbf{0}.$$

The idea is to choose $a^{(k)}$ so that (1.64) defines a GEM sequence; that is, so that (1.62) holds. It can be shown that

$$Q(\boldsymbol{\Psi}^{(k+1)};\, \boldsymbol{\Psi}^{(k)}) - Q(\boldsymbol{\Psi}^{(k)};\, \boldsymbol{\Psi}^{(k)}) = a^{(k)}\boldsymbol{S}^T(\boldsymbol{y};\, \boldsymbol{\Psi}^{(k)})\boldsymbol{A}^{(k)}\boldsymbol{S}(\boldsymbol{y};\, \boldsymbol{\Psi}^{(k)}), \tag{1.65}$$

where

$$\boldsymbol{A}^{(k)} = \boldsymbol{\mathcal{I}}_c^{-1}(\boldsymbol{\Psi}^{(k)};\, \boldsymbol{y})\{\boldsymbol{I}_d - \tfrac{1}{2}a^{(k)}\tilde{\boldsymbol{\mathcal{I}}}_c^{(k)}(\boldsymbol{y})\boldsymbol{\mathcal{I}}_c^{-1}(\boldsymbol{\Psi}^{(k)};\, \boldsymbol{y})\} \tag{1.66}$$

and where

$$\begin{aligned}\tilde{\boldsymbol{\mathcal{I}}}_c^{(k)}(\boldsymbol{y}) &= -[\partial^2 Q(\boldsymbol{\Psi};\, \boldsymbol{\Psi}^{(k)})/\partial\boldsymbol{\Psi}\partial\boldsymbol{\Psi}^T]_{\boldsymbol{\Psi}=\tilde{\boldsymbol{\Psi}}^{(k)}} \\ &= E_{\boldsymbol{\Psi}^{(k)}}\{\boldsymbol{I}_c(\tilde{\boldsymbol{\Psi}}^{(k)};\, \boldsymbol{X}) \mid \boldsymbol{y}\},\end{aligned}$$

and $\tilde{\boldsymbol{\Psi}}^{(k)}$ is a point on the line segment from $\boldsymbol{\Psi}^{(k)}$ to $\boldsymbol{\Psi}^{(k+1)}$; $\boldsymbol{I}_d$ denotes the $d \times d$ identity matrix. Thus the left-hand side of (1.65) is nonnegative if the matrix $\boldsymbol{A}^{(k)}$ is positive definite.

Typically in practice, $\boldsymbol{\mathcal{I}}_c(\boldsymbol{\Psi}^{(k)};\, \boldsymbol{y})$ is positive definite and so then we have a GEM sequence if the matrix

$$\boldsymbol{I}_d - \tfrac{1}{2}a^{(k)}\tilde{\boldsymbol{\mathcal{I}}}_c^{(k)}(\boldsymbol{y})\boldsymbol{\mathcal{I}}_c^{-1}(\boldsymbol{\Psi}^{(k)};\, \boldsymbol{y}) \tag{1.67}$$

is positive definite, which can be achieved by choosing the constant $a^{(k)}$ sufficiently small. Suppose that the sequence $\{\boldsymbol{\Psi}^{(k)}\}$ tends to some limit point as

$k \to \infty$. Then it can be seen from (1.66) that, as $k \to \infty$, $a^{(k)} < 2$ will ensure that (1.67) holds.

The derivation of (1.65) is to be given in Section 4.11, where the use of this GEM algorithm in an attempt to reduce the computation on the M-step is to be considered further.

1.5.7 EM Gradient Algorithm

The algorithm that uses one Newton–Raphson step to approximate the M-step of the EM algorithm [that is, uses (1.63) with $a^{(k)} = 1$] is referred to by Lange (1995a) as the EM gradient algorithm. It forms the basis of the quasi-Newton approach of Lange (1995b) to speed up the convergence of the EM algorithm, as to be considered in Section 4.13. But as pointed out by Lange (1995b), it is an interesting algorithm in its own right, and is to be considered further in Section 4.12.

1.5.8 EM Mapping

Any instance of the EM (GEM) algorithm as described above implicitly defines a mapping $\boldsymbol{\Psi} \to \boldsymbol{M}(\boldsymbol{\Psi})$, from the parameter space of $\boldsymbol{\Psi}$, $\boldsymbol{\Omega}$, to itself such that

$$\boldsymbol{\Psi}^{(k+1)} = \boldsymbol{M}(\boldsymbol{\Psi}^{(k)}) \quad (k = 0, 1, 2, \ldots). \tag{1.68}$$

If $\boldsymbol{\Psi}^{(k)}$ converges to some point $\boldsymbol{\Psi}^*$ and $\boldsymbol{M}(\boldsymbol{\Psi})$ is continuous, then $\boldsymbol{\Psi}^*$ must satisfy

$$\boldsymbol{\Psi}^* = \boldsymbol{M}(\boldsymbol{\Psi}^*).$$

Thus $\boldsymbol{\Psi}^*$ is a fixed point of the map $\boldsymbol{M}$.

It is easy to show that if the MLE $\hat{\boldsymbol{\Psi}}$ of $\boldsymbol{\Psi}$ is the unique global maximizer of the likelihood function, then it is a fixed point of the EM algorithm (although there is no guarantee that it is the only one). To see this, we note that the M-step of the EM algorithm (or a GEM algorithm) implies that

$$L(\boldsymbol{M}(\hat{\boldsymbol{\Psi}})) \geq L(\hat{\boldsymbol{\Psi}}). \tag{1.69}$$

Thus $\boldsymbol{M}(\hat{\boldsymbol{\Psi}}) = \hat{\boldsymbol{\Psi}}$, as otherwise (1.69) would contradict the assertion that

$$L(\hat{\boldsymbol{\Psi}}) > L(\boldsymbol{\Psi})$$

for all $\boldsymbol{\Psi}$ (not equal to $\hat{\boldsymbol{\Psi}}$) $\in \boldsymbol{\Omega}$.

1.6 EM ALGORITHM FOR MAXIMUM *A POSTERIORI* AND MAXIMUM PENALIZED LIKELIHOOD ESTIMATION

1.6.1 Maximum *a Posteriori* Estimation

The EM algorithm is easily modified to produce the maximum *a posteriori* (MAP) estimate or the maximum penalized likelihood estimate (MPLE) in incomplete-data problems. We consider first the computation of the MAP estimate in a Bayesian framework via the EM algorithm, corresponding to some prior density $p(\boldsymbol{\Psi})$ for $\boldsymbol{\Psi}$. We let the incomplete- and complete-data posterior densities for $\boldsymbol{\Psi}$ be given by $p(\boldsymbol{\Psi} \mid \boldsymbol{y})$ and $p(\boldsymbol{\Psi} \mid \boldsymbol{x})$, respectively. Then the MAP estimate of $\boldsymbol{\Psi}$ is the value of $\boldsymbol{\Psi}$ that maximizes the log (incomplete-data) posterior density which, on ignoring an additive term not involving $\boldsymbol{\Psi}$, is given by

$$\log p(\boldsymbol{\Psi} \mid \boldsymbol{y}) = \log L(\boldsymbol{\Psi}) + \log p(\boldsymbol{\Psi}). \tag{1.70}$$

Here $p(\cdot)$ is being used as a generic symbol for a p.d.f.

The EM algorithm is implemented as follows to compute the MAP estimate.

E-Step. On the $(k+1)$th iteration, calculate the conditional expectation of the log complete-data posterior density given the observed data vector $\boldsymbol{y}$, using the current MAP estimate $\boldsymbol{\Psi}^{(k)}$ of $\boldsymbol{\Psi}$. That is, calculate

$$E_{\boldsymbol{\Psi}^{(k)}}\{\log p(\boldsymbol{\Psi} \mid \boldsymbol{x}) \mid \boldsymbol{y}\} = Q(\boldsymbol{\Psi}; \boldsymbol{\Psi}^{(k)}) + \log p(\boldsymbol{\Psi}). \tag{1.71}$$

M-Step. Choose $\boldsymbol{\Psi}^{(k+1)}$ to maximize (1.71) over $\boldsymbol{\Psi} \in \boldsymbol{\Omega}$.

It can be seen that the E-step is effectively the same as for the computation of the MLE of $\boldsymbol{\Psi}$ in a frequentist framework, requiring the calculation of the Q-function, $Q(\boldsymbol{\Psi}; \boldsymbol{\Psi}^{(k)})$. The M-step differs in that the objective function for the maximization process is equal to $Q(\boldsymbol{\Psi}; \boldsymbol{\Psi}^{(k)})$ augmented by the log prior density, $\log p(\boldsymbol{\Psi})$. The presence of this latter term as the result of the imposition of a Bayesian prior for $\boldsymbol{\Psi}$ almost always makes the objective function more concave.

1.6.2 Example 1.5: A Multinomial Example (Example 1.1 Continued)

We now discuss a Bayesian version of Example 1.1. As before with this example in considering the conditional distribution of the complete-data vector $\boldsymbol{x}$ given the observed data vector $\boldsymbol{y}$, we can effectively work with the conditional distribution of the missing data vector $\boldsymbol{Z}$ given $\boldsymbol{y}$. We choose the prior distribution of Ψ to be the beta (ν_1, ν_2) distribution with density

$$p(\Psi) = \frac{\Gamma(\nu_1 + \nu_2)}{\Gamma(\nu_1)\Gamma(\nu_2)} \Psi^{\nu_1 - 1}(1 - \Psi)^{\nu_2 - 1}, \tag{1.72}$$

which is a natural conjugate of the conditional predictive distribution of the missing data. The latter is binomial with sample size y_1 and probability parameter

$$\frac{\frac{1}{4}\Psi}{\frac{1}{2} + \frac{1}{4}\Psi}.$$

From (1.22) and (1.72), we have that

$$\log p(\boldsymbol{\Psi} \mid \boldsymbol{x}) = \log L(\boldsymbol{\Psi}) + \log p(\boldsymbol{\Psi})$$
$$= (y_{12} + y_4 + \nu_1 - 1) \log \boldsymbol{\Psi} + (y_2 + y_3 + \nu_2 - 1) \log(1 - \boldsymbol{\Psi}), \tag{1.73}$$

apart from an additive constant. It can be seen from (1.73) that $p(\boldsymbol{\Psi} \mid \boldsymbol{x})$ has the beta form. The E-step is effected the same as in the computation of the MLE of $\boldsymbol{\Psi}$ in Example 1.1, with y_{12} in (1.22) replaced by its current conditional expectation

$$y_1 \frac{\frac{1}{4}\Psi^{(k)}}{\frac{1}{2} + \frac{1}{4}\Psi^{(k)}}.$$

On the M-step, the $(k+1)$th iterate for the MAP estimate is given by

$$\Psi^{(k+1)} = \frac{y_{12}^{(k)} + y_4 + \nu_1 - 1}{y_{12}^{(k)} + y_4 + y_2 + y_3 + \nu_1 + \nu_2 - 2}.$$

Note that this beta distribution is uniform on [0,1] when $\nu_1 = \nu_2 = 1$. When the prior is uniform, the MAP estimate is the same as the the MLE of Ψ in the frequentist framework. We return to related techniques for Bayesian estimation in Chapter 6.

1.6.3 Maximum Penalized Likelihood Estimation

In the case of MPL estimation, $\log p(\boldsymbol{\Psi})$ in (1.70) is taken to have the form

$$\log p(\boldsymbol{\Psi}) = -\xi K(\boldsymbol{\Psi}), \tag{1.74}$$

where $K(\boldsymbol{\Psi})$ is a roughness penalty and ξ is a smoothing parameter. Often $K(\boldsymbol{\Psi})$ is of the form

$$K(\boldsymbol{\Psi}) = \boldsymbol{\Psi}^T \boldsymbol{A} \boldsymbol{\Psi}.$$

For instance in ridge regression, $\boldsymbol{A} = \boldsymbol{I}_d$.

The EM algorithm can be applied in the same manner as above to compute the MPLE of $\boldsymbol{\Psi}$, where now ξ is an additional parameter to be estimated along with $\boldsymbol{\Psi}$. In Section 5.14, we consider a modified version of the EM algorithm, the one-step-late (OSL) algorithm as proposed by Green (1990b), that facilitates the computation of the MPLE.

1.7 BRIEF SUMMARY OF THE PROPERTIES OF THE EM ALGORITHM

In the following chapters, we shall discuss in detail various properties of the EM algorithm and some computational aspects of it, and we shall present a number of illustrations and applications of the algorithm. However, in order to give the reader a quick idea of the algorithm's potential as a useful tool in statistical estimation problems, we summarize here the reasons for its appeal. We also mention some of the criticisms leveled against the algorithm.

The EM algorithm has several appealing properties relative to other iterative algorithms such as Newton–Raphson and Fisher's scoring method for finding MLEs. Some of its advantages compared to its competitors are as follows:

1. The EM algorithm is numerically stable with each EM iteration increasing the likelihood (except at a fixed point of the algorithm).
2. Under fairly general conditions, the EM algorithm has reliable global convergence. That is, starting from an arbitrary point $\boldsymbol{\Psi}^{(0)}$ in the parameter space, convergence is nearly always to a local maximizer, barring very bad luck in the choice of $\boldsymbol{\Psi}^{(0)}$ or some local pathology in the log likelihood function.
3. The EM algorithm is typically easily implemented, because it relies on complete-data computations: the E-step of each iteration only involves taking expectations over complete-data conditional distributions and the M-step of each iteration only requires complete-data ML estimation, which is often in simple closed form.
4. The EM algorithm is generally easy to program, since no evaluation of the likelihood nor its derivatives is involved.
5. The EM algorithm requires small storage space and can generally be carried out on a small computer. For instance, it does not have to store the information matrix nor its inverse at any iteration.
6. Since the complete-data problem is likely to be a standard one, the M-step can often be carried out using standard statistical packages in situations where the complete-data MLEs do not exist in closed form. In other such situations, extensions of the EM algorithm such as the GEM and expectation–conditional maximization (ECM) algorithms often enable the M-step to be implemented iteratively in a fairly simple manner. Moreover, these extensions share the stable monotone convergence of the EM algorithm.
7. The analytical work required is much simpler than with other methods since only the conditional expectation of the log likelihood for the complete-data problem needs to be maximized. Although a certain amount of analytical work may be needed to carry out the E-step, it is not complicated in many applications.
8. The cost per iteration is generally low, which can offset the larger number of iterations needed for the EM algorithm compared to other competing procedures.
9. By watching the monotone increase in likelihood (if evaluated easily) over iterations, it is easy to monitor convergence and programming errors.

10. The EM algorithm can be used to provide estimated values of the 'missing' data.

Some of the criticisms of the EM algorithm are as follows:

1. Unlike Fisher's scoring method, it does not have a built-in procedure for producing an estimate of the covariance matrix of the parameter estimates. However, as to be discussed in Section 1.8 and to be pursued further in Chapter 4, this disadvantage can easily be removed by using appropriate methodology associated with the EM algorithm.
2. The EM algorithm may converge slowly even in some seemingly innocuous problems and in problems where there is too much 'incomplete information'.
3. The EM algorithm like the Newton-type methods does not guarantee convergence to the global maximum when there are multiple maxima. Furthermore, in this case, the estimate obtained depends upon the initial value. But, in general, no optimization algorithm is guaranteed to converge to a global or local maximum, and the EM algorithm is not magical in this regard. There are other procedures such as simulated annealing to tackle such situations. However, these are complicated to apply.
4. In some problems, the E-step may be analytically intractable, although in such situations there is the possibility of effecting it via a Monte Carlo approach, as to be discussed in Section 6.2.

1.8 HISTORY OF THE EM ALGORITHM

1.8.1 Work Before Dempster, Laird, and Rubin (1977)

The statistical literature is strewn with methods in the spirit of the EM algorithm or which are actually EM algorithms in special contexts. The formulation of the EM algorithm in its present generality is due to DLR, who also give a variety of examples of its applicability and establish its convergence and other basic properties under fairly general conditions. They identify the common thread in a number of algorithms, formulate it in a general setting, and show how some of these algorithms for special problems are special cases of their EM algorithm. They also point out new applications of the algorithm. In this subsection, we give an account of a few algorithms for special contexts, which have preceded the general formulation of DLR.

The earliest reference to literature on an EM-type of algorithm is Newcomb (1886), who considers estimation of parameters of a mixture of two univariate normals. McKendrick (1926) gives a medical application of a method in the spirit of the EM algorithm. Healy and Westmacott (1956) propose an iterative method for estimating a missing value in a randomized block design, which turns out to be an example of the EM algorithm, as to be pointed out in Examples 2.2 and 2.3 in Section 2.3.

Hartley (1958) gives a treatment of the general case of count data and enunciates the basic ideas of the EM algorithm. Buck (1960) considers the estimation of the mean vector and the covariance matrix of a p-dimensional population when some observations on all the p variables together with some observations on some of the variables only are available. He suggests a method of imputation of the missing values by regressing a missing variable on the observed variables in each case, using only cases for which observations on all variables are available. Then the parameters are estimated from the observations 'completed' in this way together with a 'correction' for the covariance matrix elements. The interesting aspect of Buck's (1960) method is that all the required regressions and correction terms can be computed by a single operation of inverting the information matrix based on the complete observations and suitable pivoting and sweeping operations. Buck's (1960) procedure gives the MLE under certain conditions. It also has the basic elements of the EM algorithm. Blight (1970) considers the problem of finding the MLEs of the parameters of an exponential family from a Type I censored sample and derives the likelihood equation. By suitably interpreting the likelihood equation, he derives an iterative method for its solution, which turns out to be an EM algorithm. He also obtains some convergence results and derives the asymptotic covariance matrix of the estimator.

In a series of papers, Baum and Petrie (1966), Baum and Eagon (1967), and Baum, Petrie, Soules, and Weiss (1970) deal with an application of the EM algorithm in a Markov model; these papers contain some convergence results, which generalize easily. Further, the algorithm developed here is the basis of present-day EM algorithms used in hidden Markov models. Orchard and Woodbury (1972) introduce the Missing Information principle, which is very much related to the spirit and basic ideas of the EM algorithm and note the general applicability of the principle. The relationship between the complete- and incomplete-data log likelihood functions established by them, leads to the fact that the MLE is a fixed point of a certain transform. This fact is also noted in many different contexts by a number of authors, who exploit it to develop quite a few forerunners to the EM algorithm. Carter and Myers (1973) consider a special type of mixture of discrete distributions, for which in the case of partially classified data, no closed-form solution exists for the MLE. They work out the likelihood equation and derive an algorithm to solve it using Hartley's (1958) method; this algorithm turns out to be an EM algorithm. Chen and Fienberg (1974) consider the problem of a two-way contingency table with some units classified both ways and some classified by only one of the ways and derive an algorithm for computation of the MLEs of the cell probabilities, which turns out to be an EM algorithm. Haberman (1974) also deals with the application of an EM-type algorithm in contingency tables with partially classified data. Efron (1967) introduces the so-called Self-Consistency principle in a nonparametric setting for a wide class of incomplete-data problems as an intuitive analog of the ML principle and introduces the Self-Consistency algorithm for right-censored problems.

Turnbull (1974) deals with nonparametric estimation of a survivorship function from doubly censored data based on the idea of self-consistency due to Efron (1967) and derives an iterative procedure. Turnbull (1976) deals with the empirical distribution with arbitrarily grouped, censored, and truncated data, derives a version of the EM algorithm, and notes that not only actually missing-data problems, but also problems such as with truncated data, can be treated as incomplete-data problems; he calls individuals who are never observed "ghosts". Thus Turnbull (1974) extends Efron's (1967) idea and shows its equivalence with the nonparametric likelihood equations for these problems. He also proves convergence of EM-like algorithms for these problems. Prior to the appearance of the DLR paper, the resolution of mixtures of distributions for a variety of situations and distributional families gave rise to a number of algorithms that can be regarded as particular applications of the EM algorithm. These papers, which are surveyed in McLachlan (1982) and McLachlan and Basford (1988), include the seminal paper of Day (1969). Also, McLachlan (1975, 1977) proposes an iterative method for forming the (normal theory-based) linear discriminant function from partially classified training data.

The basic idea of the EM algorithm is also in use in the "gene-counting method" used by geneticists in the estimation of ABO blood group gene frequencies and other genetic problems (Ceppellini, Siniscalco, and Smith, 1955; Smith, 1957), as noted in Example 2.4 of Section 2.4.

Sundberg (1974, 1976) deals with properties of the likelihood equation in the general context of incomplete-data problems from exponential families, and arrives at special forms for the likelihood equation and the information matrix, which have come to be known as Sundberg formulas. Sundberg (1976) acknowledges that his key "iteration mapping", which corresponds to the EM mapping of DLR, was suggested to him by A. Martin-Löf in a personal communication in 1967. Beale and Little (1975) develop an algorithm and the associated theory for the multivariate normal case with incomplete data.

All this work was done before DLR formulated the problem in its generality. Indeed, there are many other algorithms found in the literature before DLR which are in the spirit of the EM algorithm or are actually EM algorithms in special contexts. We have not mentioned them all here.

1.8.2 EM Examples and Applications Since Dempster, Laird, and Rubin (1977)

After the general formulation of the EM algorithm by DLR, some well-known ML estimation methods in various contexts have been shown to be EM algorithms, in DLR itself and by others. Iteratively Reweighted Least Squares (IRLS) is an iterative procedure for estimating regression coefficients, wherein each iteration is a weighted least-squares procedure with the weights changing with the iterations. It is applied to robust regression, where the estimates obtained are MLEs under suitable distributional assumptions. Dempster, Laird, and Rubin (1980) show that the IRLS procedure is an EM algorithm under distributional assumptions. Again, the well-known and standard method of estimating variance components in a mixed model, Henderson's algorithm, is shown to be an EM-type algorithm by Laird (1982). Gill (1989) observes that, in such missing-value problems, the score functions of suitably chosen parametric submodels coincide exactly with the self-consistency equations and also have an interpretation in terms of the EM algorithm. Wolynetz (1979a, 1979b, 1980) deals with the case of confined and censored data and derives the ML regression line under the assumption that residuals of the dependent variables are normally distributed, using the EM algorithm. It turns out that this line is the same as the "iterative least-squares" line derived by Schmee and Hahn (1979) in an industrial context and the "detections and bounds" regression developed by Avni and Tananbaum (1986) in the context of a problem in astronomy.

The EM algorithm has been applied to neural networks with hidden units to derive training algorithms; in the M-step, this involves a version of the iterative proportional fitting algorithm for multiway contingency tables (Byrne, 1992; Cheng and Titterington, 1994). Csiszár and Tusnády (1984), Amari, Kurata, and Nagaoka (1992), Byrne (1992), and Amari (1995a, 1995b) explore the connection between the EM algorithm and information geometry. It is pointed out that the EM algorithm is useful in the learning of hidden units in a Boltzmann machine and that the steps of the EM algorithm correspond to the e-geodesic and m-geodesic projections in a manifold of probability distributions, in the sense of statistical inference and differential geometry. The EM algorithm is also useful in the estimation of parameters in hidden Markov models, which are applicable in speech recognition (Rabiner, 1989) and image processing applications (Besag, 1986); these models can be viewed as more general versions of the classical mixture-resolution problems for which the EM algorithm has already become a standard tool (Titterington, 1990; Qian and Titterington, 1991).

The DLR paper proved to be a timely catalyst for further research into the applications of finite mixture models. This is witnessed by the subsequent stream of papers on finite mixtures in the literature, commencing with, for example, Ganesalingam and McLachlan (1978, 1979). As Aitkin and Aitkin (1994) note, almost all the post-1978 applications of mixture modeling reported in the books on mixtures by Titterington, Smith, and Makov (1985) and McLachlan and Basford (1988), use the EM algorithm.

1.8.3 Two Interpretations of EM

In the innumerable independent derivations of the EM algorithm for special problems, especially the various versions of the mixture-resolution problem, two interpretations are discernible:

1. The EM algorithm arises naturally from the particular forms taken by the derivatives of the log likelihood function. Various authors have arrived at the EM algorithm in special cases, while attempting to manipulate the likelihood equations to be able to solve them in an elegant manner.
2. Many a problem for which the MLE is complex, can be viewed as an incomplete-data problem with a corresponding complete-data problem, suitably formulated, so that the log likelihood functions of the two problems have a nice connection, which can be exploited to arrive at the EM algorithm.

The first interpretation is reflected in the following studies, all of which are on mixture resolution and which preceded DLR. Finite mixtures of univariate normal distributions are treated by Hasselblad (1966) and Behboodian (1970), arbitrary finite mixtures by Hasselblad (1969), mixtures of two multivariate normal distributions with a common covariance matrix by Day (1969), and mixtures of multivariate normal distributions with arbitrary covariance matrices by Wolfe (1967, 1970). This interpretation is also reflected in Blight (1970), who considers exponential families under Type I censoring, Tan and Chang (1972), who consider a mixture problem in genetics, in Hosmer (1973a) who carries out Monte Carlo studies on small sample sizes with mixtures of two normal distributions, in Hosmer (1973b) who extends his earlier results to the case of a partially classified sample, in the book by Duda and Hart (1973) where the use of multivariate normal mixtures in unsupervised pattern recognition is considered, in Hartley (1978) where a "switching regression" model is considered, and in Peters and Coberly (1976) who consider ML estimation of the proportions in a mixture.

The second interpretation is reflected in the works of Orchard and Woodbury (1972), who were the first to formulate the 'missing information principle' and to apply it in various problems, in Healy and Westmacott (1956) and other works on missing values in designed experiments, in Buck (1960) on the estimation of the mean and the covariance matrix of a random vector, in Baum et al. (1970) who consider the general mixture density estimation problem, in Hartley and Hocking (1971) who consider the general problem of analysis of incomplete data, in Haberman (1974, 1976, 1977) who considers log linear models for frequency tables derived by direct and indirect observations, iteratively reweighted least-squares estimation, and product models, and in the works of Ceppellini et al. (1955), Chen (1972), Goodman (1974), and Thompson (1975).

1.8.4 Developments in EM Theory, Methodology, and Applications

Dempster, Laird, and Rubin (1977) establish important fundamental properties of the algorithm. In particular, these properties imply that typically in practice the sequence of EM iterates will converge to a local maximizer of the log likelihood function $\log L(\boldsymbol{\Psi})$. If $L(\boldsymbol{\Psi})$ is unimodal in $\boldsymbol{\Omega}$ with $\boldsymbol{\Psi}^*$ being the only stationary point of $L(\boldsymbol{\Psi})$, then for any EM sequence $\{\boldsymbol{\Psi}^{(k)}\}$, $\boldsymbol{\Psi}^{(k)}$ converges to the unique maximizer $\boldsymbol{\Psi}^*$ of $L(\boldsymbol{\Psi})$. In general, if $\log L(\boldsymbol{\Psi})$ has several (local or global) maxima and stationary values, convergence of the EM sequence to either type depends on the choice of starting point. Furthermore, DLR show that convergence is linear with the rate of convergence proportional to λ_{max}, where λ_{max} is the maximal fraction of missing information. This implies that the EM algorithm can be very slow to converge, but then the intermediate values do provide very valuable statistical information. This also implies that the choice of the complete-data problem can influence the rate of convergence, since this choice will determine λ_{max}.

There have been quite a few developments in the methodology of the EM algorithm since DLR. Wu (1983) gives a detailed account of the convergence properties of the EM algorithm, addressing, in particular, the problem that the convergence of $L(\boldsymbol{\Psi}^{(k)})$ to L^* does not automatically imply the convergence of $\boldsymbol{\Psi}^{(k)}$ to a point $\boldsymbol{\Psi}^*$. On this same matter, Boyles (1983) presents an example of a generalized EM sequence that converges to the circle of the unit radius and not to a single point. Horng (1986, 1987) presents many interesting examples of sublinear convergence of the EM algorithm. Lansky, Casella, McCulloch, and Lansky (1992) establish some invariance, convergence, and rates of convergence results. The convergence properties of the EM algorithm are to be pursued further in Section 3.4.

In a series of papers, Turnbull and Mitchell (1978, 1984) and Mitchell and Turnbull (1979) discuss nonparametric ML estimation in survival/sacrifice experiments and show its self-consistency and convergence. Laird (1978) deals with nonparametric ML estimation of a mixing distribution and points out the equivalence of the Self-Consistency principle and Orchard and Woodbury's (1972) Missing Information principle. Laird (1978) also shows that in the case of parametric exponential families, these two principles have the same mathematical basis of the Sundberg formulas. She also establishes that the self-consistency algorithm is a special case of the EM algorithm.

One of the initial criticisms of the EM algorithm was that unlike Newton-type methods, it does not automatically produce an estimate of the covariance matrix of the MLE. In an important development associated with the EM methodology, Louis (1982) develops a method of finding the observed information matrix while using the EM algorithm, which is generally applicable. This method gives the observed information matrix in terms of the gradient and curvature of the complete-data log likelihood function, which is more amenable to analytical calculations than the incomplete-data analog. Fisher (1925) had observed the result that the incomplete-data score statistic is the conditional expected value of the complete-data score statistic given the incomplete observations (observed data). Efron (1977) in his comments on DLR connects Fisher's result with incompleteness. Louis (1982) makes this connection deeper by establishing it for the second moments. In other related work, Meilijson (1989) proposes a method of numerically computing the covariance matrix of the MLE, using the ingredients computed in the E- and M-steps of the algorithm, as well as a method to speed up convergence. His method of approximation avoids having to calculate second-order derivatives as with Louis' method. Meilijson (1989) shows that the single-observation scores for the incomplete-data model are obtainable as a by-product of the E-step. He also notes that the expected information matrix can be estimated consistently by the empirical covariance matrix of the individual scores. Previously, Redner and Walker (1984) in the context of the mixture resolution problem, suggested using the empirical covariance matrix of the individual scores to estimate consistently the expected information matrix.

Louis (1982) also suggests a method of speeding up convergence of the EM algorithm using the multivariate generalization of the Aitken acceleration procedure. The resulting algorithm is essentially equivalent to using the Newton–Raphson method to find a zero of the (incomplete-data) score statistic. Recently, Jamshidian and Jennrich (1993) use a generalized conjugate gradient approach to accelerate convergence of the EM algorithm. However, attempts to speed up the EM algorithm do reduce its simplicity and there is no longer any guarantee that the likelihood will always increase from one iteration to the next. These points are to be taken up in Chapter 4.

The use of the empirical information matrix as discussed above is of course applicable only in the special case of i.i.d. data. For the general case, Meng and Rubin (1991) define a procedure that obtains a numerically stable estimate of the asymptotic covariance matrix of the EM-computed estimate, using only the code for computing the complete-data covariance matrix, the code for the EM algorithm itself, and the code for standard matrix operations. In particular, neither likelihoods, nor partial derivatives of likelihoods nor log likelihoods need to be evaluated. They refer to this extension of the EM algorithm as the Supplemented EM algorithm.

As noted earlier, one of the major reasons for the popularity of the EM algorithm is that the M-step involves only complete-data ML estimation, which is often computationally simple. But, if the complete-data ML estimation is rather complicated, then the EM algorithm is less attractive because the M-step is computationally unattractive. In many cases, however, complete-data ML estimation is relatively simple if maximization is undertaken conditional on some of the parameters (or some functions of the parameters). To this end, Meng and Rubin (1993) introduce a class of generalized EM algorithms, which they call the Expectation–Conditional Maximization (ECM) algorithm. The ECM algorithm takes advantage of the simplicity of complete-data conditional maximization by replacing a complicated M-step of the EM algorithm with several computationally simpler CM-steps. Each of these CM-steps maximizes the expected complete-data log likelihood function found in the preceding E-step subject to constraints on $\boldsymbol{\Psi}$, where the collection of all constraints is such that the maximization is over the full parameter space of $\boldsymbol{\Psi}$. More recently, Liu and Rubin (1994) give a generalization of the ECM algorithm that replaces some of the CM-steps with steps that maximize the constrained actual (incomplete-data) log likelihood. They call this algorithm, the Expectation–Conditional Maximization Either (ECME) algorithm. It shares with both the EM and ECM algorithms, their stable monotone convergence and basic simplicity of implementation relative to faster converging competitors. In a further extension, Meng and van Dyk (1995) propose generalizing the ECME algorithm and the SAGE algorithm of Fessler and Hero (1994) by combining them into one algorithm, called the Alternating ECM (AECM) algorithm. It allows the specification of the complete data to be different on each CM-step.

Meng and van Dyk (1995) also consider the problem of speeding up convergence of the EM algorithm. Their approach is through the choice of the missing data in the specification of the complete-data problem in the EM framework. They introduce a working parameter in the specification of the complete data, which thus indexes a class of EM algorithms. The aim is to select a value of the working parameter that increases the speed of convergence without appreciably affecting the stability and simplicity of the resulting EM algorithm.

In other recent developments, there is the work of Lange (1995a, 1995b) on the use of the EM gradient algorithm in situations where the solution to the M-step does not exist in closed form. As discussed in Section 1.5.7 and to be considered further in Chapter 4, the EM gradient algorithm approximates the M-step by one Newton–Raphson step. Lange (1995b) subsequently uses the EM gradient algorithm to form the basis of a quasi-Newton approach to accelerate convergence of the EM algorithm. In another recent development, Heyde and Morton (1996) extend the EM algorithm to deal with estimation via general estimating functions and in particular the quasi-score.

1.9 OVERVIEW OF THE BOOK

In Chapter 2, the EM methodology presented in this chapter is illustrated in some commonly occurring situations such as missing observations in multivariate normal data sets and regression problems, the multinomial distribution with complex cell-probability structure, grouped data, data from truncated distributions, and the fitting of finite mixture models.

The basic theory of the EM algorithm is presented in Chapter 3. In particular, the convergence properties and the rates of convergence are systematically examined. Consideration is given also to the associated Missing Information principle.

In Chapter 4, two important issues associated with the use of the EM algorithm are considered, namely the provision of standard errors and the speeding up of its convergence. In so doing, we discuss several of the many modifications, extensions, and alternatives to the EM methodology that appear in the literature.

We discuss further modifications and extensions to the EM algorithm in Chapter 5. In particular, the extensions of the EM algorithm known as the the smoothed EM, ECM, multicycle ECM, ECME, and AECM algorithms are given. We also present the EM gradient algorithm and a consequent quasi-Newton approach to accelerate its convergence. Having presented the easier illustrations in Chapter 2, the more difficult problems that motivated the development of these extensions are illustrated in Chapter 5, including estimation for variance components, repeated-measures designs, and factor analysis.

In Chapter 6, further variants and extensions of the EM algorithm are considered, such as the Stochastic EM and Monte Carlo EM algorithms. Some comparisons of the EM algorithm with these methods and alternatives such as simulated annealing are also presented. The relationship of the EM algorithm to other data augmentation techniques, such as the Gibbs sampler and other Markov Chain Monte Carlo techniques, is considered. Since this area is still developing, only a short account of these developments is presented, to prepare the reader for following the ongoing developments. We conclude the book with summaries of some interesting and topical applications of the EM algorithm.

1.10 NOTATION

We now define the notation that is used consistently throughout the book. Less frequently used notation will be defined later when it is first introduced.

All vectors and matrices are in boldface. The superscript T denotes the transpose of a vector or matrix. The trace of a matrix $\boldsymbol{A}$ is denoted by $tr(\boldsymbol{A})$, while the determinant of $\boldsymbol{A}$ is denoted by $|\boldsymbol{A}|$. The null vector is denoted by $\boldsymbol{0}$. The notation $\text{diag}(a_1, \ldots, a_n)$ is used for a matrix with diagonal elements $a_1, \ldots, a_n$ and all off-diagonal elements zero.

The vector $\boldsymbol{x}$ is used to represent the so-called complete data, while the vector $\boldsymbol{y}$ represents the actual observed data (incomplete data). Where possible, a random vector is represented by the corresponding upper case of the letter used for a particular realization. In this instance, $\boldsymbol{X}$ and $\boldsymbol{Y}$ denote the complete- and incomplete-data random vectors corresponding to $\boldsymbol{x}$ and $\boldsymbol{y}$, respectively.

The incomplete-data random vector $\boldsymbol{Y}$ is taken to be of p-dimensions, having probability density function (p.d.f.) $g(\boldsymbol{y}; \boldsymbol{\Psi})$ on $\mathbb{R}^p$, where

$$\boldsymbol{\Psi} = (\Psi_1, \ldots, \Psi_d)^T$$

is the vector containing the unknown parameters in the postulated form for the p.d.f. of $\boldsymbol{Y}$. The parameter space is denoted by $\boldsymbol{\Omega}$. In the case where $\boldsymbol{Y}$ is discrete, we can still view $g(\boldsymbol{y}; \boldsymbol{\Psi})$ as a density by the adoption of counting measure.

The likelihood function for $\boldsymbol{\Psi}$ formed from the observed data $\boldsymbol{y}$ is denoted by

$$L(\boldsymbol{\Psi}) = g(\boldsymbol{y}; \boldsymbol{\Psi}),$$

while $\log L(\boldsymbol{\Psi})$ denotes the log likelihood function.

The p.d.f. of the complete-data vector $\boldsymbol{X}$ is denoted by $g_c(\boldsymbol{x}; \boldsymbol{\Psi})$, with

$$L_c(\boldsymbol{\Psi}) = g_c(\boldsymbol{x}; \boldsymbol{\Psi})$$

denoting the complete-data likelihood function for $\boldsymbol{\Psi}$ that could be formed from $\boldsymbol{x}$ if it were completely observable.

The (incomplete-data) score statistic is given by

$$S(\boldsymbol{y}; \boldsymbol{\Psi}) = \partial \log L(\boldsymbol{\Psi})/\partial \boldsymbol{\Psi},$$

while

$$S_c(\boldsymbol{x}; \boldsymbol{\Psi}) = \partial \log L_c(\boldsymbol{\Psi})/\partial \boldsymbol{\Psi}$$

denotes the corresponding complete-data score statistic.

The conditional p.d.f. of $\boldsymbol{X}$ given $\boldsymbol{y}$ is denoted by

$$k(\boldsymbol{x} \mid \boldsymbol{y}; \boldsymbol{\Psi}).$$

The so-called missing data is represented by the vector $\boldsymbol{z}$.

The sequence of EM iterates is denoted by $\{\boldsymbol{\Psi}^{(k)}\}$, where $\boldsymbol{\Psi}^{(0)}$ denotes the starting value of $\boldsymbol{\Psi}$ and $\boldsymbol{\Psi}^{(k)}$ the value of $\boldsymbol{\Psi}$ on the kth subsequent iteration of the EM algorithm. Such superfixes are also used for components of $\boldsymbol{\Psi}$ and other parameters derived from them; for instance, if μ_1 is a component of $\boldsymbol{\Psi}$ and $\sigma_{22.1}$ is a parameter defined as a function of the components of $\boldsymbol{\Psi}$, the notations $\mu_1^{(k)}$ and $\sigma_{22.1}^{(k)}$ respectively denote their kth EM iterates.

The Q-function is used to denote the conditional expectation of the complete-data log likelihood function, $\log L_c(\boldsymbol{\Psi})$, given the observed data $\boldsymbol{y}$, using the current fit for $\boldsymbol{\Psi}$. Hence on the $(k+1)$th iteration of the E-step, it is given by

$$Q(\boldsymbol{\Psi}; \boldsymbol{\Psi}^{(k)}) = E_{\boldsymbol{\Psi}^{(k)}}\{\log L_c(\boldsymbol{\Psi}) \mid \boldsymbol{y}\},$$

where the expectation operator E has the subscript $\boldsymbol{\Psi}^{(k)}$ to explicitly convey that this (conditional) expectation is being effected using $\boldsymbol{\Psi}^{(k)}$ for $\boldsymbol{\Psi}$. Concerning other moment operators, we shall use $\text{var}_{\boldsymbol{\Psi}}(W)$ for the variance of a random variable W and $\text{cov}_{\boldsymbol{\Psi}}(\boldsymbol{W})$ for the covariance matrix of a random vector $\boldsymbol{W}$, where $\boldsymbol{\Psi}$ is the parameter vector indexing the distribution of W $(\boldsymbol{W})$.

The maximum-likelihood estimate (MLE) of $\boldsymbol{\Psi}$ is denoted by $\hat{\boldsymbol{\Psi}}$.

The (incomplete-data) observed information matrix is denoted by $\boldsymbol{I}(\hat{\boldsymbol{\Psi}}; \boldsymbol{y})$, where

$$\boldsymbol{I}(\boldsymbol{\Psi}; \boldsymbol{y}) = -\partial^2 \log L(\boldsymbol{\Psi})/\partial \boldsymbol{\Psi} \partial \boldsymbol{\Psi}^T.$$

The (incomplete-data) expected information matrix is denoted by $\mathcal{I}(\boldsymbol{\Psi})$, where

$$\mathcal{I}(\boldsymbol{\Psi}) = E_{\boldsymbol{\Psi}}\{\boldsymbol{I}(\boldsymbol{\Psi};\, \boldsymbol{Y})\}.$$

For the complete data, we let

$$\boldsymbol{I}_c(\boldsymbol{\Psi};\, \boldsymbol{x}) = -\partial^2 \log L_c(\boldsymbol{\Psi})/\partial\boldsymbol{\Psi}\partial\boldsymbol{\Psi}^T,$$

while its conditional expectation given $\boldsymbol{y}$ is denoted by

$$\mathcal{I}_c(\boldsymbol{\Psi};\, \boldsymbol{y}) = E_{\boldsymbol{\Psi}}\{\boldsymbol{I}_c(\boldsymbol{\Psi};\, \boldsymbol{X}) \mid \boldsymbol{y}\}.$$

The expected information matrix corresponding to the complete data is given by

$$\mathcal{I}_c(\boldsymbol{\Psi}) = E_{\boldsymbol{\Psi}}\{\boldsymbol{I}_c(\boldsymbol{\Psi};\, \boldsymbol{X})\}.$$

The so-called missing information matrix is denoted by $\mathcal{I}_m(\boldsymbol{\Psi};\, \boldsymbol{y})$ and is defined as

$$\mathcal{I}_m(\boldsymbol{\Psi};\, \boldsymbol{y}) = -E_{\boldsymbol{\Psi}}\{\partial^2 \log k(\boldsymbol{x} \mid \boldsymbol{y};\, \boldsymbol{\Psi})/\partial\boldsymbol{\Psi}\partial\boldsymbol{\Psi}^T \mid \boldsymbol{y}\}.$$

In other notations involving I, the symbol $\boldsymbol{I}_d$ is used to denote the $d \times d$ identity matrix, while $I_A(x)$ denotes the indicator function that is 1 if x belongs to the set A and is zero otherwise.

The p.d.f. of a random vector $\boldsymbol{W}$ having a p-dimensional multivariate normal distribution with mean $\boldsymbol{\mu}$ and covariance $\boldsymbol{\Sigma}$ is denoted by $\phi(\boldsymbol{w};\, \boldsymbol{\mu},\, \boldsymbol{\Sigma})$, where

$$\phi(\boldsymbol{w};\, \boldsymbol{\mu},\, \boldsymbol{\Sigma}) = (2\pi)^{-p/2} \mid \boldsymbol{\Sigma} \mid^{-\frac{1}{2}} \exp\{-\tfrac{1}{2}(\boldsymbol{w} - \boldsymbol{\mu})^T \boldsymbol{\Sigma}^{-1}(\boldsymbol{w} - \boldsymbol{\mu})\}.$$

The notation $\phi(w;\, \mu,\, \sigma^2)$ is used to denote the p.d.f. of a univariate normal distribution with mean μ and variance σ^2.

CHAPTER 2

Examples of the EM Algorithm

2.1 INTRODUCTION

Before we proceed to present the basic theory underlying the EM algorithm, we give in this chapter a variety of examples to demonstrate how the EM algorithm can be conveniently applied to find the MLE in some commonly occurring situations.

The first three examples concern the application of the EM algorithm to problems where there are missing data in the conventional sense. The other examples relate to problems where an incomplete-data formulation and the EM algorithm derived thereby result in an elegant way of computing MLEs. As to missing-data problems, although they are one of the most important classes of problems where the EM algorithm is profitably used, we do not discuss them further in this book except as continuation of these examples in view of the availability of an excellent book by Little and Rubin (1987), which is devoted to the topic and covers the application of the EM algorithm in that context. Discussions of the missing-value problem in a general setting and in the multivariate normal case may also be found in Little (1983a, 1983b, 1993, 1994) and in Little and Rubin (1989, 1990).

2.2 MULTIVARIATE DATA WITH MISSING VALUES

2.2.1 Example 2.1: Bivariate Normal Data with Missing Values

Let $\boldsymbol{W} = (W_1, W_2)^T$ be a bivariate random vector having a normal distribution

$$\boldsymbol{W} \sim N(\boldsymbol{\mu}, \boldsymbol{\Sigma}) \tag{2.1}$$

with mean $\boldsymbol{\mu} = (\mu_1, \mu_2)^T$ and covariance matrix

$$\boldsymbol{\Sigma} = \begin{pmatrix} \sigma_{11} & \sigma_{12} \\ \sigma_{12} & \sigma_{22} \end{pmatrix}.$$

The bivariate normal density is given by

$$\phi(\boldsymbol{w}; \boldsymbol{\Psi}) = (2\pi)^{-1} \mid \boldsymbol{\Sigma} \mid^{-\frac{1}{2}} \exp\{-\tfrac{1}{2}(\boldsymbol{w} - \boldsymbol{\mu})^T \boldsymbol{\Sigma}^{-1}(\boldsymbol{w} - \boldsymbol{\mu})\}, \tag{2.2}$$

where the vector of parameters $\boldsymbol{\Psi}$ is given by

$$\boldsymbol{\Psi} = (\mu_1, \mu_2, \sigma_{11}, \sigma_{12}, \sigma_{22})^T.$$

Suppose we wish to find the MLE of $\boldsymbol{\Psi}$ on the basis of a random sample of size n taken on $\boldsymbol{W}$, where the data on the ith variate W_i are missing in m_i of the units $(i = 1, 2)$. We label the data so that $\boldsymbol{w}_j = (w_{1j}, w_{2j})^T$ $(j = 1, \ldots, m)$ denote the fully observed data points, where $m = n - m_1 - m_2$, $\boldsymbol{w}_{2j}$ $(j = m+1, \ldots, m+m_1)$ denote the m_1 observations with the values of the first variate w_{1j} missing, and $\boldsymbol{w}_{1j}$ $(j = m+m_1+1, \ldots, n)$ denote the m_2 observations with the values of the second variate w_{2j} missing.

It is supposed that the "missingness" can be considered to be completely random, so that the observed data can be regarded as a random sample of size m from the bivariate normal distribution (2.1) and an independent pair of independent random samples of size m_i from the univariate normal distributions

$$W_i \sim N(\mu_i, \sigma_{ii}) \tag{2.3}$$

for $i = 1, 2$.

The observed data are therefore given by

$$\boldsymbol{y} = (\boldsymbol{w}_1^T, \ldots, \boldsymbol{w}_m^T, \boldsymbol{v}^T)^T,$$

where the vector $\boldsymbol{v}$ is given by

$$\boldsymbol{v} = (w_{2,m+1}, \ldots, w_{2,m+m_1}, w_{1,m+m_1+1}, \ldots, w_{1,n})^T.$$

The log likelihood function for $\boldsymbol{\Psi}$ based on the observed data $\boldsymbol{y}$ is

$$\begin{aligned}\log L(\boldsymbol{\Psi}) = &-n\log(2\pi) - \tfrac{1}{2}m\log \mid \boldsymbol{\Sigma} \mid - \tfrac{1}{2}\sum_{j=1}^{m}(\boldsymbol{w}_j - \boldsymbol{\mu})^T \boldsymbol{\Sigma}^{-1}(\boldsymbol{w}_j - \boldsymbol{\mu}) \\ &- \tfrac{1}{2}\sum_{i=1}^{2} m_i \log \sigma_{ii} - \tfrac{1}{2}\left\{\sigma_{11}^{-1}\sum_{j=m+m_1+1}^{n}(w_{1j} - \mu_1)^2 \right. \\ &\left. + \sigma_{22}^{-1}\sum_{j=m+1}^{m+m_1}(w_{2j} - \mu_2)^2\right\}. \end{aligned} \tag{2.4}$$

An obvious choice for the complete data here are the n bivariate observations. The complete-data vector $\boldsymbol{x}$ is then given by

$$\boldsymbol{x} = (\boldsymbol{w}_1^T, \ldots, \boldsymbol{w}_n^T)^T,$$

for which the missing-data vector z is

$$z = (w_{1,m+1}, \ldots, w_{1,m+m_1}, w_{2,m+m_1+1}, \ldots, w_{2,n})^T.$$

The complete-data log likelihood function for $\boldsymbol{\Psi}$ is

$$\begin{aligned}\log L_c(\boldsymbol{\Psi}) &= -n\log(2\pi) - \tfrac{1}{2}n\log|\boldsymbol{\Sigma}| - \tfrac{1}{2}\sum_{j=1}^{n}(\boldsymbol{w}_j - \boldsymbol{\mu})^T\boldsymbol{\Sigma}^{-1}(\boldsymbol{w}_j - \boldsymbol{\mu}) \\ &= -n\log(2\pi) - \tfrac{1}{2}n\log\xi \\ &\quad -\tfrac{1}{2}\xi^{-1}[\sigma_{22}T_{11} + \sigma_{11}T_{22} - 2\sigma_{12}T_{12} \\ &\qquad -2\{T_1(\mu_1\sigma_{22} - \mu_2\sigma_{12}) + T_2(\mu_2\sigma_{11} - \mu_1\sigma_{12})\} \\ &\qquad + n(\mu_1^2\sigma_{22} + \mu_2^2\sigma_{11} - 2\mu_1\mu_2\sigma_{12})],\end{aligned} \tag{2.5}$$

where

$$T_i = \sum_{j=1}^{n} w_{ij} \quad (i = 1, 2), \tag{2.6}$$

$$T_{hi} = \sum_{j=1}^{n} w_{hj}w_{ij} \quad (h, i = 1, 2), \tag{2.7}$$

and where

$$\xi = \sigma_{11}\sigma_{22}(1 - \rho^2)$$

and

$$\rho = \sigma_{12}/(\sigma_{11}\sigma_{22})^{\frac{1}{2}}$$

is the correlation between W_1 and W_2.

It can be seen that $L_c(\boldsymbol{\Psi})$ belongs to the regular exponential family with sufficient statistic

$$\boldsymbol{T} = (T_1, T_2, T_{11}, T_{12}, T_{22})^T.$$

If the complete-data vector $\boldsymbol{x}$ were available, then the (complete-data) MLE of $\boldsymbol{\Psi}$, $\hat{\hat{\boldsymbol{\Psi}}}$, would be easily computed. From the usual results for (complete-data) ML estimation for the bivariate normal distribution, $\hat{\hat{\boldsymbol{\Psi}}}$ is given by

$$\hat{\hat{\mu}}_i = T_i/n \qquad (i = 1, 2), \tag{2.8}$$

$$\hat{\hat{\sigma}}_{hi} = (T_{hi} - n^{-1}T_hT_i)/n \quad (h, i = 1, 2). \tag{2.9}$$

We now consider the E-step on the $(k+1)$th iteration of the EM algorithm, where $\boldsymbol{\Psi}^{(k)}$ denotes the value of $\boldsymbol{\Psi}$ after the kth EM iteration. It can be seen from (2.5) that in order to compute the current conditional expectation of the complete-data log likelihood,

$$Q(\boldsymbol{\Psi};\ \boldsymbol{\Psi}^{(k)}) = E_{\boldsymbol{\Psi}^{(k)}}\{\log L_c(\boldsymbol{\Psi}) \mid \boldsymbol{y}\},$$

we require the current conditional expectations of the sufficient statistics T_i and $T_{hi}(h, i = 1, 2)$. Thus in effect we require

$$E_{\boldsymbol{\Psi}^{(k)}}(W_{1j} \mid w_{2j})$$

and

$$E_{\boldsymbol{\Psi}^{(k)}}(W_{1j}^2 \mid w_{2j})$$

for $j = m+1, \ldots, m+m_1$, and

$$E_{\boldsymbol{\Psi}^{(k)}}(W_{2j} \mid w_{1j})$$

and

$$E_{\boldsymbol{\Psi}^{(k)}}(W_{2j}^2 \mid w_{1j})$$

for $j = m+m_1+1, \ldots, n$.

From the well-known properties of the bivariate normal distribution, the conditional distribution of W_2 given $W_1 = w_1$ is normal with mean

$$\mu_2 + \sigma_{12}\sigma_{11}^{-1}(w_1 - \mu_1)$$

and variance

$$\sigma_{22.1} = \sigma_{22}(1 - \rho^2).$$

Thus,

$$E_{\boldsymbol{\Psi}^{(k)}}(W_{2j} \mid w_{1j}) = w_{2j}^{(k)}, \tag{2.10}$$

where

$$w_{2j}^{(k)} = \mu_2^{(k)} + (\sigma_{12}^{(k)}/\sigma_{11}^{(k)})(w_{1j} - \mu_1^{(k)}), \tag{2.11}$$

and

$$E_{\boldsymbol{\Psi}^{(k)}}(W_{2j}^2 \mid w_{1j}) = w_{2j}^{(k)^2} + \sigma_{22.1}^{(k)} \tag{2.12}$$

for $j = m+m_1+1, \ldots, n$. Similarly, $E_{\boldsymbol{\Psi}^{(k)}}(W_{1j} \mid w_{2j})$ and $E_{\boldsymbol{\Psi}^{(k)}}(W_{1j}^2 \mid w_{2j})$ are obtained by interchanging the subscripts 1 and 2 in (2.10) to (2.12).

Note that if we were to simply impute the $w_{2j}^{(k)}$ for the missing w_{2j} in the complete-data log likelihood (and likewise for the missing w_{1j}), we would not get the same expression as the Q-function yielded by the E-step, because of the omission of the term $\sigma_{22.1}^{(k)}$ on the right-hand side of (2.12).

The M-step on the $(k+1)$th iteration is implemented simply by replacing T_i and T_{hi} by $T_i^{(k)}$ and $T_{hi}^{(k)}$, respectively, where the latter are defined by replacing the missing w_{ij} and w_{ij}^2 with their current conditional expectations as specified by (2.10) and (2.12) for $i = 2$ and by their corresponding forms for $i = 1$. Accordingly, $\boldsymbol{\Psi}^{(k+1)}$ is given by

$$\mu_i^{(k+1)} = T_i^{(k)}/n \qquad (i = 1, 2), \tag{2.13}$$

$$\sigma_{hi}^{(k+1)} = (T_{hi}^{(k)} - n^{-1}T_h^{(k)}T_i^{(k)})/n \quad (h, i = 1, 2). \tag{2.14}$$

2.2.2 Numerical Illustration

To illustrate the application of the EM algorithm to this type of problem, we now apply it to the data set below from Rubin (1987). Here, there are $n = 10$ bivariate observations in $m_2 = 2$ of which the value of the second variate w_2 is missing (indicated by ?), but there are no cases of values of the first variate w_1 missing; thus $m_1 = 0$.

Variate 1:	8	11	16	18	6	4	20	25	9	13
Variate 2:	10	14	16	15	20	4	18	22	?	?

In this case, where only one of the variates is incompletely observed, explicit expressions exist for the (incomplete-data) MLEs of the components of $\boldsymbol{\Psi}$. They are obtained by factoring the (incomplete-data) likelihood function $L(\boldsymbol{\Psi})$ into one factor corresponding to the marginal density of the variate completely observed (here W_1) and a factor corresponding to the conditional density of W_2 given w_1; see, for example, Little and Rubin (1987, p. 100). The explicit formulas for the MLE $\hat{\boldsymbol{\Psi}}$ so obtained are

$$\begin{aligned}
\hat{\mu}_1 &= \sum_{j=1}^{n} w_{1j}/n, \\
\hat{\sigma}_{11} &= \sum_{j=1}^{n} (w_{ij} - \hat{\mu}_1)^2/n, \\
\hat{\mu}_2 &= \overline{w}_2 + \hat{\beta}(\hat{\mu}_1 - \overline{w}_1), \\
\hat{\sigma}_{22} &= s_{22} + \hat{\beta}^2(\hat{\sigma}_{11} - s_{11}), \\
\hat{\beta} &= s_{12}/s_{11},
\end{aligned}$$

where

$$\overline{w}_i = \sum_{j=1}^{m} w_{ij}/m \quad (i = 1, 2)$$

and

$$s_{hi} = \sum_{j=1}^{m} (w_{hj} - \bar{y}_h)(w_{ij} - \bar{y}_i)/m \quad (h, i = 1, 2).$$

Results of the EM algorithm so applied to the bivariate data above are given in Table 2.1, where the entries for $k = \infty$ correspond to the MLEs computed from their explicit formulas. Since the first variate is completely observed in this illustration, $T_1^{(k)} = T_1$ and $T_{11}^{(k)} = T_{11}$ for each k, and so $\mu_1^{(k+1)}$ and $\sigma_{11}^{(k+1)}$ remain the same over the EM iterations.

It can be seen that in this simple illustration, the EM algorithm has essentially converged after a few iterations. The starting value for $\boldsymbol{\Psi}^{(0)}$ here was obtained by computing the estimates of the parameters on the basis of the m completely recorded data points.

2.2.3 Multivariate Data: Buck's Method

The spirit of the EM algorithm above is captured in a method proposed by Buck (1960) for the estimation of the mean $\boldsymbol{\mu}$ and the covariance matrix $\boldsymbol{\Sigma}$ of a p-dimensional random vector $\boldsymbol{W}$ based on a random sample with observations missing on some variables in some data points. Buck's method is as follows:

Step 1. Estimate $\boldsymbol{\mu}$ and $\boldsymbol{\Sigma}$ by the sample mean $\overline{\boldsymbol{w}}$ and the sample covariance matrix $\boldsymbol{S}$ using only completely observed cases.

Table 2.1. Results of the EM Algorithm for Example 2.1 (Missing Data on One Variate).

Iteration	$\mu_1^{(k)}$	$\mu_2^{(k)}$	$\sigma_{11}^{(k)}$	$\sigma_{12}^{(k)}$	$\sigma_{22}^{(k)}$	$-2\log L(\boldsymbol{\Psi}^{(k)})$
1	13.000	14.87500	40.200	32.37500	28.85938	1019.642
2	13.000	14.55286	40.200	21.23856	24.57871	210.9309
3	13.000	14.59924	40.200	20.92417	26.28657	193.3312
4	13.000	14.61165	40.200	20.89313	26.66079	190.5504
5	13.000	14.61444	40.200	20.88690	26.73555	190.0147
6	13.000	14.61506	40.200	20.88552	26.75037	189.9080
7	13.000	14.61519	40.200	20.88525	26.75337	189.8866
8	13.000	14.61523	40.200	20.88514	26.75388	189.8823
9	13.000	14.61523	40.200	20.88513	26.75398	189.8816
10	13.000	14.61523	40.200	20.88516	26.75408	189.8814
11	13.000	14.61523	40.200	20.88516	26.75405	189.8814
12	13.000	14.61523	40.200	20.88516	26.75405	189.8814
∞	13.000	14.61523	40.200	20.88516	26.75405	189.8814

Step 2. Estimate each missing observation by using the linear regression of the variable concerned on the variables on which observations have been made for that case. For this purpose, compute the required linear regressions by using $\overline{\boldsymbol{w}}$ and $\boldsymbol{S}$. Thus

$$\hat{w}_{ij} = \begin{cases} w_{ij}, & \text{if } w_{ij} \text{ is observed,} \\ \overline{w}_i + \sum_l \hat{\beta}_{il}^j (w_{lj} - \overline{w}_l), & \text{if } w_{ij} \text{ is missing,} \end{cases}$$

where the $\hat{\beta}_{il}^j$ are the coefficients in the regression as explained above.

Step 3. Compute the mean vector $\overline{\boldsymbol{w}}^*$ and corrected sum-of-squares and products matrix $\boldsymbol{A}$ using the completed observations $\hat{w}_{ij}$.

Step 4. Calculate the final estimates as follows:

$$\hat{\boldsymbol{\mu}} = \overline{\boldsymbol{w}}^*; \quad \hat{\sigma}_{hi} = \left(A_{hi} + \sum_{j=1}^{n} c_{hij} \right) \bigg/ n,$$

where the correction term c_{hij} is the residual covariance of w_h and w_i given the variables observed in the jth data point, if both w_{hj} and w_{ij} are missing and zero otherwise. This can be calculated from $\boldsymbol{S}$.

The EM algorithm applied under the assumption of multivariate normality for this problem is just an iterative version of Buck's (1960) method.

2.3 LEAST SQUARES WITH MISSING DATA

2.3.1 Healy–Westmacott Procedure

Suppose we are interested in estimating by least squares the parameters of a linear model

$$y_j = \beta_0 + \boldsymbol{\beta}^T \boldsymbol{x}_j + e_j \quad (j = 1, \ldots, n) \tag{2.15}$$

when some of the observations $y_1, \ldots, y_n$ are missing. In (2.15), $\boldsymbol{x}_j$ is a p-dimensional vector containing the values of the design variables corresponding to the jth response variable y_j, and e_j denotes the error term which has mean zero and variance σ^2 $(j = 1, \ldots, n)$.

The model parameters may be regression coefficients or various effects in a designed experiment. In designed experiments, the $\boldsymbol{x}_j$-values at which the experiment is conducted are fixed and the response y_j is observed at these points. Here the model assumed is on the conditional distribution of Y_j, the random variable corresponding to y_j, given $\boldsymbol{x}_j$ for $j = 1, \ldots, n$ and they are assumed to be independent; the missing cases do not contribute information on the model parameters unlike in the previous example. However, in designed experiments, the experimental values of $\boldsymbol{x}_j$ are chosen to make the computations simple. If some y_j values are missing, then least-squares computations get complicated. An approach of the following sort was suggested by Healy and Westmacott (1956), which exploits the simplicity of the least-squares computations in the complete-data case, in carrying out least-squares computations for the incomplete case:

1. Select trial values for all missing values.
2. Perform the complete-data analysis, that is, compute least-squares estimates of model parameters by the complete-data method.
3. Predict missing values using the estimates obtained above.
4. Substitute predicted values for missing values.
5. Go to Step 2 and continue until convergence of missing values or the residual sum of squares (RSS) is reached.

2.3.2 Example 2.2: Linear Regression with Missing Dependent Values

Let us apply this method to the analysis of a 3^2 experiment with missing values. For $j = 1, \ldots, n$, the response variable y_j is the number of lettuce plants at the jth combined level of two factors, nitrogen (x_{1j}) and phosphorus (x_{2j}). The three levels of both nitrogen and phosphorus are denoted by $-1, 0, 1$. Responses corresponding to $(-1, -1)^T$ and $(0, 1)^T$ are missing. A data set adapted from Cochran and Cox (1957) is given in Table 2.2. Suppose we wish to estimate the parameters of the linear regression

$$y_j = \beta_0 + \beta_1 x_{1j} + \beta_2 x_{2j} + e_j,$$

where the common variance σ^2 of the error terms e_j is to be estimated by the residual mean square.

From the data of a full 3^2 experiment, least-squares estimates are easily computed as follows:

$$\hat{\beta}_0 = \bar{y}; \quad \hat{\beta}_(1) = \frac{y_{+(1)} - y_{-(1)}}{6}; \quad \hat{\beta}_2 = \frac{y_{+(2)} - y_{-(2)}}{6},$$

where $\bar{y}$ is the mean of the nine observations, $y_{+(i)}$ and $y_{-(i)}$ are the total of the responses at which x_{ij} is +1 and -1, respectively, for $i = 1, 2$. The least-

Table 2.2. Number of Lettuce Plants Emerging in a Partial 3^2 Experiment (Example 2.2).

Nitrogen Level (x_{1j})	Phosphorus Level (x_{2j})	No. of Lettuce Plants (y_j)
−1	0	409
−1	1	341
0	−1	413
0	0	358
1	−1	326
1	0	291
1	1	312

Source: Adapted from Cochran and Cox (1957).

squares estimates in the incomplete-data case are somewhat more inelegant and complicated. Starting from initial estimates of 358, −51, and −55 for β_0, β_1, and β_2, respectively, computed from the data points $(0,0)^T, (-1,0)^T$, and $(0,-1)^T$, the Healy–Westmacott procedure yields results given in Table 2.3.

Table 2.3. Results of the Healy–Westmacott Algorithm on the Data of Example 2.2.

Iteration	β_0	β_1	β_2	$\hat{y}(-1,-1)$	$\hat{y}(0,-1)$	RSS
1	357.4445	−47.50000	−41.16667	464.0000	303.0000	1868.158
2	356.9321	−44.51854	−35.97223	446.1111	316.2778	1264.418
3	356.4870	−43.07048	−33.74382	437.4229	320.9599	1045.056
4	356.2272	−42.38354	−32.75969	433.3013	322.7431	955.8019
5	356.0931	−42.06175	−32.31716	431.3704	323.4675	917.2295
6	356.0276	−41.91201	−32.11602	430.4720	323.7759	900.0293
7	355.9964	−41.84261	−32.02402	430.0556	323.9115	892.2330
8	355.9817	−41.81050	−31.98178	429.8630	323.9724	888.6705
9	355.9749	−41.79567	−31.96235	429.7740	323.9999	887.0355
10	355.9717	−41.78882	−31.95339	429.7329	324.0125	886.2831
11	355.9703	−41.78566	−31.94928	429.7139	324.0183	885.9361
12	355.9696	−41.78422	−31.94739	429.7052	324.0210	885.7762
13	355.9693	−41.78355	−31.94651	429.7012	324.0222	885.7018
14	355.9691	−41.78322	−31.94608	429.6993	324.0228	885.6691
15	355.9691	−41.78308	−31.94590	429.6984	324.0230	885.6527
16	355.9690	−41.78302	−31.94582	429.6980	324.0232	885.6473
17	355.9690	−41.78298	−31.94578	429.6979	324.0232	885.6418
18	355.9690	−41.78298	−31.94577	429.6978	324.0233	885.6418
19	355.9690	−41.78298	−31.94577	429.6978	324.0233	885.6418

The final estimates, of course, coincide with the estimates found by direct least squares, which are obtained upon solving the normal equations

$$\bar{y} = \beta_0 + \frac{\beta_1}{7}; \quad y_{+(1)} - y_{-(1)} = \beta_0 + 5\beta_1 - \beta_2; \quad y_{+(2)} - y_{-(2)} = \beta_1 + 4\beta_2.$$

2.3.3 Example 2.3: Missing Values in a Latin Square Design

Consider the data in Table 2.4 (adapted from Cochran and Cox, 1957) on errors in shoot heights (of wheat in centimeters) of six samplers (treatments) arranged in a Latin square, the columns being six areas and the rows being the order in which the areas were sampled. Two of the observations (indicated by ??) are missing.

Assuming the standard additive model of row, column, and treatment effects, in a complete Latin square, the estimates of the row, column, and treatment differences are simply those obtained from the corresponding means. Thus the expected value of the observation in row i and column the j with treatment k is $(3R_i + 3C_j + 3T_k - G)/18$, where R_i, C_j, and T_k are the totals of the ith row, the jth column, and the kth treatment, respectively, and G is the total of all the observations. Thus, this formula will be used for predicting a missing value in the Healy–Westmacott algorithm. The residual sum of squares is computed by the standard analysis of variance resolving the total sum of squares into its components of row, column, treatment, and residual sums of squares. In Table 2.5, we present the results obtained by the Healy–Westmacott procedure started from initial estimates based on the effect means calculated from the available data.

The final estimates of row, column, and treatment totals are:

Row:	28.9,	28.23,	27.3,	37.9,	20.91,	28.4
Column:	42.0,	34.8,	16.63,	18.2,	25.1,	34.0
Treatment:	36.4,	33.5,	36.7,	40.43,	16.5,	7.2

Table 2.4. Sampler's Errors in Shoot Heights (cm) (6 × 6 Latin Square).

	Areas					
Order	1	2	3	4	5	6
I	F 3.5	B 4.2	A 6.7	D 6.6	C 4.1	E 3.8
II	B 8.9	F 1.9	D ??	A 4.5	E 2.4	C 5.8
III	C 9.6	E 3.7	F−2.7	B 3.7	D 6.0	A 7.0
IV	D 10.5	C 10.2	B 4.6	E 3.7	A 5.1	F 3.8
V	E ??	A 7.2	C 4.0	F−3.3	B 3.5	D 5.0
VI	A 5.9	D 7.6	E−0.7	C 3.0	F 4.0	B 8.6

Source: Adapted from Cochran and Cox (1957).

Table 2.5. Results of the Healy–Westmacott Algorithm on the Data of Example 2.3.

Iteration	$\hat{y}_{23}$	$\hat{y}_{51}$	Residual Sum of Squares
1	4.73	3.598	65.847
2	4.73	3.598	65.847

2.3.4 Healy–Westmacott Procedure as an EM Algorithm

The Healy–Westmacott method can be looked upon as an application of the EM algorithm. Suppose we assume that the conditional distribution of the jth response variable y_j, given the vector $\boldsymbol{x}_j$, is normal with mean $\boldsymbol{\beta}^T\boldsymbol{x}_j$, where $\boldsymbol{x}_j$ is a row of the design matrix $\boldsymbol{X}$, and variance σ^2 (not depending upon $\boldsymbol{X}$). Then the least-squares estimate of $\boldsymbol{\beta}$ and the estimate obtained in the least-squares analysis for σ^2 are also the MLEs. Hence the problem of estimation of the parameter vector,

$$\boldsymbol{\Psi} = (\boldsymbol{\beta}, \sigma^2)^T,$$

can be looked upon as a ML estimation problem, and thus in the case of missing values, the EM algorithm can be profitably applied, if the complete-data problem is simpler. Let us consider the EM algorithm in this role. Suppose that the data are labeled so that the first m observations are complete with the last $(n - m)$ cases having the response y_j missing. Note that we are using here the universal notation $\boldsymbol{X}$ to denote the design matrix. Elsewhere in this book, $\boldsymbol{X}$ denotes the random vector corresponding to the complete-data vector $\boldsymbol{x}$. Also, the use in this section of $\boldsymbol{x}_j$ as the jth row of $\boldsymbol{X}$ is not to be confused as being a replication of $\boldsymbol{x}$, which is used elsewhere in the book to denote the complete-data vector.

The incomplete-data log likelihood function is given by

$$\begin{aligned}\log L(\boldsymbol{\Psi}) = &-\tfrac{1}{2}m\log(2\pi) - \tfrac{1}{2}m\log\sigma^2 - \tfrac{1}{2}\sum_{j=1}^{m} y_j^2/\sigma^2 \\ &- \sum_{j=1}^{m}(\boldsymbol{\beta}^T\boldsymbol{x}_j)^2/\sigma^2 + \sum_{j=1}^{m} y_j(\boldsymbol{\beta}^T\boldsymbol{x}_j)/\sigma^2. \qquad (2.16)\end{aligned}$$

The complete-data log likelihood is the same except that the summation in (2.16) ranges up to n. Concerning the E-step on the $(k+1)$th iteration, it follows from (2.16) that in order to compute the current conditional expectation of the complete-data log likelihood function, we require the conditional expectations of the first two moments of the missing responses Y_j given the vector

$$\boldsymbol{y} = (y_1, \ldots, y_m)^T$$

of the observed responses and the design matrix $\boldsymbol{X}$. They are given by

$$E_{\boldsymbol{\Psi}^{(k)}}(Y_j \mid \boldsymbol{y}, \boldsymbol{X}) = \boldsymbol{x}_j^T \boldsymbol{\beta}^{(k)}$$

and

$$E_{\boldsymbol{\Psi}^{(k)}}(Y_j^2 \mid \boldsymbol{y}, \boldsymbol{X}) = (\boldsymbol{x}_j^T \boldsymbol{\beta}^{(k)})^2 + \sigma^{(k)^2} \tag{2.17}$$

for $j = m+1, \ldots, n$.

Implementation of the M-step at the $(k+1)$th iteration is straightforward, since we use the least-squares computation of the complete-data design to find $\boldsymbol{\beta}^{(k+1)}$, and $\sigma^{(k+1)^2}$ is given by the current residual mean square. We thus obtain

$$\boldsymbol{\beta}^{(k+1)} = (\boldsymbol{X}^T \boldsymbol{X})^{-1} \boldsymbol{X}^T \boldsymbol{Y}^{(k)}$$

and

$$\sigma^{(k+1)^2} = \frac{1}{n} \left\{ \sum_{j=1}^{m} (y_j - \boldsymbol{\beta}^{(k)^T} \boldsymbol{x}_j)^2 + (n-m)\sigma^{(k)^2} \right\},$$

where

$$\boldsymbol{Y}^{(k)} = (y_1, \ldots, y_m, y_{m+1}^{(k)}, \ldots, y_n^{(k)})^T,$$

and

$$y_j^{(k)} = \boldsymbol{x}_j^T \boldsymbol{\beta}^{(k)}.$$

The first equation does not involve $\sigma^{(k)^2}$, and at convergence,

$$\sigma^{(k+1)^2} = \sigma^{(k)^2} = \hat{\sigma}^2.$$

Thus

$$\hat{\sigma}^2 = \frac{1}{n} \left\{ \sum_{j=1}^{m} (y_j - \hat{\boldsymbol{\beta}}^T \boldsymbol{x}_j)^2 + (n-m)\hat{\sigma}^2 \right\},$$

or

$$\hat{\sigma}^2 = \frac{1}{m} \sum_{j=1}^{m} (y_j - \hat{\boldsymbol{\beta}}^T \boldsymbol{x}_j)^2.$$

Hence the EM algorithm can omit the M-step for $\sigma^{(k)^2}$ and the calculation of (2.17) on the E-step, and compute $\hat{\sigma}^2$ after the convergence of the sequence $\{\boldsymbol{\beta}^{(k)}\}$. It can be easily seen that this is exactly the Healy–Westmacott method described above, confirming that the Healy–Westmacott method is an example of the EM algorithm; see Little and Rubin (1987, p. 152–155) for further discussion.

2.4 EXAMPLE 2.4: MULTINOMIAL WITH COMPLEX CELL STRUCTURE

In Section 1.4.2, we gave an introductory example on the EM algorithm applied to a multinomial distribution with cell probabilities depending on a single unknown parameter. The EM algorithm can handle more complicated problems of this type that occur in genetics. To illustrate this, we consider a multinomial problem with cell probabilities depending on two unknown parameters.

Suppose we have $n = 435$ observations on a multinomial with four cells with cell probability structure given in Table 2.6. Also given in the table are observed frequencies in these cells. The observed data are therefore given by the vector of cell frequencies

$$\boldsymbol{y} = (n_O, n_A, n_B, n_{AB})^T.$$

The vector of unknown parameters is

$$\boldsymbol{\Psi} = (p, q)^T,$$

since $r = 1 - p - q$. The object is to find the MLE of $\boldsymbol{\Psi}$ on the basis of $\boldsymbol{y}$. This is a well-known problem of gene frequency estimation in genetics, and is discussed, for example, in Kempthorne (1957), Elandt-Johnson (1971), and Rao (1973).

The log likelihood function for $\boldsymbol{\Psi}$, apart from an additive constant, is

$$\log L(\boldsymbol{\Psi}) = 2n_0 \log r + n_A \log(p^2 + 2pr) + n_B \log(q^2 + 2qr) + 2n_{AB} \log(pq),$$

which does not admit a closed-form solution for $\hat{\boldsymbol{\Psi}}$, the MLE of $\boldsymbol{\Psi}$.

In applying the EM algorithm to this problem, a natural choice for the complete-data vector is

$$\boldsymbol{x} = (n_O, \boldsymbol{z}^T)^T, \tag{2.18}$$

where

$$\boldsymbol{z} = (n_{AA}, n_{AO}, n_{BB}, n_{BO})^T$$

represents the unobservable or 'missing' data, taken to be the frequencies

Table 2.6. Observed Multinomial Data for Example 2.4.

Category (Cell)	Cell Probability	Observed Frequency
O	r^2	$n_0 = 176$
A	$p^2 + 2pr$	$n_A = 182$
B	$q^2 + 2qr$	$n_B = 60$
AB	$2pq$	$n_{AB} = 17$

n_{AA}, n_{AO}, n_{BB}, and n_{BO} corresponding to the middle cells as given in Table 2.7. Notice that since the total frequency n is fixed, the five cell frequencies in (2.18) are sufficient to represent the complete data. If we take the distribution of $\boldsymbol{x}$ to be multinomial with n draws with respect to the six cells with probabilities as specified in Table 2.7, then it is obvious that the vector $\boldsymbol{y}$ of observed frequencies has the required multinomial distribution, as specified in Table 2.6.

The complete-data log likelihood function for $\boldsymbol{\Psi}$ can be written in the form (apart from an additive constant) as

$$\log L_c(\boldsymbol{\Psi}) = 2\,n_A^+ \log p + 2n_B^+ \log q + 2n_O^+ \log r, \tag{2.19}$$

where

$$n_A^+ = n_{AA} + \tfrac{1}{2}n_{AO} + \tfrac{1}{2}n_{AB},$$

$$n_B^+ = n_{BB} + \tfrac{1}{2}n_{BO} + \tfrac{1}{2}n_{AB},$$

and

$$n_O^+ = n_O + \tfrac{1}{2}n_{AO} + \tfrac{1}{2}n_{BO}.$$

It can be seen that (2.19) has the form of a multinomial log likelihood for the frequencies $2n_A^+$, $2n_B^+$, and $2n_O^+$ with respect to three cells with probabilities p, q, and r, respectively. Thus the complete-data MLEs of these probabilities obtained by maximization of (2.19) are given by

$$\hat{\hat{p}} = \frac{n_A^+}{n}; \quad \hat{\hat{q}} = \frac{n_B^+}{n}. \tag{2.20}$$

As seen in Section 1.5.3, the E- and M-steps simplify when the complete-data likelihood function belongs to the regular exponential family, as in this example. The E-step requires just the calculation of the current conditional expectation of the sufficient statistic for $\boldsymbol{\Psi}$, which here is $(n_A^+, n_B^+)^T$. The M-

Table 2.7. Complete-Data Structure for Example 2.4.

Category (Cell)	Cell Probability	Notation for Frequency
O	r^2	n_O
AA	p^2	n_{AA}
AO	$2pr$	n_{AO}
BB	q^2	n_{BB}
BO	$2qr$	n_{BO}
AB	$2pq$	n_{AB}

step then obtains $\boldsymbol{\Psi}^{(k+1)}$ as a solution of the equation obtained by equating this expectation equal to $\boldsymbol{\Psi}$. In effect for this problem, $\boldsymbol{\Psi}^{(k+1)}$ is given by replacing n_A^+ and n_B^+ in the right-hand side of (2.20) by their current conditional expectations given the observed data.

To compute these conditional expectations of n_A^+ and n_B^+ (the E-step), we require the conditional expectation of the unobservable data $\boldsymbol{z}$ for this problem. Consider the first element of $\boldsymbol{z}$, which from (2.18) is n_{AA}. Then it is easy to verify that conditional on $\boldsymbol{y}$, effectively n_A, n_{AA} has a binomial distribution with sample size n_A and probability parameter

$$p^{(k)^2}/(p^{(k)^2} + 2p^{(k)}r^{(k)}),$$

with $\boldsymbol{\Psi}^{(k)}$ used in place of the unknown parameter vector $\boldsymbol{\Psi}$ on the $(k+1)$th iteration. Thus the current conditional expectation of n_{AA} given $\boldsymbol{y}$ is obtained by

$$E_{\boldsymbol{\Psi}^{(k)}}(n_{AA}) = n_{AA}^{(k)},$$

where

$$n_{AA}^{(k)} = n_A\, p^{(k)^2}/(p^{(k)^2} + 2p^{(k)}r^{(k)}). \tag{2.21}$$

Similarly, the current conditional expectations of n_{AO}, n_{BB}, and n_{BO}, given $\boldsymbol{y}$, can be calculated.

Execution of the M-step gives

$$p^{(k+1)} = (n_{AA}^{(k)} + \tfrac{1}{2}n_{AO}^{(k)} + \tfrac{1}{2}n_{AB})/n \tag{2.22}$$

and

$$q^{(k+1)} = (n_{BB}^{(k)} + \tfrac{1}{2}n_{BO}^{(k)} + \tfrac{1}{2}n_{AB})/n. \tag{2.23}$$

Results of the EM algorithm for this problem are given in Table 2.8. The MLE of $\boldsymbol{\Psi}$ can be taken to be the value of $\boldsymbol{\Psi}^{(k)}$ on iteration $k = 4$. Although not formulated in the language of the EM algorithm, this idea was current in the genetics literature as the gene-counting method (see Ceppellini et al., 1955; Smith, 1957).

Table 2.8. Results of the EM Algorithm for Example 2.4.

Iteration	$p^{(k)}$	$q^{(k)}$	$r^{(k)}$	$-\log L(\boldsymbol{\Psi}^{(k)})$
0	0.26399	0.09299	0.643016	2.5619001
1	0.26436	0.09316	0.64248	2.5577875
2	0.26443	0.09317	0.64240	2.5577729
3	0.26444	0.09317	0.64239	2.5577726
4	0.26444	0.09317	0.64239	2.5577726

2.5 EXAMPLE 2.5: ANALYSIS OF PET AND SPECT DATA

The EM algorithm has been employed in ML estimation of parameters in a computerized image reconstruction process such as SPECT (single-photon emission computed tomography) or PET (positron emission tomography). An excellent account of the statistical aspects of emission tomography may be found in the papers by Kay (1994) and McColl, Holmes, and Ford (1994), which are contained in the special issue of the journal *Statistical Methods in Medical Research* on the topic of emission tomography. In both PET and SPECT a radioactive tracer is introduced into the organ under study of the human or animal patient. The radioisotope is incorporated into a molecule that is absorbed into the tissue of the organ and resides there in concentrations that indicate levels of metabolic activity and blood flow. As the isotope decays, it emits either single photons (SPECT) or positrons (PET), which are counted by bands of gamma detectors strategically placed around the patient's body. The method of collection of the data counts differs quite considerably for PET and SPECT because of the fundamental differences in their underlying physical processes. SPECT isotopes are gamma emitters that tend to have long half lives. PET isotopes emit positrons which annihilate with nearby electrons in the tissue to generate pairs of photons that fly off on paths almost 180° apart. In both imaging modalities, not all emissions are counted because the photons may travel along directions that do not cross the detectors or because they may be attenuated by the body's tissues.

With PET and SPECT, the aim is to estimate the spatial distribution of the isotope concentration in the organ on the basis of the projected counts recorded at the detectors. In a statistical framework, it is assumed that the emissions occur according to a spatial Poisson point process in the region under study with an unknown intensity function, which is usually referred to as the emission density. In order to estimate the latter, the process is discretized as follows. The space over which the reconstruction is required is finely divided into a number n of rectangular pixels (or voxels in three-dimensions), and it is assumed that the (unknown) emission density is a constant λ_i for the ith pixel $(i = 1, \ldots, n)$. Let y_j denote the number of counts recorded by the jth detector $(j = 1, \ldots, d)$, where d denotes the number of detectors. As it is common to work with arrays of 128×128 or 256×256 square pixels, the detectors move within the plane of measurement during a scan in order to record more observations than parameters. They count for only a short time at each position.

After this discretization of the problem, reconstruction aims to infer the vector of emission densities

$$\boldsymbol{\lambda} = (\lambda_1, \ldots, \lambda_n)^T$$

from the vector of the observable counts $\boldsymbol{y} = (y_1, \ldots, y_d)^T$. In both PET and SPECT, it leads to a Poisson regression model for the counts. Given the vector

$\boldsymbol{\lambda}$ of isotope densities, the counts $y_1, \ldots, y_d$, are conditionally independent according to a Poisson distribution, namely

$$Y_j \sim P(\mu_j), \tag{2.24}$$

where the mean μ_j of Y_j is modeled as

$$\mu_j = \sum_{i=1}^{n} \lambda_i p_{ij} \quad (j = 1, \ldots, d), \tag{2.25}$$

and p_{ij} is the conditional probability that a photon/positron is counted by the jth detector given that it was emitted from within the ith pixel.

The advantage of this statistical approach is that it takes into account the Poisson variation underlying the data, unlike the class of deterministic methods based on a filtered back-projection approach. The conditional probabilities p_{ij}, which appear as the coefficients in the additive Poisson regression model (2.25), depend on the geometry of the detection system, the activity of the isotope and exposure time, and the extent of attenuation and scattering between source and detector. The specification of the p_{ij} is much more complicated with SPECT than PET since attenuation can be neglected for the latter (see, for example, Green, 1990b, McColl et al., 1994 and Kay, 1994).

We now outline the application of the EM algorithm to this problem of tomography reconstruction. This approach was pioneered independently by Shepp and Vardi (1982) and Lange and Carson (1984), and developed further by Vardi, Shepp, and Kaufman (1985).

An obvious choice of the complete-data vector in this example is $\boldsymbol{x} = (\boldsymbol{y}^T, \boldsymbol{z}^T)^T$, where the vector $\boldsymbol{z}$ consists of the unobservable data counts z_{ij} with z_{ij} defined to be the number of photons/positrons emitted within pixel i and recorded at the jth detector ($i = 1, \ldots, n$; $j = 1, \ldots, d$). It is assumed that given $\boldsymbol{\lambda}$, the $\{Z_{ij}\}$ are conditionally independent, with each Z_{ij} having a Poisson distribution specified as

$$Z_{ij} \sim P(\lambda_i p_{ij}) \quad (i = 1, \ldots, d; \; j = 1, \ldots, n).$$

Since

$$y_j = \sum_{i=1}^{n} z_{ij} \quad (j = 1, \ldots, d),$$

it is obvious that these assumptions for the unobservable data $\{z_{ij}\}$ imply the model (2.24) for the incomplete data $\{y_j\}$ in $\boldsymbol{y}$.

The complete-data log likelihood is given by

$$\log L_c(\boldsymbol{\lambda}) = \sum_{i=1}^{n} \sum_{j=1}^{d} \{-\lambda_i p_{ij} + z_{ij} \log(\lambda_i p_{ij}) - \log z_{ij}!\}. \tag{2.26}$$

As the Poisson distribution belongs to the linear exponential family, (2.26) is linear in the unobservable data z_{ij}. Hence the E-step (on the $(k+1)$th iteration) simply requires the calculation of the conditional expectation of Z_{ij} given the observed data $\boldsymbol{y}$, using the current fit $\boldsymbol{\lambda}^{(k)}$ for $\boldsymbol{\lambda}$. From a standard probability result, the conditional distribution of Z_{ij} given $\boldsymbol{y}$ and $\boldsymbol{\lambda}^{(k)}$ is binomial with sample size parameter y_j and probability parameter

$$\lambda_i p_{ij} \Big/ \sum_{h=1}^{n} \lambda_h p_{hj} \quad (i = 1, \ldots, n; \; j = 1, \ldots, d). \tag{2.27}$$

Using (2.27), it follows that

$$E_{\boldsymbol{\lambda}^{(k)}}(Z_{ij} \mid \boldsymbol{y}) = z_{ij}^{(k)},$$

where

$$z_{ij}^{(k)} = y_j \lambda_i^{(k)} p_{ij} \Big/ \sum_{h=1}^{n} \lambda_h^{(k)} p_{hj}. \tag{2.28}$$

With z_{ij} replaced by $z_{ij}^{(k)}$ in (2.26), application of the M-step on the $(k+1)$th iteration gives

$$\begin{aligned} \lambda_i^{(k+1)} &= q_i^{-1} \sum_{j=1}^{d} E_{\boldsymbol{\lambda}^{(k)}}(Z_{ij} \mid \boldsymbol{y}) \\ &= \lambda_i^{(k)} q_i^{-1} \sum_{j=1}^{d} \left\{ y_j p_{ij} \Big/ \sum_{h=1}^{n} \lambda_h^{(k)} p_{hj} \right\} \quad (i = 1, \ldots, n), \end{aligned} \tag{2.29}$$

where

$$q_i = \sum_{j=1}^{d} p_{ij}$$

is the probability that an emission from the ith pixel is recorded by one of the d detectors $(i = 1, \ldots, n)$. As noted by Vardi et al. (1985), it is much simpler and involves no loss of generality to take $q_i = 1$. This is because we can work with the density of emission counts that are actually detected; that is, we let λ_i represent the mean number of emissions from the ith pixel that are detected.

It has been shown (Vardi et al., 1985) that $\boldsymbol{\lambda}^{(k)}$ converges to a global maximizer of the incomplete-data likelihood function $L(\boldsymbol{\lambda})$, but it will not be unique if the number of pixels n is greater than the number of detectors d; see Byrne (1993). An attractive feature of (2.29) is that the positivity constraints on the estimates of the λ_i are satisfied, providing the initial estimates $\lambda_i^{(0)}$ are all positive.

After a certain number of iterations, the EM algorithm produces images that begin to deteriorate. The point at which deterioration begins depends on the number of the projected counts y_j ($j = 1, \ldots, d$). The more the counts, the later it begins. The reason for this deterioration in the reconstructed image is that, for a low number of counts, increasing the value of the likelihood function increases the distortion due to noise. As a consequence, various modifications to the EM algorithm have been proposed to produce more useful final images. Although these modified versions of the EM algorithm are algorithms that can be applied to a variety of problems, the motivation for their development in the first instance came in the context of improving the quality of images reconstructed from PET and SPECT data. Hence it is in this context that we shall outline in Chapter 5, some of these modifications of the EM algorithm; for a review and a bibliography of the EM algorithm in tomography, see Krishnan (1995). Further modifications of the EM algorithm in its application to this problem have been given recently by Fessler and Hero (1994) and Meng and van Dyk (1995).

2.6 EXAMPLE 2.6: MULTIVARIATE t-DISTRIBUTION WITH KNOWN DEGREES OF FREEDOM

2.6.1 ML Estimation of Multivariate t-Distribution

The multivariate t-distribution has many potential applications in applied statistics. A p-dimensional random variable $\boldsymbol{W}$ is said to have a multivariate t-distribution $t_p(\boldsymbol{\mu}, \boldsymbol{\Sigma}, \nu)$ with location $\boldsymbol{\mu}$, positive definite inner product matrix $\boldsymbol{\Sigma}$, and ν degrees of freedom if given the weight u,

$$\boldsymbol{W} \mid u \sim N(\boldsymbol{\mu}, \boldsymbol{\Sigma}/u), \tag{2.30}$$

where the random variable U corresponding to the weight u is distributed as

$$U \sim \text{gamma}\,(\tfrac{1}{2}\nu, \tfrac{1}{2}\nu). \tag{2.31}$$

The gamma (α, β) density function $f(u; \alpha, \beta)$ is given by

$$f(u; \alpha, \beta) = \{\beta^{\alpha} u^{\alpha-1}/\Gamma(\alpha)\} \exp(-\beta u) I_{(0,\infty)}(u); \quad (\alpha, \beta > 0).$$

On integrating out u from the joint density function of $\boldsymbol{W}$ and U that can be formed from (2.30) and (2.31), the density function of $\boldsymbol{W}$ is given by

$$f_p(\boldsymbol{w}; \boldsymbol{\mu}, \boldsymbol{\Sigma}, \nu) = \frac{\Gamma(\frac{\nu+p}{2})\, |\boldsymbol{\Sigma}|^{-1/2}}{(\pi\nu)^{\frac{1}{2}p}\Gamma(\frac{\nu}{2})\{1 + \delta(\boldsymbol{w}, \boldsymbol{\mu}; \boldsymbol{\Sigma})/\nu\}^{\frac{1}{2}(\nu+p)}}, \tag{2.32}$$

where

$$\delta(\boldsymbol{w}, \boldsymbol{\mu}; \boldsymbol{\Sigma}) = (\boldsymbol{w} - \boldsymbol{\mu})^T \boldsymbol{\Sigma}^{-1} (\boldsymbol{w} - \boldsymbol{\mu})$$

denotes the Mahalanobis squared distance between $\boldsymbol{w}$ and $\boldsymbol{\mu}$ (with $\boldsymbol{\Sigma}$ as the covariance matrix). As ν tends to infinity, U converges to one with probability one, and so $\boldsymbol{W}$ becomes marginally multivariate normal with mean $\boldsymbol{\mu}$ and covariance matrix $\boldsymbol{\Sigma}$.

We consider now the application of the EM algorithm for ML estimation of the parameters $\boldsymbol{\mu}$ and $\boldsymbol{\Sigma}$ in the t-density (2.32) in the case where the degrees of freedom ν is known. For instance ν can be assumed to be known in statistical analyses where different specified degrees of freedom ν are used for judging the robustness of the analyses; see Lange, Little, and Taylor (1989) and Lange and Sinsheimer (1993). For $\nu < \infty$, ML estimation of $\boldsymbol{\mu}$ is robust in the sense that observations with large Mahalanobis distances are downweighted. This can be clearly seen from the form of the equation (2.42) to be derived for the MLE of $\boldsymbol{\mu}$. As ν decreases, the degree of downweighting of outliers increases.

The EM algorithm for known ν is given in Rubin (1983), and is extended to the case with missing data in $\boldsymbol{W}$ in Little (1988) and in Little and Rubin (1987, pp. 209–216). More recently, Liu and Rubin (1994, 1995) have shown how the MLE can be found much more efficiently by using the ECME algorithm. The use of the latter algorithm for this problem is to be described in Chapter 5, where the general case of unknown ν is to be considered. As cautioned by Liu and Rubin (1995), care must be taken especially with small or unknown ν, because the likelihood function can have many spikes with very high likelihood values but little associated posterior mass under any reasonable prior. In that case, the associated parameter estimates may be of limited practical interest by themselves, even though formally they are local or even global maxima of the likelihood function. It is, nevertheless, important to locate such maxima because they can critically influence the behavior of iterative simulation algorithms designed to summarize the entire posterior distribution; see Gelman and Rubin (1992).

Suppose that $\boldsymbol{w}_1, \ldots, \boldsymbol{w}_n$ denote an observed random sample from the $t_p(\boldsymbol{\mu}, \boldsymbol{\Sigma}, \nu)$ distribution. That is,

$$\boldsymbol{y} = (\boldsymbol{w}_1^T, \ldots, \boldsymbol{w}_n^T)^T$$

denotes the observed data vector. The problem is to find the MLE of $\boldsymbol{\Psi}$ on the basis of $\boldsymbol{y}$, where $\boldsymbol{\Psi}$ contains the elements of $\boldsymbol{\mu}$ and the distinct elements of $\boldsymbol{\Sigma}$. From (2.32), the log likelihood function for $\boldsymbol{\Psi}$ that can be formed from $\boldsymbol{y}$ is

$$\begin{aligned}\log L(\boldsymbol{\Psi}) &= \sum_{j=1}^{n} \log f_p(\boldsymbol{w}_j;\, \boldsymbol{\mu},\, \boldsymbol{\Sigma},\, \nu), \\ &= -\tfrac{1}{2}np\log(\pi\nu) + n\{\log\Gamma(\tfrac{\nu+p}{2}) - \log\Gamma(\tfrac{1}{2}\nu)\} \\ &\qquad -\tfrac{1}{2}n\log|\boldsymbol{\Sigma}| + \tfrac{1}{2}n(\nu+p)\log\nu \\ &\qquad -\tfrac{1}{2}(\nu+p)\sum_{j=1}^{n}\log\{\nu + \delta(\boldsymbol{w}_j,\, \boldsymbol{\mu};\boldsymbol{\Sigma})\}, \end{aligned} \tag{2.33}$$

which does not admit a closed-form solution for the MLE of $\boldsymbol{\Psi}$.

In the light of the definition (2.30) of this t-distribution, it is convenient to view the observed data $\boldsymbol{y}$ as incomplete. The complete-data vector $\boldsymbol{x}$ is taken to be

$$\boldsymbol{x} = (\boldsymbol{y}^T,\, \boldsymbol{z}^T)^T,$$

where

$$\boldsymbol{z} = (u_1,\, \ldots,\, u_n)^T.$$

The missing variables $u_1, \ldots, u_n$ are defined so that

$$\boldsymbol{W}_j \mid u_j \sim N(\boldsymbol{\mu},\, \boldsymbol{\Sigma}/u_j), \tag{2.34}$$

independently for $j = 1, \ldots, n$, and

$$U_1,\, \ldots,\, U_n \overset{\text{i.i.d.}}{\sim} \text{gamma}\,(\tfrac{1}{2}\nu,\, \tfrac{1}{2}\nu). \tag{2.35}$$

Here in this example, the missing data vector $\boldsymbol{z}$ consists of variables that would never be observable as data in the usual sense.

Because of the conditional structure of the complete-data model specified by (2.30) and (2.31), the complete-data likelihood function can be factored into the product of the conditional density of $\boldsymbol{W}$ given $\boldsymbol{z}$ and the marginal density of $\boldsymbol{Z}$. Accordingly, the complete-data log likelihood can be written as

$$\log L_c(\boldsymbol{\Psi}) = \log L_{1c}(\boldsymbol{\Psi}) + a(\boldsymbol{z}),$$

where

$$\begin{aligned}\log L_{1c}(\boldsymbol{\Psi}) &= -\tfrac{1}{2}np\log(2\pi) - \tfrac{1}{2}n\log|\boldsymbol{\Sigma}| + \tfrac{1}{2}p\sum_{j=1}^{n}\log\, u_j \\ &\qquad -\tfrac{1}{2}\sum_{j=1}^{n}u_j(\boldsymbol{w}_j - \boldsymbol{\mu})^T\boldsymbol{\Sigma}^{-1}(\boldsymbol{w}_j - \boldsymbol{\mu}) \end{aligned} \tag{2.36}$$

and

$$a(z) = -n \log \Gamma(\tfrac{1}{2}\nu) + \tfrac{1}{2}n\nu \log(\tfrac{1}{2}\nu) + \tfrac{1}{2}\nu \sum_{j=1}^{n} (\log u_j - u_j) - \sum_{j=1}^{n} \log u_j. \tag{2.37}$$

Now the E-step on the $(k+1)$th iteration of the EM algorithm requires the calculation of $Q(\boldsymbol{\Psi}; \boldsymbol{\Psi}^{(k+1)})$, the current conditional expectation of the complete-data log likelihood function $\log L_c(\boldsymbol{\Psi})$. In the present case of known ν, we need focus only on the first term $\log L_{1c}(\boldsymbol{\Psi})$ in the expression (2.36) for $\log L_c(\boldsymbol{\Psi})$ (since the other term does not involve unknown parameters). As this term is linear in the unobservable data u_j, the E-step is effected simply by replacing u_j with its current conditional expectation given w_j.

Since the gamma distribution is the conjugate prior distribution for U, it is not difficult to show that the conditional distribution of U given $\boldsymbol{W} = \boldsymbol{w}$ is

$$U \mid \boldsymbol{w} \sim \text{gamma}\,(m_1, m_2), \tag{2.38}$$

where

$$m_1 = \tfrac{1}{2}(\nu + p)$$

and

$$m_2 = \tfrac{1}{2}\{\nu + \delta(\boldsymbol{w}, \boldsymbol{\mu}; \boldsymbol{\Sigma})\}. \tag{2.39}$$

From (2.38), we have that

$$E(U \mid \boldsymbol{w}) = \frac{\nu + p}{\nu + \delta(\boldsymbol{w}, \boldsymbol{\mu}; \boldsymbol{\Sigma})}. \tag{2.40}$$

Thus from (2.40),

$$E_{\boldsymbol{\Psi}^{(k)}}(U_j \mid \boldsymbol{w}_j) = u_j^{(k)},$$

where

$$u_j^{(k)} = \frac{\nu + p}{\nu + \delta(\boldsymbol{w}_j, \boldsymbol{\mu}^{(k)}; \boldsymbol{\Sigma}^{(k)})}. \tag{2.41}$$

The M-step is easily implemented on noting that $L_{1c}(\boldsymbol{\Psi})$ corresponds to the likelihood function formed from n independent observations $\boldsymbol{w}_1, \ldots, \boldsymbol{w}_n$ with common mean $\boldsymbol{\mu}$ and covariance matrices $\boldsymbol{\Sigma}/u_1, \ldots, \boldsymbol{\Sigma}/u_n$, respectively. After execution of the E-step, each u_j is replaced by $u_j^{(k)}$, and so the M-step is equivalent to computing the weighted sample mean and sample covariance matrix of $\boldsymbol{w}_1, \ldots, \boldsymbol{w}_n$ with weights $u_1^{(k)}, \ldots, u_n^{(k)}$. Hence

$$\boldsymbol{\mu}^{(k+1)} = \sum_{j=1}^{n} u_j^{(k)} \boldsymbol{y}_j \bigg/ \sum_{j=1}^{n} u_j^{(k)} \tag{2.42}$$

and

$$\boldsymbol{\Sigma}^{(k+1)} = \frac{1}{n} \sum_{j=1}^{n} u_j^{(k)} (\boldsymbol{w}_j - \boldsymbol{\mu}^{(k+1)})(\boldsymbol{w}_j - \boldsymbol{\mu}^{(k+1)})^T. \tag{2.43}$$

It can be seen in this case of known ν that the EM algorithm is equivalent to iteratively reweighted least squares. The E-step updates the weights $u_j^{(k)}$, while the M-step effectively chooses $\boldsymbol{\mu}^{(k+1)}$ and $\boldsymbol{\Sigma}^{(k+1)}$ by weighted least-squares estimation.

Liu and Rubin (1995) note that the above results can be easily extended to linear models where the mean $\boldsymbol{\mu}$ of $\boldsymbol{W}_j$ is replaced by $\boldsymbol{X}_j\boldsymbol{\beta}$, where $\boldsymbol{X}_j$ is a matrix containing the observed values of some covariates associated with $\boldsymbol{W}_j$.

2.6.2 Numerical Example: Stack Loss Data

As a numerical example of the ML fitting of the t-distribution via the EM algorithm, we present an example from Lange et al. (1989), who analyzed the stack-loss data set of Brownlee (1965), which has been subjected to many robust analyses by various authors. Table 2.9 contains the data and Table 2.10 shows the slope of the regression of stack loss (y_j) on air flow (x_{1j}), temperature (x_{2j}), and acid (x_{3j}) for the linear regression model

$$y_j = \beta_0 + \sum_{i=1}^{3} \beta_i x_{ij} + e_j,$$

with t-distributed errors e_j for values of the degrees of freedom ν ranging from $\nu = 0.5$ to $\nu = \infty$ (normal). Also, included in Table 2.10, are the values of four other estimators from Ruppert and Carroll (1980) of which two are trimmed least-squares ($\hat{\boldsymbol{\Psi}}_{KB}$, $\hat{\boldsymbol{\Psi}}_{PE}$), and two are the M-estimates of Huber (1964) and Andrews (1974). Also, the results are given for the ML estimate of ν, which was found to be $\hat{\nu} = 1.1$. The values of the log likelihood, $\log L(\hat{\boldsymbol{\Psi}})$, at the solutions are given in the second column for various values of ν. Twice the difference between the best fitting t and normal model log likelihoods is 5.44 which, on reference to the chi-squared distribution with one degree of freedom, suggests asymptotically a significant improvement in fit. As noted by Lange et al. (1989), the results of the fit for the ML case of $\hat{\nu} = 1.1$ are similar to those of Andrews (1974), which Ruppert and Carroll (1980) favored based on the closeness of the fit to the bulk of the data.

Table 2.9. Stack Loss Data.

Air Flow x_1	Temperature x_2	Acid x_3	Stack Loss y
80	27	89	42
80	27	88	37
75	25	90	37
62	24	87	28
62	22	87	18
62	23	87	18
62	24	93	19
62	24	93	20
58	23	87	15
58	18	80	14
58	17	89	14
58	17	88	13
58	18	82	11
58	19	93	12
50	18	89	8
50	18	86	7
50	19	72	8
50	19	79	8
50	20	80	9
56	20	82	15
70	20	91	15

Source: Adapted from Brownlee (1965).

2.7 FINITE NORMAL MIXTURES

2.7.1 Example 2.7: Univariate Component Densities

We now extend the problem in Example 1.2 of Section 1.4.3 to the situation where the component densities in the mixture model (1.27) are not completely specified.

In this example, the g component densities are taken to be univariate normal with unknown means $\mu_1, \ldots, \mu_g$ and common unknown variance σ^2. We henceforth write $f_i(w)$ as $f_i(w; \boldsymbol{\theta}_i)$, where $\boldsymbol{\theta}_i = (\mu_i, \sigma^2)^T$ and

$$\boldsymbol{\theta} = (\mu_1, \ldots, \mu_g, \sigma^2)^T$$

contains the distinct unknown parameters in these g normal component den-

Table 2.10. Estimates of Regression Coefficients with *t*-Distributed Errors and by Other Methods.

Method	Log Likelihood	Intercept (β_0)	Air Flow (β_1)	Temperature (β_2)	Acid (β_3)
Normal ($t, \nu = \infty$)	−33.0	−39.92	0.72	1.30	−0.15
$t, \nu = 8$	−32.7	−40.71	0.81	0.97	−0.13
$t, \nu = 4$	−32.1	−40.07	0.86	0.75	−0.12
$t, \nu = 3$	−31.8	−39.13	0.85	0.66	−0.10
$t, \nu = 2$	−31.0	−38.12	0.85	0.56	−0.09
$t, \hat{\nu} = 1.1$	−30.3	−38.50	0.85	0.49	−0.07
$t, \nu = 1$	−30.3	−38.62	0.85	0.49	−0.04
$t, \nu = 0.5$	−31.2	−40.82	0.84	0.54	−0.04
Normal minus four outliers		−37.65	0.80	0.58	−0.07
$\hat{\boldsymbol{\Psi}}_{\text{KB}}$		−42.83	0.93	0.63	−0.10
$\hat{\boldsymbol{\Psi}}_{\text{PE}}$		−40.37	0.72	0.96	−0.07
Huber		−41.00	0.83	0.91	−0.13
Andrews		−37.20	0.82	0.52	−0.07

Source: Adapted from Lange et al. (1989), with permission from the Journal of the American Statistical Association.

sities. The vector $\boldsymbol{\Psi}$ containing all the unknown parameters is now

$$\boldsymbol{\Psi} = (\boldsymbol{\theta}^T, \pi_1, \ldots, \pi_{g-1})^T.$$

The normal mixture model to be fitted is thus

$$f(w; \boldsymbol{\Psi}) = \sum_{i=1}^{g} \pi_i f_i(w; \boldsymbol{\theta}_i), \tag{2.44}$$

where

$$\begin{aligned} f_i(w; \boldsymbol{\theta}_i) &= \phi(w; \mu_i, \sigma^2) \\ &= (2\pi\sigma^2)^{-1/2} \exp\{-\tfrac{1}{2}(w - \mu_i)^2/\sigma^2\}. \end{aligned}$$

The estimation of $\boldsymbol{\Psi}$ on the basis of $\boldsymbol{y}$ is only meaningful if $\boldsymbol{\Psi}$ is identifiable; that is, distinct values of $\boldsymbol{\Psi}$ determine distinct members of the family

$$\{f(w; \boldsymbol{\Psi}) : \quad \boldsymbol{\Psi} \in \boldsymbol{\Omega}\},$$

where $\boldsymbol{\Omega}$ is the specified parameter space. This is true for normal mixtures in that we can determine $\boldsymbol{\Psi}$ up to a permutation of the component labels. For example, for $g = 2$, we cannot distinguish $(\pi_1, \mu_1, \mu_2, \sigma^2)^T$ from $(\pi_2, \mu_2, \mu_1, \sigma^2)^T$, but this lack of identifiability is of no concern in practice, as it can be easily overcome by the imposition of the constraint $\pi_1 \leq \pi_2$ (see McLachlan and Basford, 1988, Section 1.5). The reader is further referred to Titterington et al. (1985, Section 3.1) for a lucid account of the concept of identifiability for mixtures.

As in the case of Example 1.2 of Section 1.4.3 where only the mixing proportions were unknown, we take the complete-data vector $\boldsymbol{x}$ to be

$$\boldsymbol{x} = (\boldsymbol{y}^T, \boldsymbol{z}^T)^T,$$

where the unobservable vector $\boldsymbol{z}$ is as defined by (1.30). The complete-data log likelihood function for $\boldsymbol{\Psi}$ is given by (1.33), but where now

$$\begin{aligned} C &= \sum_{i=1}^{g} \sum_{j=1}^{n} z_{ij} \log f_i(w_j;\, \boldsymbol{\theta}_i) \\ &= -\tfrac{1}{2} n \log(2\pi) \\ &\quad -\tfrac{1}{2} \sum_{i=1}^{g} \sum_{j=1}^{n} z_{ij} \{\log \sigma^2 + (w_j - \mu_i)^2/\sigma^2\}. \end{aligned}$$

The E-step is the same as before, requiring the calculation of (1.35). The M-step now requires the computation of not only (1.37), but also the values $\mu_1^{(k+1)}, \ldots, \mu_g^{(k+1)}$ and $\sigma^{(k+1)^2}$ that, along with $\pi_1^{(k)}, \ldots, \pi_{g-1}^{(k)}$, maximize $Q(\boldsymbol{\Psi};\, \boldsymbol{\Psi}^{(k)})$. Now

$$\sum_{j=1}^{n} z_{ij} w_j \Big/ \sum_{j=1}^{n} z_{ij} \tag{2.45}$$

and

$$\sum_{i=1}^{g} \sum_{j=1}^{n} z_{ij} (w_j - \mu)^2 / n \tag{2.46}$$

are the MLE's of μ_i and σ^2, respectively, if the z_{ij} were observable. As $\log L_c(\boldsymbol{\Psi})$ is linear in the z_{ij}, it follows that the z_{ij} in (2.45) and (2.46) are replaced by their current conditional expectations $z_{ij}^{(k)}$, which here are the current estimates $\tau_i(w_j;\, \boldsymbol{\Psi}^{(k)})$ of the posterior probabilities of membership of the components of the mixture, given by

$$\tau_i(w_j;\, \boldsymbol{\Psi}^{(k)}) = \pi_i^{(k)} f_i(w_j;\, \boldsymbol{\theta}_i^{(k)}) / f(w_j;\, \boldsymbol{\Psi}^{(k)}) \quad (i = 1, \ldots, g).$$

This yields

$$\mu_i^{(k+1)} = \sum_{j=1}^{n} z_{ij}^{(k)} w_j \Big/ \sum_{j=1}^{n} z_{ij}^{(k)} \quad (i = 1, \ldots, g) \tag{2.47}$$

and

$$\sigma^{(k+1)^2} = \sum_{i=1}^{g}\sum_{j=1}^{n} z_{ij}^{(k)}(w_j - \mu_i^{(k+1)})^2/n, \tag{2.48}$$

and $\pi_i^{(k+1)}$ is given by (1.37).

The (vector) likelihood equation for $\hat{\mu}_i$ and $\hat{\sigma}^2$ through direct differentiation of the incomplete-data log likelihood function, $\log L(\boldsymbol{\Psi})$, is quite easily manipulated to show that it is identifiable with the iterative solutions (2.47) and (2.48) produced by application of the EM algorithm. For example, on differentiating (2.44) with respect to μ_i, we obtain that $\hat{\mu}_i$ satisfies the equation

$$\sum_{j=1}^{n} \hat{\pi}_i\{f_i(w_j;\, \hat{\boldsymbol{\theta}}_i)/f(w_j;\, \hat{\boldsymbol{\Psi}})\}(w_j - \hat{\mu}_i) = 0 \quad (i = 1, \ldots, g),$$

which can be written as

$$\hat{\mu}_i = \sum_{j=1}^{n} \tau_i(w_j;\, \hat{\boldsymbol{\Psi}})w_j \bigg/ \sum_{j=1}^{n} \tau_i(w_j;\, \hat{\boldsymbol{\Psi}}), \tag{2.49}$$

and which is identifiable with (2.47).

2.7.2 Example 2.8: Multivariate Component Densities

The results above for univariate normal component densities easily generalize to the multivariate case. On the M-step at the $(k+1)$th iteration, the updated estimates of the ith component mean $\boldsymbol{\mu}_i$ and the common covariance matrix $\boldsymbol{\Sigma}$ are given by

$$\boldsymbol{\mu}_i^{(k+1)} = \sum_{j=1}^{n} z_{ij}^{(k)}\boldsymbol{w}_j \bigg/ \sum_{j=1}^{n} z_{ij}^{(k)} \quad (i = 1, \ldots, g) \tag{2.50}$$

and

$$\boldsymbol{\Sigma}^{(k+1)} = \sum_{i=1}^{g}\sum_{j=1}^{n} z_{ij}^{(k)}(\boldsymbol{w}_j - \boldsymbol{\mu}_i^{(k+1)})(\boldsymbol{w}_j - \boldsymbol{\mu}_i^{(k+1)})^T/n. \tag{2.51}$$

The essence of the EM algorithm for this problem is already there in the work of Day (1969).

In the case of normal components with arbitrary covariance matrices, equation (2.51) is replaced by

$$\boldsymbol{\Sigma}_i^{(k+1)} = \sum_{j=1}^{n} z_{ij}^{(k)}(\boldsymbol{w}_j - \boldsymbol{\mu}_i^{(k+1)})(\boldsymbol{w}_j - \boldsymbol{\mu}_i^{(k+1)})^T \bigg/ \sum_{j=1}^{n} z_{ij}^{(k)} \quad (i = 1, \ldots, g). \tag{2.52}$$

The likelihood function $L(\boldsymbol{\Psi})$ tends to have multiple local maxima for normal mixture models. In this case of unrestricted component covariance matrices, $L(\boldsymbol{\Psi})$ is unbounded, as each data point gives rise to a singularity on the edge of the parameter space; see, for example, McLachlan and Basford (1988, Chapter 2); Gelman, Carlin, Stern, and Rubin (1995), and Lindsay, 1995). It suffices to direct attention to local maxima in the interior of the parameter space, as under essentially the usual regularity conditions (Kiefer, 1978, and Peters and Walker, 1978), there exists a sequence of roots of the likelihood equation that is consistent and asymptotically efficient. With probability tending to one, these roots correspond to local maxima in the interior of the parameter space. In practice, however, consideration has to be given to the problem of relatively large local maxima that occur as a consequence of a fitted component having a very small (but nonzero) variance for univariate data or generalized variance (the determinant of the covariance matrix) for multivariate data. Such a component corresponds to a cluster containing a few data points either relatively close together or almost lying in a lower dimensional subspace in the case of multivariate data. There is thus a need to monitor the relative size of the component variances for univariate observations and of the generalized component variances for multivariate data in an attempt to identify these spurious local maximizers. Hathaway (1983, 1985, 1986) considers a constrained formulation of this problem in order to avoid singularities and to reduce the number of spurious local maximizers. Of course the possibility here that for a given starting point the EM algorithm may converge to a spurious local maximizer or may not converge at all is not a failing of this algorithm. Rather it is a consequence of the properties of the likelihood function for the normal mixture model with unrestricted component matrices in the case of ungrouped data.

2.7.3 Numerical Example: Red Blood Cell Volume Data

We consider here a data set from McLachlan and Jones (1988) in their study of the change in the red blood cell populations of cows exposed to the tick-borne parasite *Anaplasma marginale* in a laboratory trial. The data collected were in the form of red blood cell volume distributions obtained from a Coulter counter. The observed cell counts so obtained from a cow 21 days after inoculation is listed in Table 2.11, along with the expected cell frequencies as obtained by McLachlan and Jones (1988) for this data set under the normal mixture model below. The lower and upper truncation values for these red blood cell volume counts are 28.8 fl and 158.4 fl, respectively, and the grouping is into 18 intervals of equal width of 7.2 fl. A cursory inspection of the observed cell counts displayed in histogram form on the logarithmic scale in Figure 2.1 suggests that the red blood cell volume is bimodal at 21 days after inoculation. Accordingly, McLachlan and Jones (1988) modeled the log of the red blood cell volume distribution by a two-component normal density,

Table 2.11. Doubly Truncated Red Blood Cell Volume Data for a Cow 21 Days after Inoculation.

Group No.	Lower Endpoint of Cell Volume Interval	Frequencies	
		Observed	Expected
1	28.8	10	6.5
2	36.0	21	27.0
3	43.2	51	54.8
4	50.4	77	70.0
5	57.6	70	66.4
6	64.8	50	53.9
7	72.0	44	44.2
8	79.2	40	42.1
9	86.4	46	45.9
10	93.6	54	50.8
11	100.8	53	53.2
12	108.0	54	51.4
13	115.2	44	45.8
14	122.4	36	38.2
15	129.6	29	29.9
16	136.8	21	22.3
17	144.0	16	15.9
18	151.2	13	10.9

Source: Adapted from McLachlan and Jones (1988), with permission of the Biometric Society.

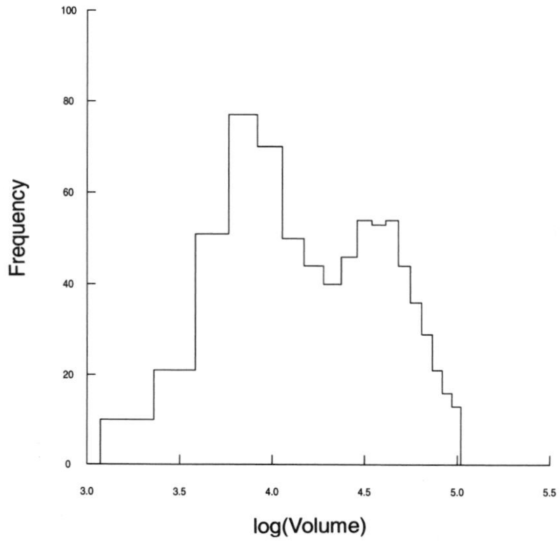

Figure 2.1. Plot of histogram of red blood cell volume data.

Table 2.12. Results of EM Algorithm Applied to Midpoints of the Class Intervals for Red Blood Cell Volume Data.

Parameter	Initial Value	MLE
π_1	0.45	0.5192
μ_1	4.00	4.1103
μ_2	4.40	4.7230
σ_1^2	0.08	0.0685
σ_2^2	0.05	0.0286

$$f(w;\ \boldsymbol{\Psi}) = \pi_1 \phi(w;\ \mu_1,\ \sigma_1^2) + \pi_2 \phi(w;\ \mu_2,\ \sigma_2^2). \tag{2.53}$$

McLachlan and Jones (1988) fitted this model to the data in their original grouped and truncated form, and their results are to be reported in Section 2.8.7, which is devoted to ML estimation from data grouped into intervals. But to illustrate the fitting of a normal mixture model to individual data points, we fit this model by taking the observations within an interval to be at its midpoint and also ignoring the extreme intervals. The adequacy of this approximation depends on the ratio of the width of the intervals relative to the variance; see Heitjan (1989).

From the initial value of $\boldsymbol{\Psi}^{(0)}$ as specified in Table 2.12, the EM algorithm converged after 65 iterations. The stopping criterion was based on the relative change in the log likelihood, using a tolerance value of 10^{-8}.

2.8 EXAMPLE 2.9: GROUPED AND TRUNCATED DATA

2.8.1 Introduction

We consider now the application of the EM algorithm to continuous data that are grouped into intervals and may also be truncated. The results are then specialized to the case of univariate normal data. In general with ML estimation for grouped data, Dempster and Rubin (1983) show how Shepherd's corrections arise through application of the EM algorithm; see Heitjan (1989). Specific examples of ML estimation for grouped discrete data may be found, for instance, in Schader and Schmid (1985) and Adamidis and Loukas (1993), who consider the fitting of the negative binomial distribution and Poisson mixtures in binomial proportions, respectively.

2.8.2 Specification of Complete Data

Let W be a random variable with p.d.f. $f(w;\ \boldsymbol{\Psi})$ specified up to a vector $\boldsymbol{\Psi}$ of unknown parameters. Suppose that the sample space $\mathcal{W}$ of W is partitioned into v mutually exclusive intervals $\mathcal{W}_j \ (j = 1, \ldots, v)$. Independent observations are made on W, but only the number n_j falling in $\mathcal{W}_j \ (j = 1, \ldots, r)$ is recorded, where $r \leq v$. That is, individual observations are not recorded but only the class intervals $\mathcal{W}_j$ in which they fall are recorded; further, even such observations are made only if the W value falls in one of the intervals $\mathcal{W}_j$ $(j = 1, \ldots, r)$.

For given

$$n = \sum_{j=1}^{r} n_j,$$

the observed data

$$\boldsymbol{y} = (n_1, \ldots, n_r)^T$$

has a multinomial distribution, consisting of n draws on r categories with probabilities $P_j(\boldsymbol{\Psi})/P(\boldsymbol{\Psi})$, $j = 1, \ldots, r$, where

$$P_j(\boldsymbol{\Psi}) = \int_{\mathcal{W}_j} f(w;\, \boldsymbol{\Psi})dw \tag{2.54}$$

and

$$P(\boldsymbol{\Psi}) = \sum_{j=1}^{r} P_j(\boldsymbol{\Psi}).$$

Thus the (incomplete-data) log likelihood is given by

$$\log L(\boldsymbol{\Psi}) = \sum_{j=1}^{r} n_j \log\{P_j(\boldsymbol{\Psi})/P(\boldsymbol{\Psi})\} + c_1, \tag{2.55}$$

where

$$c_1 = \log\left\{n! \Big/ \prod_{j=1}^{r} n_j!\right\}.$$

We can solve this problem within the EM framework by introducing the vectors

$$\boldsymbol{u} = (n_{r+1}, \ldots, n_v)^T \tag{2.56}$$

and

$$\boldsymbol{w}_j = (w_{j1}, \ldots, w_{j,n_j})^T \quad (j = 1, \ldots, v) \tag{2.57}$$

as the missing data. The vector $\boldsymbol{u}$ contains the unobservable frequencies in the case of truncation $(r < v)$, while $\boldsymbol{w}_j$ contains the n_j unobservable individual observations in the jth interval $\mathcal{W}_j$ $(j = 1, \ldots, v)$.

The complete-data vector $\boldsymbol{x}$ corresponding to the missing data defined by (2.56) and (2.57) is

$$\boldsymbol{x} = (\boldsymbol{y}^T, \boldsymbol{u}^T, \boldsymbol{w}_1^T, \ldots, \boldsymbol{w}_v^T)^T.$$

These missing data have been introduced so that the complete-data log likelihood function for $\boldsymbol{\Psi}$, $\log L_c(\boldsymbol{\Psi})$, is equivalent to

$$\sum_{j=1}^{v} \sum_{l=1}^{n_j} \log f(w_{jl};\, \boldsymbol{\Psi}), \tag{2.58}$$

which is the log likelihood function formed under the assumption that the observations in $\boldsymbol{w}_1, \ldots, \boldsymbol{w}_v$ constitute an observed random sample of size $n+m$ from $f(w; \boldsymbol{\Psi})$ on $\mathcal{W}$, where

$$m = \sum_{j=r+1}^{v} n_j.$$

We now consider the specification of the distributions of the missing data $\boldsymbol{u}$ and $\boldsymbol{w}_1, \ldots, \boldsymbol{w}_v$, so that $\log L_c(\boldsymbol{\Psi})$ is equivalent to (2.58) and hence implies the incomplete-data log likelihood, $\log L(\boldsymbol{\Psi})$, as given by (2.55). We shall proceed by considering $L_c(\boldsymbol{\Psi})$ as the product of the density of $\boldsymbol{y}$, the conditional density of $\boldsymbol{u}$ given $\boldsymbol{y}$, and the conditional density of $\boldsymbol{w}_1, \ldots, \boldsymbol{w}_v$ given $\boldsymbol{y}$ and $\boldsymbol{u}$. Concerning the latter, $\boldsymbol{w}_1, \ldots, \boldsymbol{w}_v$ are taken to be independent with the n_j observations w_{jk} $(k = 1, \ldots, n_j)$ in $\boldsymbol{w}_j$, constituting an observed random sample of size n_j from the density

$$h_j(w; \boldsymbol{\Psi}) = f(w; \boldsymbol{\Psi})/P_j(\boldsymbol{\Psi}) \quad (j = 1, \ldots, v). \tag{2.59}$$

It remains now to specify the conditional distribution of $\boldsymbol{u}$ given $\boldsymbol{y}$, which we shall write as $d(\boldsymbol{u} \mid \boldsymbol{y}; \boldsymbol{\Psi})$. This is one instance with the application of the EM algorithm where the specification of the distribution of the missing data is perhaps not straightforward. Hence we shall leave it unspecified for the moment and proceed with the formation of $L_c(\boldsymbol{\Psi})$ in the manner described above. It follows that

$$\log L_c(\boldsymbol{\Psi}) = \log L(\boldsymbol{\Psi}) + \log d(\boldsymbol{u} \mid \boldsymbol{y}; \boldsymbol{\Psi}) + \sum_{j=1}^{v} \sum_{l=1}^{n_j} \log h_j(w_{jl}; \boldsymbol{\Psi})$$

which, on using (2.55) and (2.59), equals

$$\begin{aligned}
\log L_c(\boldsymbol{\Psi}) &= \sum_{j=1}^{r} n_j \log\{P_j(\boldsymbol{\Psi})/P(\boldsymbol{\Psi})\} + c_1 \\
&\quad + \log d(\boldsymbol{u} \mid \boldsymbol{y}; \boldsymbol{\Psi}) + \sum_{j=1}^{v} \sum_{l=1}^{n_j} \log\{f(w_{jl}; \boldsymbol{\Psi})/P_j(\boldsymbol{\Psi})\} \\
&= \sum_{j=1}^{v} \sum_{l=1}^{n_j} \log f(w_{jl}; \boldsymbol{\Psi}) \\
&\quad + \log d(\boldsymbol{u} \mid \boldsymbol{y}; \boldsymbol{\Psi}) - \log \left[\{P(\boldsymbol{\Psi})\}^n \prod_{j=r+1}^{v} P_j\{(\boldsymbol{\Psi})\}^{n_j} \right] + c_1.
\end{aligned} \tag{2.60}$$

Thus (2.60) is equivalent to (2.58), and so implies the incomplete-data log likelihood (2.55), if the conditional distribution of $\boldsymbol{u}$ given $\boldsymbol{y}$ is specified to be

$$d(\boldsymbol{u} \mid \boldsymbol{y};\, \boldsymbol{\Psi}) = c_2\{P(\boldsymbol{\Psi})\}^n \prod_{j=r+1}^{v} \{P_j(\boldsymbol{\Psi})\}^{n_j}, \tag{2.61}$$

where c_2 is a normalizing constant that does not depend on $\boldsymbol{\Psi}$. It can be shown that the right-hand side of (2.61) defines a proper probability function if

$$c_2 = (m+n-1)! \Bigg/ \left\{ (n-1)! \prod_{j=r+1}^{v} n_j! \right\}.$$

The use of (2.61) can be viewed as simply a device to produce the desired form (2.58) for the complete-data likelihood $L_c(\boldsymbol{\Psi})$. The distribution (2.6.1) reduces to the negative binomial in the case $r = v - 1$; see Achuthan and Krishnan (1992) for an example of its use in a genetics problem with truncation.

2.8.3 E-Step

On working with (2.58) as the complete-data log likelihood function for $\boldsymbol{\Psi}$, the E-step on the $(k+1)$th iteration is effected by first taking the expectation of $\log L_c(\boldsymbol{\Psi})$ conditional also on $\boldsymbol{w}_1, \ldots, \boldsymbol{w}_v$ and $\boldsymbol{u}$, as well as $\boldsymbol{y}$. On taking the expectation of $\log L_c(\boldsymbol{\Psi})$ over $\boldsymbol{w}_1, \ldots, \boldsymbol{w}_v$ and finally $\boldsymbol{u}$, it follows that on the $(k+1)$th iteration, the expectation of (2.58) conditional on $\boldsymbol{y}$ is given by

$$Q(\boldsymbol{\Psi};\, \boldsymbol{\Psi}^{(k)}) = \sum_{j=1}^{v} n_j^{(k)} Q_j(\boldsymbol{\Psi};\, \boldsymbol{\Psi}^{(k)}) \tag{2.62}$$

where

$$Q_j(\boldsymbol{\Psi};\, \boldsymbol{\Psi}^{(k)}) = E_{\boldsymbol{\Psi}^{(k)}}\{\log f(W;\, \boldsymbol{\Psi}) \mid W \in \mathcal{W}_j\} \tag{2.63}$$

and where

$$\begin{aligned} n_j^{(k)} &= n_j & (j = 1, \ldots, r) \\ &= E_{\boldsymbol{\Psi}^{(k)}}(n_j \mid \boldsymbol{y}) & (j = r+1, \ldots, v). \end{aligned}$$

With the conditional distribution of $\boldsymbol{u}$ given $\boldsymbol{y}$ specified by (2.61), it can be shown that

$$E_{\boldsymbol{\Psi}^{(k)}}(n_j \mid \boldsymbol{y}) = nP_j(\boldsymbol{\Psi}^{(k)})/P(\boldsymbol{\Psi}^{(k)}) \quad (j = r+1, \ldots, v). \tag{2.64}$$

We have seen that a crude approach to estimation from grouped data is to form the likelihood function with the n_j observations in the jth interval $\mathcal{W}_j$ taken to be at the midpoint

$$\overline{w}_j = \tfrac{1}{2}(w_{j-1} + w_j)$$

with (or without) an appropriate adjustment to handle truncation or the extreme intervals in the case of infinite endpoints. With this approach the jth interval contributes

$$n_j \log f(\overline{w};\ \boldsymbol{\Psi})$$

to the log likelihood. We see from (2.62) that with the EM algorithm the corresponding contribution to the log likelihood is

$$n_j E_{\boldsymbol{\Psi}^{(k)}}\{\log f(W;\ \boldsymbol{\Psi}) \mid W \in \mathcal{W}_j\}.$$

Thus the EM algorithm uses the current conditional expectation of the log density over the jth interval. The midpoint approach simply approximates this expectation by evaluating the log density at the midpoint of the interval.

2.8.4 M-Step

On the M-step at the $(k+1)$th iteration, we have on differentiation of $Q(\boldsymbol{\Psi};\ \boldsymbol{\Psi}^{(k)})$ with respect to $\boldsymbol{\Psi}$ that $\boldsymbol{\Psi}^{(k+1)}$ is a root of the equation

$$\partial Q(\boldsymbol{\Psi};\ \boldsymbol{\Psi}^{(k)})/\partial \boldsymbol{\Psi} = \sum_{j=1}^{v} n_j^{(k)} \partial Q_j(\boldsymbol{\Psi};\ \boldsymbol{\Psi}^{(k)})/\partial \boldsymbol{\Psi}, \tag{2.65}$$

where

$$\partial Q_j(\boldsymbol{\Psi};\ \boldsymbol{\Psi}^{(k)})/\partial \boldsymbol{\Psi} = E_{\boldsymbol{\Psi}^{(k)}}\{\partial \log f(W;\ \boldsymbol{\Psi})/\partial \boldsymbol{\Psi} \mid W \in \mathcal{W}_j\}, \tag{2.66}$$

on interchanging the operations of differentiation and expectation.

2.8.5 Confirmation of Incomplete-Data Score Statistic

It is of interest in this example to check that the expression

$$[\partial Q(\boldsymbol{\Psi};\ \boldsymbol{\Psi}^{(k)})/\partial \boldsymbol{\Psi}]_{\boldsymbol{\Psi}^{(k)}=\boldsymbol{\Psi}}$$

for the incomplete-data score statistic $\boldsymbol{S}(\boldsymbol{y};\ \boldsymbol{\Psi})$ in terms of the complete-data specification agrees with what would be obtained by direct differentiation of the incomplete-data log likelihood $\log L(\boldsymbol{\Psi})$ as given by (2.55).

Now on evaluating the right-hand side of (2.66) at $\boldsymbol{\Psi}^{(k)} = \boldsymbol{\Psi}$, it follows that

$$\begin{aligned}[\partial Q_j(\boldsymbol{\Psi};\ \boldsymbol{\Psi}^{(k)})/\partial \boldsymbol{\Psi}]_{\boldsymbol{\Psi}^{(k)}=\boldsymbol{\Psi}} &= \{\partial P_j(\boldsymbol{\Psi})/\partial \boldsymbol{\Psi}\}/P_j(\boldsymbol{\Psi}) \\ &= \partial \log P_j(\boldsymbol{\Psi})/\partial \boldsymbol{\Psi}. \end{aligned} \tag{2.67}$$

Considering the summation in (2.65) over the truncated intervals, we have from (2.64) and (2.67) that

$$\left[\sum_{j=r+1}^{v} n_j^{(k)} \partial Q_j(\boldsymbol{\Psi};\, \boldsymbol{\Psi}^{(k)})/\partial \boldsymbol{\Psi}\right]_{\boldsymbol{\Psi}^{(k)}=\boldsymbol{\Psi}}$$

$$= \{n/P(\boldsymbol{\Psi})\} \sum_{j=r+1}^{v} \partial P_j(\boldsymbol{\Psi})/\partial \boldsymbol{\Psi} \tag{2.68}$$

$$= -\{n/P(\boldsymbol{\Psi})\}\partial P(\boldsymbol{\Psi})/\partial \boldsymbol{\Psi} \tag{2.69}$$

$$= -n\,\partial \log P(\boldsymbol{\Psi})/\partial \boldsymbol{\Psi}. \tag{2.70}$$

The result (2.69) follows from (2.68) since

$$\sum_{j=r+1}^{v} P_j(\boldsymbol{\Psi}) = 1 - P(\boldsymbol{\Psi}).$$

On using (2.67) and (2.70) in (2.65), we have that

$$\begin{aligned} \boldsymbol{S}(\boldsymbol{y};\, \boldsymbol{\Psi}) &= [\partial Q(\boldsymbol{\Psi};\, \boldsymbol{\Psi}^{(k)})/\partial \boldsymbol{\Psi}]_{\boldsymbol{\Psi}^{(k)}=\boldsymbol{\Psi}} \\ &= \sum_{j=1}^{r} n_j \partial \log P_j(\boldsymbol{\Psi})/\partial \boldsymbol{\Psi} - n\,\partial \log P(\boldsymbol{\Psi})/\partial \boldsymbol{\Psi}, \end{aligned} \tag{2.71}$$

which obviously agrees with the expression for $\boldsymbol{S}(\boldsymbol{y};\, \boldsymbol{\Psi})$ obtained by differentiating $\log L(\boldsymbol{\Psi})$ directly with respect to $\boldsymbol{\Psi}$.

We can express $\partial \log P(\boldsymbol{\Psi})/\partial \boldsymbol{\Psi}$ in (2.71) as

$$n\,\partial \log P(\boldsymbol{\Psi})/\partial \boldsymbol{\Psi} = \sum_{j=1}^{r} \{nP_j(\boldsymbol{\Psi})/P(\boldsymbol{\Psi})\}\partial \log P_j(\boldsymbol{\Psi})/\partial \boldsymbol{\Psi},$$

and so $\boldsymbol{S}(\boldsymbol{y};\, \boldsymbol{\Psi})$ can be expressed in the form

$$\boldsymbol{S}(\boldsymbol{y};\, \boldsymbol{\Psi}) = \sum_{j=1}^{r} \left\{ n_j - \frac{nP_j(\boldsymbol{\Psi})}{P(\boldsymbol{\Psi})} \right\} \partial \log P_j(\boldsymbol{\Psi})/\partial \boldsymbol{\Psi}. \tag{2.72}$$

We shall make use of this result in Section 4.4.

2.8.6 M-Step for Grouped Normal Data

To discuss the M-step, we consider the case where

$$f(w;\, \boldsymbol{\Psi}) = \phi(w;\, \mu,\, \sigma^2)$$

denotes the normal density with mean μ and variance σ^2 and

$$\boldsymbol{\Psi} = (\mu,\ \sigma^2)^T.$$

In this case,

$$Q(\boldsymbol{\Psi};\ \boldsymbol{\Psi}^{(k)}) = -\tfrac{1}{2}\left(n + \sum_{j=r+1}^{v} n_j^{(k)}\right)\{\log(2\pi) + \log\sigma^2\}$$
$$-\tfrac{1}{2}\sigma^{-2}\sum_{j=1}^{v} n_j^{(k)} E_{\boldsymbol{\Psi}^{(k)}}\{(W-\mu)^2 \mid W \in \mathcal{W}_j\}.$$

It can now be shown that the value $\boldsymbol{\Psi}^{(k+1)}$ of $\boldsymbol{\Psi}$ that maximizes $Q(\boldsymbol{\Psi},\ \boldsymbol{\Psi}^{(k)})$ is given by

$$\boldsymbol{\Psi}^{(k+1)} = (\mu^{(k+1)},\ \sigma^{(k+1)^2})^T,$$

where

$$\mu^{(k+1)} = \sum_{j=1}^{v} n_j^{(k)}\{E_{\boldsymbol{\Psi}^{(k)}}(W \mid W \in \mathcal{W}_j)\} \Big/ \sum_{j=1}^{v} n_j^{(k)} \tag{2.73}$$

and

$$\sigma^{(k+1)^2} = \sum_{j=1}^{v} n_j^{(k)}[E_{\boldsymbol{\Psi}^{(k)}}\{(W-\mu^{(k+1)})^2 \mid W \in \mathcal{W}_j\}] \Big/ \sum_{j=1}^{v} n_j^{(k)}. \tag{2.74}$$

We see from (2.73) and (2.74) that in computing $\mu^{(k+1)}$ and $\sigma^{(k+1)^2}$, an unobservable frequency n_j in the case of truncation is replaced by

$$n_j^{(k)} = nP_j(\boldsymbol{\Psi}^{(k)})/P(\boldsymbol{\Psi}^{(k)}).$$

For the computation of the mean $\mu^{(k+1)}$, the individual values of W in the interval $\mathcal{W}_j$ are replaced by the current conditional expectation of W given W falls in $\mathcal{W}_j$. Similarly, for the computation of the variance $\sigma^{(k+1)^2}$, the individual values of $(W-\mu^{(k+1)})^2$ in the interval $\mathcal{W}_j$ are replaced by the current conditional expectation of $(W-\mu^{(k+1)})^2$ given W falls in $\mathcal{W}_j$. Jones and McLachlan (1990) have developed an algorithm for fitting finite mixtures of normal distributions to data that are grouped into intervals and that may also be truncated.

2.8.7 Numerical Example: Grouped Log Normal Data

In the previous section, we fitted on the log scale a two-component normal mixture model (2.53) to the midpoints of the class intervals in Table 2.8 into which the red blood cell volume have been grouped. We now report the

Table 2.13. Results of EM Algorithm Applied to Red Blood Cell Volume Data in Grouped and Truncated Form on the Log Scale.

	MLE	
Parameter	Midpoint Approximation	Exact Result
π_1	0.5192	0.4521
	(0.0417)	(0.0521)
μ_1	4.1103	4.0728
	(0.0326)	(0.0384)
μ_2	4.7230	4.7165
	(0.0175)	(0.0242)
σ_1^2	0.0685	0.0575
	(0.0105)	(0.0107)
σ_2^2	0.0286	0.0438
	(0.0047)	(0.0099)

results of McLachlan and Jones (1988) who fitted this model to the data in their original grouped and truncated form. In Section 2.7, we described the fitting of a normal mixture density, while in Section 2.8 we described the fitting of a single normal density to grouped and truncated data. These two M-steps can be combined to give the M-step for the fitting of a normal mixture density to data that are grouped and truncated, as detailed in McLachlan and Jones (1988). In Table 2.13, we display the results of the fit that they obtained for the present data set after 117 iterations of the EM algorithm started from the same value for $\boldsymbol{\Psi}$ as in Table 2.12. Also displayed there is the corresponding fit obtained in the previous section for the midpoint approximation method. The entries in parentheses refer to the standard errors of the above estimates and they were obtained using methodology discussed in Chapter 4. The standard errors for the estimates from the midpoint approximation method were obtained by using the inverse of the empirical information matrix (4.41) to approximate the covariance matrix of the MLE, while for the exact MLE method, the corresponding form (4.45) of the empirical information matrix for grouped data was used.

It can be seen from Table 2.13 that the midpoint approximation technique overestimates the mixing parameter. The estimates of the two component means are similar for the midpoint approximation and exact ML methods, while the former overestimates the first component variance and underestimates the second component variance. The approximate standard errors of the parameter estimates are lower for the midpoint approximation method.

The expected frequencies for this model fit are displayed alongside the observed frequencies in Table 2.11. The chi-squared goodness-of-fit statistic is equal to 5.77 on twelve degrees of freedom, confirming that the two-component log normal mixture provides an adequate fit to the observed frequency counts.

CHAPTER 3

Basic Theory of the EM Algorithm

3.1 INTRODUCTION

Having illustrated the EM algorithm in a variety of problems and having presented a general formulation of it in earlier chapters, we now begin a systematic account of its theory. In this chapter, we show that the likelihood increases with each EM or GEM iteration, and under fairly general conditions, the likelihood values converge to stationary values. However, in order to caution the reader that the EM algorithm will not always lead to at least a local if not a global maximum of the likelihood, we present two examples where this is not the case and convergence is to a saddle point in one case and to a local *minimum* in the other case, of the likelihood. The principles of Self-Consistency and Missing Information are explained in the sequel. The notion that the rate at which the EM algorithm converges depends upon the amount of information missing in the incomplete data *vis a vis* the formulated complete data, is made explicit by deriving results regarding the rate of convergence in terms of information matrices for the incomplete- and complete-data problems. Theoretical results are illustrated mostly with examples discussed in earlier chapters.

Our intent here is to collect in one place and present in a unified manner the available results concerning the basic theory of EM—aspects of convergence and rates of convergence. The treatment in this chapter is inevitably mathematical. Readers not interested in mathematical details may skip proofs of theorems.

3.2 MONOTONICITY OF THE EM ALGORITHM

Dempster, Laird, and Rubin (1977) show that the (incomplete-data) likelihood function $L(\boldsymbol{\Psi})$ is not decreased after an EM iteration; that is,

$$L(\boldsymbol{\Psi}^{(k+1)}) \geq L(\boldsymbol{\Psi}^{(k)}) \tag{3.1}$$

for $k = 0, 1, 2, \ldots$. To see this, let

$$k(\boldsymbol{x} \mid \boldsymbol{y};\, \boldsymbol{\Psi}) = g_c(\boldsymbol{x};\, \boldsymbol{\Psi})/g(\boldsymbol{y};\, \boldsymbol{\Psi}) \tag{3.2}$$

be the conditional density of $\boldsymbol{X}$ given $\boldsymbol{Y} = \boldsymbol{y}$. Then the log likelihood is given by

$$\begin{aligned}\log L(\boldsymbol{\Psi}) &= \log g(\boldsymbol{y};\, \boldsymbol{\Psi}) \\ &= \log g_c(\boldsymbol{x};\, \boldsymbol{\Psi}) - \log k(\boldsymbol{x} \mid \boldsymbol{y};\, \boldsymbol{\Psi}) \\ &= \log L_c(\boldsymbol{\Psi}) - \log k(\boldsymbol{x} \mid \boldsymbol{y};\, \boldsymbol{\Psi}).\end{aligned} \tag{3.3}$$

On taking the expectations of both sides of (3.3) with respect to the conditional distribution of $\boldsymbol{X}$ given $\boldsymbol{Y} = \boldsymbol{y}$, using the fit $\boldsymbol{\Psi}^{(k)}$ for $\boldsymbol{\Psi}$, we have that

$$\begin{aligned}\log L(\boldsymbol{\Psi}) &= E_{\boldsymbol{\Psi}^{(k)}}\{\log L_c(\boldsymbol{\Psi}) \mid \boldsymbol{y}\} - E_{\boldsymbol{\Psi}^{(k)}}\{\log k(\boldsymbol{X} \mid \boldsymbol{y};\, \boldsymbol{\Psi}) \mid \boldsymbol{y}\} \\ &= Q(\boldsymbol{\Psi};\, \boldsymbol{\Psi}^{(k)}) - H(\boldsymbol{\Psi};\, \boldsymbol{\Psi}^{(k)}),\end{aligned} \tag{3.4}$$

where

$$H(\boldsymbol{\Psi};\, \boldsymbol{\Psi}^{(k)}) = E_{\boldsymbol{\Psi}^{(k)}}\{\log k(\boldsymbol{X} \mid \boldsymbol{y};\, \boldsymbol{\Psi}) \mid \boldsymbol{y}\}. \tag{3.5}$$

From (3.4), we have that

$$\begin{aligned}&\log L(\boldsymbol{\Psi}^{(k+1)}) - \log L(\boldsymbol{\Psi}^{(k)}) \\ &\quad = \{Q(\boldsymbol{\Psi}^{(k+1)};\, \boldsymbol{\Psi}^{(k)}) - Q(\boldsymbol{\Psi}^{(k)};\, \boldsymbol{\Psi}^{(k)})\} \\ &\qquad -\{H(\boldsymbol{\Psi}^{(k+1)};\, \boldsymbol{\Psi}^{(k)}) - H(\boldsymbol{\Psi}^{(k)};\, \boldsymbol{\Psi}^{(k)})\}.\end{aligned} \tag{3.6}$$

The first difference on the right-hand side of (3.6) is nonnegative since $\boldsymbol{\Psi}^{(k+1)}$ is chosen so that

$$Q(\boldsymbol{\Psi}^{(k+1)};\, \boldsymbol{\Psi}^{(k)}) \geq Q(\boldsymbol{\Psi};\, \boldsymbol{\Psi}^{(k)}) \tag{3.7}$$

for all $\boldsymbol{\Psi} \in \boldsymbol{\Omega}$. Hence (3.1) holds if the second difference on the right-hand side of (3.6) is nonpositive; that is, if

$$H(\boldsymbol{\Psi}^{(k+1)};\, \boldsymbol{\Psi}^{(k)}) - H(\boldsymbol{\Psi}^{(k)};\, \boldsymbol{\Psi}^{(k)}) \leq 0. \tag{3.8}$$

Now for any $\boldsymbol{\Psi}$,

$$\begin{aligned}&H(\boldsymbol{\Psi};\, \boldsymbol{\Psi}^{(k)}) - H(\boldsymbol{\Psi}^{(k)};\, \boldsymbol{\Psi}^{(k)}) \\ &= E_{\boldsymbol{\Psi}^{(k)}}[\log\{k(\boldsymbol{X} \mid \boldsymbol{y};\, \boldsymbol{\Psi})/k(\boldsymbol{X} \mid \boldsymbol{y};\, \boldsymbol{\Psi}^{(k)})\} \mid \boldsymbol{y}] \\ &\leq \log[E_{\boldsymbol{\Psi}^{(k)}}\{k(\boldsymbol{X} \mid \boldsymbol{y};\, \boldsymbol{\Psi})/k(\boldsymbol{X} \mid \boldsymbol{y};\, \boldsymbol{\Psi}^{(k)})\} \mid \boldsymbol{y}] \qquad (3.9) \\ &= \log \int_{\mathcal{X}(\boldsymbol{y})} k(\boldsymbol{x} \mid \boldsymbol{y};\, \boldsymbol{\Psi})\, d\boldsymbol{x} \\ &= 0, \qquad (3.10)\end{aligned}$$

where the inequality in (3.9) is a consequence of Jensen's inequality and the concavity of the logarithmic function.

This establishes (3.8), and hence the inequality (3.1), showing that the likelihood $L(\boldsymbol{\Psi})$ is not decreased after an EM iteration. The likelihood will be increased if the inequality (3.7) is strict. Thus for a bounded sequence of likelihood values $\{L(\boldsymbol{\Psi}^{(k)}\}$, $L(\boldsymbol{\Psi}^{(k)})$ converges monotonically to some L^*.

A consequence of (3.1) is the self-consistency of the EM algorithm. For if the MLE $\hat{\boldsymbol{\Psi}}$ globally maximizes $L(\boldsymbol{\Psi})$, it must satisfy

$$Q(\hat{\boldsymbol{\Psi}}; \hat{\boldsymbol{\Psi}}) \geq Q(\boldsymbol{\Psi}; \hat{\boldsymbol{\Psi}}) \tag{3.11}$$

for all $\boldsymbol{\Psi}$. Otherwise

$$Q(\hat{\boldsymbol{\Psi}}; \hat{\boldsymbol{\Psi}}) < Q(\boldsymbol{\Psi}_o; \hat{\boldsymbol{\Psi}})$$

for some $\boldsymbol{\Psi}_o$, implying that

$$L(\boldsymbol{\Psi}_o) > L(\hat{\boldsymbol{\Psi}}),$$

which would contradict the fact that $\hat{\boldsymbol{\Psi}}$ is the global maximizer of $L(\boldsymbol{\Psi})$.

It will be seen that the differential form of (3.11) is that $\hat{\boldsymbol{\Psi}}$ is a root of the equation

$$[\partial Q(\boldsymbol{\Psi}; \hat{\boldsymbol{\Psi}})/\partial \boldsymbol{\Psi}]_{\boldsymbol{\Psi}=\hat{\boldsymbol{\Psi}}} = \mathbf{0}.$$

This equation will be established shortly in the next section. As noted by Efron (1982), the self-consistency property of the MLE has been rediscovered in many different contexts since Fisher's (1922, 1925, 1934) original papers. Here it forms the basis of the EM algorithm.

3.3 MONOTONICITY OF A GENERALIZED EM ALGORITHM

From the definition of a GEM algorithm given in Section 1.5.5, $\boldsymbol{\Psi}^{(k+1)}$ is chosen not to globally maximize $Q(\boldsymbol{\Psi}; \boldsymbol{\Psi}^{(k)})$ with respect to $\boldsymbol{\Psi}$, but rather to satisfy

$$Q(\boldsymbol{\Psi}^{(k+1)}; \boldsymbol{\Psi}^{(k)}) \geq Q(\boldsymbol{\Psi}^{(k)}; \boldsymbol{\Psi}^{(k)}). \tag{3.12}$$

We have seen in the previous section that this condition (3.12) is sufficient to ensure that (3.1) holds for an iterative sequence $\{\boldsymbol{\Psi}^{(k)}\}$. Thus the likelihood is not decreased after a GEM iteration.

3.4 CONVERGENCE OF AN EM SEQUENCE TO A STATIONARY VALUE

3.4.1 Introduction

As shown in the last section, for a sequence of likelihood values $\{L(\boldsymbol{\Psi}^{(k)})\}$

bounded above, $L(\boldsymbol{\Psi}^{(k)})$ converges monotonically to some value L^*. In almost all applications, L^* is a stationary value. That is, $L^* = L(\boldsymbol{\Psi}^*)$ for some point $\boldsymbol{\Psi}^*$ at which

$$\partial L(\boldsymbol{\Psi})/\partial\boldsymbol{\Psi} = \mathbf{0},$$

or equivalently,

$$\partial \log L(\boldsymbol{\Psi})/\partial\boldsymbol{\Psi} = \mathbf{0}.$$

Moreover, in many practical applications, L^* will be a local maximum. In any event, if an EM sequence $\{\boldsymbol{\Psi}^{(k)}\}$ is trapped at some stationary point $\boldsymbol{\Psi}^*$ that is not a local or global maximizer of $L(\boldsymbol{\Psi})$ (for example, a saddle point), a small random perturbation of $\boldsymbol{\Psi}$ away from the saddle point $\boldsymbol{\Psi}^*$ will cause the EM algorithm to diverge from the saddle point.

In general, if $L(\boldsymbol{\Psi})$ has several stationary points, convergence of the EM sequence to either type (local or global maximizers, saddle points) depends on the choice of starting point $\boldsymbol{\Psi}^{(0)}$. In the case where the likelihood function $L(\boldsymbol{\Psi})$ is unimodal in $\boldsymbol{\Omega}$ (and a certain differentiability condition is satisfied), any EM sequence converges to the unique MLE, irrespective of its starting point $\boldsymbol{\Psi}^{(0)}$. This result is to be given at the end of the next section as a corollary to the convergence theorems to be presented in the sequel.

On differentiating both sides of (3.4), we have that

$$\partial \log L(\boldsymbol{\Psi})/\partial\boldsymbol{\Psi} = \partial Q(\boldsymbol{\Psi}; \boldsymbol{\Psi}^{(k)})/\partial\boldsymbol{\Psi} - \partial H(\boldsymbol{\Psi}; \boldsymbol{\Psi}^{(k)})/\partial\boldsymbol{\Psi}. \tag{3.13}$$

The inequality (3.10) implies that

$$H(\boldsymbol{\Psi}; \boldsymbol{\Psi}^{(k)}) \leq H(\boldsymbol{\Psi}^{(k)}; \boldsymbol{\Psi}^{(k)})$$

for all $\boldsymbol{\Psi} \in \boldsymbol{\Omega}$, and so

$$[\partial H(\boldsymbol{\Psi}; \boldsymbol{\Psi}^{(k)})/\partial\boldsymbol{\Psi}]_{\boldsymbol{\Psi}=\boldsymbol{\Psi}^{(k)}} = \mathbf{0}. \tag{3.14}$$

Let $\boldsymbol{\Psi}_o$ be some arbitrary value of $\boldsymbol{\Psi}$. On putting $\boldsymbol{\Psi}^{(k)} = \boldsymbol{\Psi}_o$ in (3.13), we have from (3.14) that

$$\partial \log L(\boldsymbol{\Psi}_o)/\partial\boldsymbol{\Psi} = [\partial Q(\boldsymbol{\Psi}; \boldsymbol{\Psi}_o)/\partial\boldsymbol{\Psi}]_{\boldsymbol{\Psi}=\boldsymbol{\Psi}_o}, \tag{3.15}$$

where $\partial \log L(\boldsymbol{\Psi}_o)/\partial\boldsymbol{\Psi}$ denotes $\partial \log L(\boldsymbol{\Psi})/\partial\boldsymbol{\Psi}$ evaluated at $\boldsymbol{\Psi} = \boldsymbol{\Psi}_o$.

Suppose that $\boldsymbol{\Psi} = \boldsymbol{\Psi}^*$, where $\boldsymbol{\Psi}^*$ is a stationary point of $L(\boldsymbol{\Psi})$. Then from (3.15),

$$\begin{aligned}\partial \log L(\boldsymbol{\Psi}^*)/\partial\boldsymbol{\Psi} &= [\partial Q(\boldsymbol{\Psi}; \boldsymbol{\Psi}^*)/\partial\boldsymbol{\Psi}]_{\boldsymbol{\Psi}=\boldsymbol{\Psi}^*} \\ &= \mathbf{0}. \end{aligned} \tag{3.16}$$

It can be seen from (3.16) that the EM algorithm can converge to a saddle point $\boldsymbol{\Psi}^*$ if $Q(\boldsymbol{\Psi}; \boldsymbol{\Psi}^*)$ is globally maximized over $\boldsymbol{\Psi} \in \boldsymbol{\Omega}$ at $\boldsymbol{\Psi}^*$.

This led Wu (1983) to propose the condition

$$\sup_{\boldsymbol{\Psi}\in\boldsymbol{\Omega}} Q(\boldsymbol{\Psi}; \boldsymbol{\Psi}^*) > Q(\boldsymbol{\Psi}^*; \boldsymbol{\Psi}^*) \tag{3.17}$$

for any stationary point $\boldsymbol{\Psi}^*$ that is not a local maximizer of $L(\boldsymbol{\Psi})$. This condition in conjunction with his regularity conditions to be given in the next subsection will ensure that all limit points of any instance of the EM algorithm are local maximizers of $L(\boldsymbol{\Psi})$ and that $L(\boldsymbol{\Psi}^{(k)})$ converges monotonically to $L^* = L(\boldsymbol{\Psi}^*)$ for some local maximizer $\boldsymbol{\Psi}^*$. However, the utility of condition (3.17) is limited given that it is typically hard to verify.

In Section 3.5, we shall give an example of an EM sequence converging to a saddle point, and also an example of an EM sequence that converges to a local minimizer if started from some isolated initial values. A reader not concerned with these esoteric issues of the EM algorithm may wish to proceed directly to Section 3.7.

3.4.2 Regularity Conditions of Wu

As can be seen from (1.54), the M-step of the EM algorithm involves the point-to-set map

$$\mathcal{M}(\boldsymbol{\Psi}^{(k)}) = \arg\max_{\boldsymbol{\Psi}\in\boldsymbol{\Omega}} Q(\boldsymbol{\Psi}; \boldsymbol{\Psi}^{(k)});$$

that is, $\mathcal{M}(\boldsymbol{\Psi}^{(k)})$ is the set of values of $\boldsymbol{\Psi}$ that maximize $Q(\boldsymbol{\Psi}; \boldsymbol{\Psi}^{(k)})$ over $\boldsymbol{\Psi} \in \boldsymbol{\Omega}$. For a GEM algorithm, $\mathcal{M}(\boldsymbol{\Psi}^{(k)})$ is specified by the choice of $\boldsymbol{\Psi}^{(k+1)}$ on the M-step so that

$$Q(\boldsymbol{\Psi}^{(k+1)}; \boldsymbol{\Psi}^{(k)}) \geq Q(\boldsymbol{\Psi}^{(k)}; \boldsymbol{\Psi}^{(k)}).$$

Wu (1983) makes use of existing results in the optimization literature for point-to-set maps to establish conditions to ensure the convergence of a sequence of likelihood values $\{L(\boldsymbol{\Psi}^{(k)})\}$ to a stationary value of $L(\boldsymbol{\Psi})$. The point-to-set map $\mathcal{M}(\boldsymbol{\Psi})$ is said to be closed at $\boldsymbol{\Psi} = \boldsymbol{\Psi}_o$ if $\boldsymbol{\Psi}_m \rightarrow \boldsymbol{\Psi}_o$, $\boldsymbol{\Psi}_m \in \boldsymbol{\Omega}$, and $\boldsymbol{\phi}_m \rightarrow \boldsymbol{\phi}_o$, $\boldsymbol{\phi}_m \in \mathcal{M}(\boldsymbol{\Psi}_m)$, implies that $\boldsymbol{\phi}_o \in \mathcal{M}(\boldsymbol{\Psi}_o)$. For a point-to-point map, continuity implies closedness.

Wu (1983) supposes that the following assumptions hold.

$$\boldsymbol{\Omega} \text{ is a subset in } d\text{-dimensional Euclidean space } \mathbb{R}^d. \tag{3.18}$$

$$\boldsymbol{\Omega}_{\boldsymbol{\Psi}_o} = \{\boldsymbol{\Psi} \in \boldsymbol{\Omega} : L(\boldsymbol{\Psi}) \geq L(\boldsymbol{\Psi}_o)\} \text{ is compact for any } L(\boldsymbol{\Psi}_o) > -\infty. \tag{3.19}$$

$$L(\boldsymbol{\Psi}) \text{ is continuous in } \boldsymbol{\Omega} \text{ and differentiable in the interior of } \boldsymbol{\Omega}. \tag{3.20}$$

A consequence of the conditions (3.18) to (3.20) is that any sequence $\{L(\boldsymbol{\Psi}^{(k)})\}$ is bounded above for any $\boldsymbol{\Psi}^{(0)} \in \boldsymbol{\Omega}$, where, to avoid trivialities, it is assumed that the starting point $\boldsymbol{\Psi}^{(0)}$ satisfies $L(\boldsymbol{\Psi}^{(0)}) > -\infty$. As Wu (1983) acknowledges, the compactness assumption (3.19) can be restrictive when no realistic compactification of the original parameter space is available. The regularity conditions (3.18) to (3.20) are assumed to hold for the remainder of this section and the next. It is also assumed there that each $\boldsymbol{\Psi}^{(k)}$ is in the interior of $\boldsymbol{\Omega}$. That is, $\boldsymbol{\Psi}^{(k+1)}$ is a solution of the equation

$$\partial Q(\boldsymbol{\Psi}; \boldsymbol{\Psi}^{(k)})/\partial \boldsymbol{\Psi} = \mathbf{0}. \tag{3.21}$$

Wu (1983) notes that this assumption will be implied, for example, by

$$\Omega_{\boldsymbol{\Psi}_o} \text{ is in the interior of } \boldsymbol{\Omega} \text{ for any } \boldsymbol{\Psi}_o \in \boldsymbol{\Omega}. \tag{3.22}$$

An example where the compactness regularity condition (3.19) is not satisfied is a mixture of univariate normals with means μ_1 and μ_2 and unrestricted variances σ_1^2 and σ_2^2. Each data point w_j gives rise to a singularity in $L(\boldsymbol{\Psi})$ on the edge of the parameter space $\boldsymbol{\Omega}$. More specifically, if $\mu_1(\mu_2)$ is set equal to w_j, then $L(\boldsymbol{\Psi})$ tends to infinity as $\sigma_1^2(\sigma_2^2)$ tends to zero. Thus in this example, if $\boldsymbol{\Psi}_o$ is any point in $\boldsymbol{\Omega}$ with μ_1 or μ_2 set equal to w_j, then clearly $\boldsymbol{\Omega}_{\boldsymbol{\Psi}_o}$ is not compact. This space can be made compact by imposing the constraint $\sigma_i^2 \geq \epsilon$ $(i = 1, 2)$. However, the condition (3.21) will not hold for EM sequences started sufficiently close to the boundary of the modified parameter space.

3.4.3 Main Convergence Theorem for a Generalized EM Sequence

We now state without proof the main convergence theorem given by Wu (1983) for a GEM algorithm. It also applies to the EM algorithm, since the latter is a special case of a GEM algorithm.

Theorem 3.1. *Let $\{\boldsymbol{\Psi}^{(k)}\}$ be an instance of a GEM algorithm generated by $\boldsymbol{\Psi}^{(k+1)} \in \mathcal{M}(\boldsymbol{\Psi}^{(k)})$. Suppose that (i) $\mathcal{M}(\boldsymbol{\Psi}^{(k)})$ is closed over the complement of $\mathcal{S}$, the set of stationary points in the interior of $\boldsymbol{\Omega}$ and that (ii)*

$$L(\boldsymbol{\Psi}^{(k+1)}) > L(\boldsymbol{\Psi}^{(k)}) \text{ for all } \boldsymbol{\Psi}^{(k)} \notin \mathcal{S}.$$

Then all the limit points of $\{\boldsymbol{\Psi}^{(k)}\}$ are stationary points and $L(\boldsymbol{\Psi}^{(k)})$ converges monotonically to $L^* = L(\boldsymbol{\Psi}^*)$ for some stationary point $\boldsymbol{\Psi}^* \in \mathcal{S}$.

Condition (ii) holds for an EM sequence. For consider a $\boldsymbol{\Psi}^{(k)} \notin \mathcal{S}$. Then from (3.15),

$$[\partial Q(\boldsymbol{\Psi}; \boldsymbol{\Psi}^{(k)})/\partial \boldsymbol{\Psi}]_{\boldsymbol{\Psi}=\boldsymbol{\Psi}^{(k)}} = \partial \log L(\boldsymbol{\Psi}^{(k)})/\partial \boldsymbol{\Psi}$$
$$\neq \mathbf{0},$$

since $\boldsymbol{\Psi}^{(k)} \notin \mathcal{S}$. Hence, $Q(\boldsymbol{\Psi}; \boldsymbol{\Psi}^{(k)})$ is not maximized at $\boldsymbol{\Psi} = \boldsymbol{\Psi}^{(k)}$, and so by the definition of the M-step of the EM algorithm and (3.11),

$$Q(\boldsymbol{\Psi}^{(k+1)}; \boldsymbol{\Psi}^{(k)}) > Q(\boldsymbol{\Psi}^{(k)}; \boldsymbol{\Psi}^{(k)}).$$

This implies

$$L(\boldsymbol{\Psi}^{(k+1)}) > L(\boldsymbol{\Psi}^{(k)},$$

thus establishing condition (ii) of Theorem 3.1.

For the EM algorithm, Wu (1983) notes that a sufficient condition for the closedness of the EM map is that

$$Q(\boldsymbol{\Psi}; \boldsymbol{\phi}) \text{ is continuous in both } \boldsymbol{\Psi} \text{ and } \boldsymbol{\phi}. \tag{3.23}$$

This condition is very weak and should hold in most practical situations. It has been shown to be satisfied for the case of a curved exponential family

$$g_c(\boldsymbol{x}; \boldsymbol{\Psi}) = b(\boldsymbol{x}) \exp\{\boldsymbol{\Psi}^T \boldsymbol{t}(\boldsymbol{x})\}/a(\boldsymbol{\Psi}),$$

where $\boldsymbol{\Psi}$ lies in a compact submanifold $\boldsymbol{\Omega}_0$ of the d-dimensional region $\boldsymbol{\Omega}$, as defined by (1.55).

This leads to the following theorem of Wu (1983) for an EM sequence.

3.4.4 A Convergence Theorem for an EM Sequence

Theorem 3.2. *Suppose that* $Q(\boldsymbol{\Psi}; \boldsymbol{\phi})$ *satisfies the continuity condition* (3.23). *Then all the limit points of any instance* $\{\boldsymbol{\Psi}^{(k)}\}$ *of the EM algorithm are stationary points of* $L(\boldsymbol{\Psi})$, *and* $L(\boldsymbol{\Psi}^{(k)})$ *converges monotonically to some value* $L^* = L(\boldsymbol{\Psi}^*)$ *for some stationary point* $\boldsymbol{\Psi}^*$.

Theorem 3.2 follows immediately from Theorem 3.1, since its condition (i) is implied by the continuity assumption (3.23), while its condition (ii) is automatically satisfied for an EM sequence.

According to Wu (1983), Theorem 3.1 is the most general result for EM and GEM algorithms. Theorem 3.2 was obtained by Baum et al. (1970) and Haberman (1977) for two special models. Boyles (1983) gives a similar result for general models, but under stronger regularity conditions. One key condition in Baum et al. (1970) and Boyles (1983) is that $\mathcal{M}(\boldsymbol{\Psi}^{(k)})$ is a continuous point-to-point map over $\boldsymbol{\Omega}$, which is stronger than a closed point-to-point map over the complement of $\mathcal{S}$, as assumed in Theorem 3.1.

From the viewpoint of the user, Theorem 3.2 provides the most useful result, since it only requires conditions that are easy to verify.

3.5 CONVERGENCE OF AN EM SEQUENCE OF ITERATES

3.5.1 Introduction

The convergence of the sequence of likelihood values $\{L(\boldsymbol{\Psi}^{(k)})\}$ to some value L^* does not automatically imply the convergence of the corresponding sequence of iterates $\{\boldsymbol{\Psi}^{(k)}\}$ to a point $\boldsymbol{\Psi}^*$. But as Wu (1983) stresses, from a numerical viewpoint, the convergence of $\{\boldsymbol{\Psi}^{(k)}\}$ is not as important as the convergence of $\{L(\boldsymbol{\Psi}^{(k)})\}$ to a stationary value, in particular to a local maximum.

In this section, we present some results of Wu (1983) on the convergence of an EM sequence of iterates. As we shall see below, convergence of the EM iterates usually requires more stringent regularity conditions than for the convergence of the likelihood values.

3.5.2 Two Convergence Theorems of Wu

Define

$$\mathcal{S}(a) = \{\boldsymbol{\Psi} \in \boldsymbol{\Omega} : L(\boldsymbol{\Psi}) = a\}$$

to be the subset of the set $\mathcal{S}$ of stationary points in the interior of $\boldsymbol{\Omega}$ at which $L(\boldsymbol{\Psi})$ equals a.

Theorem 3.3. *Let $\{\boldsymbol{\Psi}^{(k)}\}$ be an instance of a GEM algorithm satisfying conditions (i) and (ii) of Theorem 3.1. Suppose $\mathcal{S}(L^*)$ consists of the single point $\boldsymbol{\Psi}^*$ (that is, there cannot be two different stationary points with the same value L^*), where L^* is the limit of $L(\boldsymbol{\Psi}^{(k)})$. Then $\boldsymbol{\Psi}^{(k)}$ converges to $\boldsymbol{\Psi}^*$.*

This theorem follows immediately from Theorem 3.1 given that $\mathcal{S}(L^*)$ is a singleton. Wu (1983) notes that the above assumption $\mathcal{S}(L^*) = \{\boldsymbol{\Psi}^*\}$ can be greatly relaxed if we assume

$$||\boldsymbol{\Psi}^{(k+1)} - \boldsymbol{\Psi}^{(k)}|| \rightarrow 0, \text{ as } k \rightarrow \infty, \tag{3.24}$$

as a necessary condition for $\boldsymbol{\Psi}^{(k)}$ to tend to $\boldsymbol{\Psi}^*$.

Theorem 3.4. *Let $\{\boldsymbol{\Psi}^{(k)}\}$ be an instance of a GEM algorithm satisfying conditions (i) and (ii) of Theorem 3.1. If* (3.24) *holds, then all the limit points of $\{\boldsymbol{\Psi}^{(k)}\}$ are in a connected and compact subset of $\mathcal{S}(L^*)$. In particular, if $\mathcal{S}(L^*)$ is discrete, then $\boldsymbol{\Psi}^{(k)}$ converges to some $\boldsymbol{\Psi}^*$ in $\mathcal{S}(L^*)$.*

Proof. From condition (3.19), $\{\boldsymbol{\Psi}^{(k)}\}$ is a bounded sequence. According to Theorem 28.1 of Ostrowski (1966), the set of limit points of a bounded sequence $\{\boldsymbol{\Psi}^{(k)}\}$ with (3.24) satisfied is connected and compact. From Theorem 3.1, all limit points of $\{\boldsymbol{\Psi}^{(k)}\}$ are in $\mathcal{S}(L^*)$. □

This theorem was proved by Boyles (1983) under different regularity conditions, while Theorem 3.2 was obtained by Hartley and Hocking (1971) for a special model.

Convergence of $\boldsymbol{\Psi}^{(k)}$ to a stationary point can be proved without recourse to Theorem 3.1. For any value L_o, let

$$\mathcal{L}(L_o) = \{\boldsymbol{\Psi} \in \boldsymbol{\Omega} : L(\boldsymbol{\Psi}) = L_o\}.$$

Theorem 3.5. *Let $\{\boldsymbol{\Psi}^{(k)}\}$ be an instance of a GEM algorithm with the additional property that*

$$[\partial Q(\boldsymbol{\Psi};\, \boldsymbol{\Psi}^{(k)})/\partial \boldsymbol{\Psi}]_{\boldsymbol{\Psi}=\boldsymbol{\Psi}^{(k+1)}} = \mathbf{0}. \tag{3.25}$$

Suppose $\partial Q(\boldsymbol{\Psi};\, \boldsymbol{\phi})/\partial \boldsymbol{\Psi}$ is continuous in $\boldsymbol{\Psi}$ and $\boldsymbol{\phi}$. Then $\boldsymbol{\Psi}^{(k)}$ converges to a stationary point $\boldsymbol{\Psi}^*$ with $L(\boldsymbol{\Psi}^*) = L^*$, the limit of $L(\boldsymbol{\Psi}^{(k)})$, if either

$$\mathcal{L}(L^*) = \{\boldsymbol{\Psi}^*\} \tag{3.26}$$

or

$$\|\boldsymbol{\Psi}^{(k+1)} - \boldsymbol{\Psi}^{(k)}\| \rightarrow 0, \text{ as } k \rightarrow \infty, \text{ and } \mathcal{L}(L^*) \text{ is discrete.} \tag{3.27}$$

Proof. As noted earlier, the assumed regularity conditions (3.18) to (3.20) imply that $\{L(\boldsymbol{\Psi}^{(k)})\}$ is bounded above, and so it converges to some point L^*. In the case of (3.26), where $\mathcal{L}(L^*)$ consists of the single point $\boldsymbol{\Psi}^*$, $\boldsymbol{\Psi}^{(k)}$ obviously converges to $\boldsymbol{\Psi}^*$.

For $\mathcal{L}(L^*)$ discrete but not a singleton, condition (3.24) is sufficient to ensure that $\boldsymbol{\Psi}^{(k)}$ converges to a point $\boldsymbol{\Psi}^*$ in $\mathcal{L}(L^*)$, as seen in the proof of Theorem 3.4.

From (3.15),

$$\begin{aligned}
\partial \log L(\boldsymbol{\Psi}^*)/\partial \boldsymbol{\Psi} &= [\partial Q(\boldsymbol{\Psi};\, \boldsymbol{\Psi}^*)/\partial \boldsymbol{\Psi}]_{\boldsymbol{\Psi}=\boldsymbol{\Psi}^*} && (3.28)\\
&= \lim_{k\rightarrow\infty} [\partial Q(\boldsymbol{\Psi};\, \boldsymbol{\Psi}^{(k)})/\partial \boldsymbol{\Psi}]_{\boldsymbol{\Psi}=\boldsymbol{\Psi}^{(k+1)}} && (3.29)\\
&= \mathbf{0}, && (3.30)
\end{aligned}$$

establishing that the limit point $\boldsymbol{\Psi}^*$ is a stationary point of $L(\boldsymbol{\Psi})$. In the above, (3.29) follows from (3.28) by the continuity of $\partial Q(\boldsymbol{\Psi};\, \boldsymbol{\phi})/\partial \boldsymbol{\Psi}$ in $\boldsymbol{\Psi}$ and $\boldsymbol{\phi}$, while (3.30) follows from (3.29) by (3.25). □

It should be noted that condition (3.25) is satisfied by any EM sequence under the regularity conditions assumed here. Since $\mathcal{S}(L)$ is a subset of $\mathcal{L}(L)$, conditions (3.26) and (3.27) of Theorem 3.5 are stronger than the corresponding ones in Theorems 3.3 and 3.4, respectively. The advantage of Theorem 3.5 is that it does not require the conditions (i) and (ii) of Theorem 3.1.

3.5.3 Convergence of an EM Sequence to a Unique Maximum Likelihood Estimate

An important special case of Theorem 3.5 in the previous subsection is the following corollary for a unimodal likelihood function $L(\boldsymbol{\Psi})$ with only one stationary point in the interior of $\boldsymbol{\Omega}$.

Corollary. *Suppose that* $L(\boldsymbol{\Psi})$ *is unimodal in* $\boldsymbol{\Omega}$ *with* $\boldsymbol{\Psi}^*$ *being the only stationary point and that* $\partial Q(\boldsymbol{\Psi}; \boldsymbol{\phi})/\partial \boldsymbol{\Psi}$ *is continuous in* $\boldsymbol{\Psi}$ *and* $\boldsymbol{\phi}$. *Then any EM sequence* $\{\boldsymbol{\Psi}^{(k)}\}$ *converges to the unique maximizer* $\boldsymbol{\Psi}^*$ *of* $L(\boldsymbol{\Psi})$; *that is, it converges to the unique MLE of* $\boldsymbol{\Psi}$.

3.6 EXAMPLES OF NONTYPICAL BEHAVIOR OF AN EM (GEM) SEQUENCE

3.6.1 Example 3.1: Convergence to a Saddle Point

In Example 2.1 in Section 2.2.1, we considered the fitting of a bivariate normal density to a random sample with some observations missing (completely at random) on each of the variates W_1 and W_2. We now use this example in the particular case where the mean vector $\boldsymbol{\mu}$ is specified to be zero to illustrate the possible convergence of the EM algorithm to a saddle point and not a local maximum. The data set taken from Murray (1977) consists of $n = 12$ pairs, four of which are complete, four incomplete without observations on Variate 1, and four incomplete without observations on Variate 2, making $m_1 = m_2 = 4$ incomplete pairs, as given below, where ? indicates a missing data point.

Variate 1:	1	1	−1	−1	?	?	?	?	2	2	−2	−2
Variate 2:	1	−1	1	−1	2	2	−2	−2	?	?	?	?

As $\boldsymbol{\mu}$ is taken to be zero, the vector of unknown parameters is given by

$$\boldsymbol{\Psi} = (\sigma_{11}, \sigma_{22}, \rho)^T.$$

In Figure 3.1, we have plotted $\log L(\boldsymbol{\Psi})$ against ρ and σ^2 under the imposition of the constraint

$$\sigma_{11} = \sigma_{22} = \sigma^2.$$

Murray (1977) reports that for the observed data $\boldsymbol{y}$ as listed above, the likelihood function $L(\boldsymbol{\Psi})$ has a saddle point at

$$\boldsymbol{\Psi}_1 = (5/2, 5/2, 0)^T, \tag{3.31}$$

and two maxima at

$$\boldsymbol{\Psi}_2 = (8/3, 8/3, 1/2)^T$$

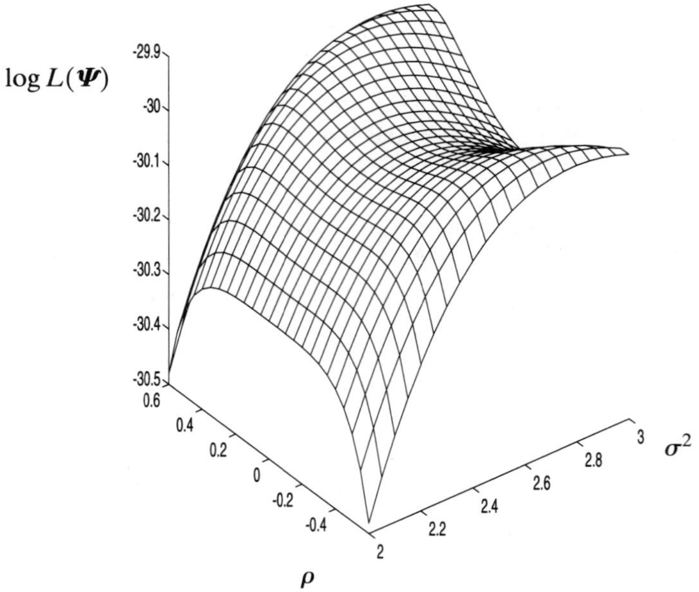

Figure 3.1. Plot of $\log L(\Psi)$ in the case $\sigma_{11} = \sigma_{22} = \sigma^2$.

and

$$\boldsymbol{\Psi}_3 = (8/3,\ 8/3,\ -1/2)^T.$$

He found that the EM sequence $\{\boldsymbol{\Psi}^{(k)}\}$ converged to $\boldsymbol{\Psi}_1$ if the EM algorithm were started from any point with $\rho = 0$. Otherwise, it converged to either $\boldsymbol{\Psi}_2$ or $\boldsymbol{\Psi}_3$.

We now examine this further. From (2.5) with $\boldsymbol{\mu}$ set equal to zero, the complete-data log likelihood is given by

$$\begin{aligned}\log L_c(\boldsymbol{\Psi}) &= -n\log(2\pi) - \tfrac{1}{2}n\log\xi \\ &\quad -\tfrac{1}{2}\xi^{-1}(\sigma_{22}T_{11} + \sigma_{11}T_{22} - 2\sigma_{12}T_{12}),\end{aligned} \tag{3.32}$$

where

$$\xi = \sigma_{11}\sigma_{22}(1-\rho^2)$$

and where the sufficient statistics T_{hi} $(h, i = 1, 2)$ are defined by (2.7). To evaluate $Q(\boldsymbol{\Psi}; \boldsymbol{\Psi}^{(k)})$, the current conditional expectation of $\log L_c(\Psi)$ given $\boldsymbol{y}$, we need to calculate

$$E_{\boldsymbol{\Psi}^{(k)}}(W_{ij} \mid \boldsymbol{y}) \quad \text{and} \quad E_{\boldsymbol{\Psi}^{(k)}}(W_{ij}^2 \mid \boldsymbol{y}).$$

From the previous results (2.10) and (2.12) on the conditional moments of the components of a normal random vector, we have for a missing w_{2j} that

$$E_{\boldsymbol{\Psi}^{(k)}}(W_{2j} \mid \boldsymbol{y}) = w_{1j}^{(k)},$$

where

$$w_{1j}^{(k)} = \rho^{(k)}(\sigma_{22}^{(k)}/\sigma_{11}^{(k)})^{1/2} w_{1j}, \tag{3.33}$$

and

$$E_{\boldsymbol{\Psi}^{(k)}}(W_{2j}^2 \mid \boldsymbol{y}) = \sigma_{22}^{(k)}(1 - \rho^{(k)^2}). \tag{3.34}$$

The corresponding conditional moments for missing w_{1j} are obtained by interchanging the subscripts 1 and 2 in the right-hand side of (3.33) and of (3.34).

Let $\boldsymbol{\Psi}_o^{(k)}$ be any value of $\boldsymbol{\Psi}^{(k)}$ with $\rho^{(k)} = 0$. Then on using (3.33) and (3.34) and noting that

$$\sum_{j=1}^{4} w_{1j} w_{2j} = 0,$$

it follows that

$$\begin{aligned} Q(\boldsymbol{\Psi}; \boldsymbol{\Psi}_o^{(k)}) = & -\log(2\pi) - 6\log(1 - \rho^2) \\ & -6\log\sigma_{11} - 6\log\sigma_{22} \\ & -\frac{1}{2}\left(\sum_{j=1}^{4} w_{1j}^2 + \sum_{j=9}^{12} w_{1j}^2 + 4\sigma_{11}^{(k)}\right) \Big/ \sigma_{11} \\ & -\frac{1}{2}\left(\sum_{j=1}^{8} w_{2j}^2 + 4\sigma_{22}^{(k)}\right) \Big/ \sigma_{22} \\ = & -\log(2\pi) - 6\log(1 - \rho^2) \\ & -6\log\sigma_{11} - 6\log\sigma_{22} \\ & -(10 + 2\sigma_{11}^{(k)})/\sigma_{11} \\ & -(10 + 2\sigma_{22}^{(k)})/\sigma_{22}. \end{aligned} \tag{3.35}$$

On equating the derivative of (3.35) with respect to $\boldsymbol{\Psi}$ to zero, it can be seen that the value $\boldsymbol{\Psi}^{(k+1)}$ of $\boldsymbol{\Psi}$ that maximizes $Q(\boldsymbol{\Psi}; \boldsymbol{\Psi}_o^{(k)})$ is given by

$$\boldsymbol{\Psi}^{(k+1)} = (\sigma_{11}^{(k+1)}, \sigma_{22}^{(k+1)}, \rho^{(k+1)})^T,$$

where

$$\rho^{(k+1)} = 0$$

and

$$\sigma_{ii}^{(k+1)} = (5 + \sigma_{ii}^{(k)})/3 \quad (i = 1, 2). \tag{3.36}$$

As $k \to \infty$, $\rho^{(k+1)}$ remains at zero, while on equating the right-hand side of (3.36) to $\sigma_{ii}^{(k+1)}$, we have that

$$\sigma_{ii}^{(k+1)} \rightarrow 5/2 \quad (i = 1, 2),$$

that is, $\boldsymbol{\Psi}^{(k+1)}$ tends to $\boldsymbol{\Psi}_1$. This establishes the fact that the sequence $\{\boldsymbol{\Psi}^{(k+1)}\}$ converges to $\boldsymbol{\Psi}_1$ if started from any point with $\rho = 0$.

As noted earlier, the EM algorithm can converge to a saddle point $\boldsymbol{\Psi}_s$ of the (incomplete-data) log likelihood $L(\boldsymbol{\Psi})$ if $Q(\boldsymbol{\Psi}; \boldsymbol{\Psi}_s)$ is maximized at $\boldsymbol{\Psi} = \boldsymbol{\Psi}_s$. This is what happens in this example if the EM algorithm is started from any point with $\rho = 0$. It then converges to the saddle point $\boldsymbol{\Psi}_s = \boldsymbol{\Psi}_1$, as given by (3.31), with $Q(\boldsymbol{\Psi}; \boldsymbol{\Psi}_1)$ being maximized at $\boldsymbol{\Psi} = \boldsymbol{\Psi}_1$. In fact, $L(\boldsymbol{\Psi})$ is globally maximized at $\boldsymbol{\Psi}_1$ when $\rho = 0$. However, if the EM algorithm is slightly perturbed from $\boldsymbol{\Psi}_1$, then it will diverge from $\boldsymbol{\Psi}_1$.

3.6.2 Example 3.2: Convergence to a Local Minimum

The fixed points of the EM algorithm include all local maxima of the likelihood, and sometimes saddle points and local minima. We have given an example where a fixed point is a saddle point. We now give an example from Arslan, Constable, and Kent (1993), where the EM algorithm can converge to a local minimum. When the likelihood has multiple local maxima, the parameter space can be partitioned into domains of convergence, one for each maximum. The prime reason for Arslan et al. (1993) presenting their examples was to demonstrate that these domains need not be connected sets.

For simplicity, Arslan et al. (1993) considered a special univariate case of the t-distribution (2.32) where only the location parameter μ is unknown and $\boldsymbol{\Sigma}$ (a scalar) is set equal to 1. In this special case, by taking the degrees of freedom ν very small, the likelihood function $L(\mu)$ can be made to have several local maxima.

From (2.41) and (2.42),

$$\mu^{(k+1)} = \sum_{j=1}^{n} u_j^{(k)} w_j \Big/ \sum_{j=1}^{n} u_j^{(k)},$$

where

$$u_j^{(k)} = \frac{\nu + 1}{\nu + (w_j - \mu^{(k)})^2}.$$

The mapping M induced here by the EM algorithm is therefore defined by

$$\mu^{(k+1)} = M(\mu^{(k)}),$$

where

$$M(\mu) = \sum_{j=1}^{n} u_j(\mu) w_j \Big/ \sum_{j=1}^{n} u_j(\mu)$$

and

$$u_j(\mu) = \frac{\nu + 1}{\nu + (w_j - \mu)^2} \quad (j = 1, \ldots, n).$$

Arslan et al. (1993) took $\nu = 0.05$ and $n = 4$ and the observed data vector as $\boldsymbol{y} = (-20, 1, 2, 3)^T$. Thus ν is very small and one of the data points, $w_1 = -20$, might be regarded as an outlier.

The log likelihood function is plotted in Figure 3.2, where it can be seen that $\log L(\mu)$ has four local maxima at

$$\mu_1 = 19.993, \quad \mu_2 = 1.086, \quad \mu_3 = 1.997, \quad \text{and} \quad \mu_4 = 2.906,$$

and three local minima at

$$\mu_1^* = -14.516, \quad \mu_2^* = 1.373, \quad \text{and} \quad \mu_3^* = 2.647.$$

(Some of these values above as obtained in our analysis are slightly different from those reported in Arslan et al., 1993.)

The associated EM map $M(\mu)$ is plotted in Figure 3.3. It can be seen that it has seven fixed points as given above.

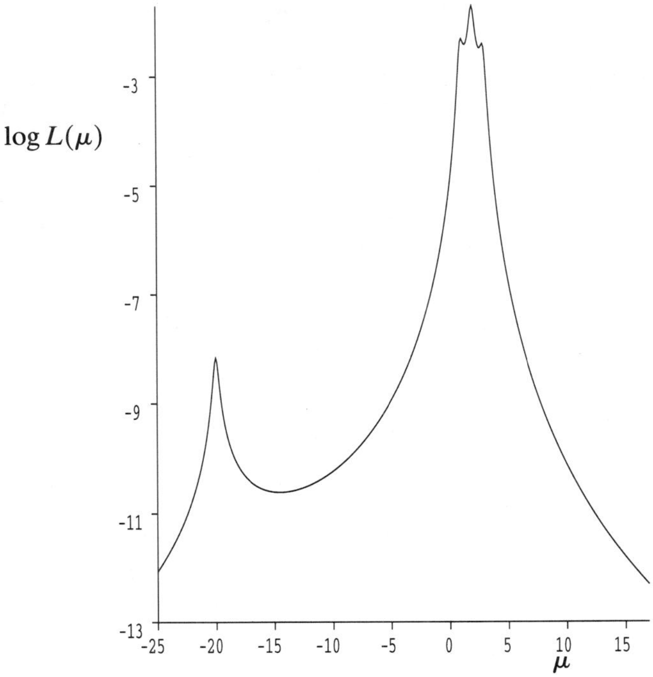

Figure 3.2. Plot of log likelihood function $\log L(\mu)$ versus μ.

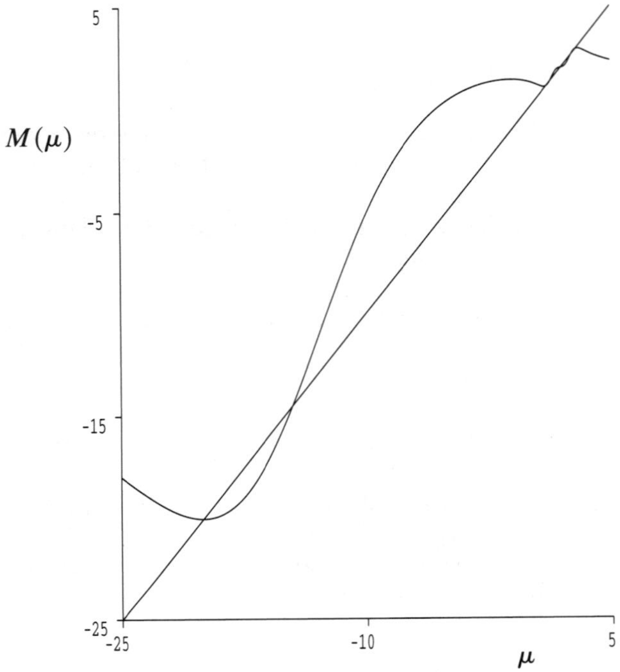

Figure 3.3. Plot of EM mapping $M(\mu)$ versus μ.

It can be seen from Figure 3.4, which is a blow-up of a portion of Figure 3.3, that there are two values of μ other than the fixed point μ_2^* for which

$$M(\mu) = \mu_2^*.$$

We shall label these two points α_1 and α_2, with $\alpha_1 = -1.874$ and $\alpha_2 = -0.330$. It follows that if the EM algorithm is started with $\mu^{(0)} = \alpha_1$ or α_2, then

$$\begin{aligned} \mu^{(1)} &= M(\mu^{(0)}) \\ &= \mu_2^*, \end{aligned} \tag{3.37}$$

and so

$$\mu^{(k)} = \mu_2^* \qquad (k = 1, 2, \ldots).$$

That is, $\mu^{(k)}$ converges to the fixed point μ_2^* of the EM algorithm, corresponding to a local minimum of the likelihood function $L(\mu)$. Of course this curious behavior can happen only from these isolated starting points.

It is of interest to see here what happens if the EM algorithm is started from a point a long way from the data. As μ tends to $\pm\infty$,

$$u_j \approx \frac{\nu + 1}{\nu + \mu^2}$$

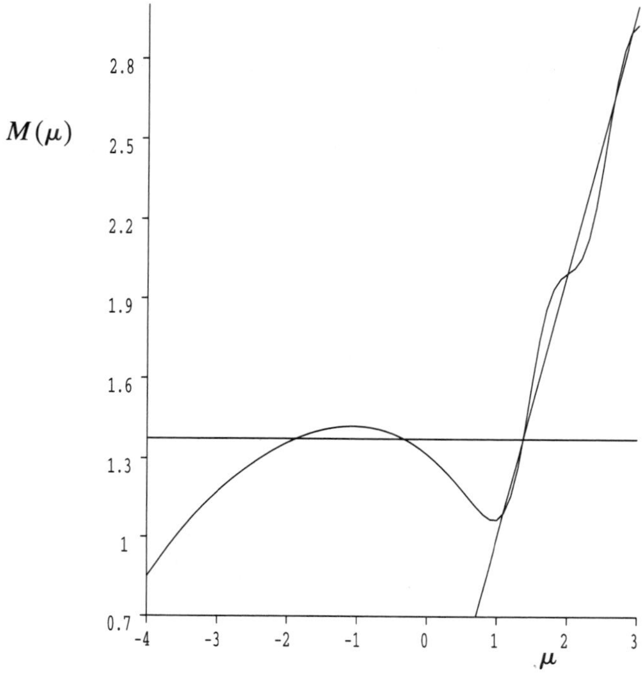

Figure 3.4. Blow-up of Figure 3.3.

for all j, and so

$$\begin{aligned} M(\mu) &\approx \sum_{j=1}^{n} w_j/n \\ &= \overline{w}. \end{aligned} \tag{3.38}$$

Thus when the starting value $\mu^{(0)}$ is taken very far away from the data, $\mu^{(1)}$ is somewhere near the sample mean.

It should be stressed that since ν is very small, the example given here is artificial. However, as argued by Arslan et al. (1993), it provides a useful warning because estimation for the one-parameter t-distribution is so straightforward. If erratic convergence behavior can occur in this example, who knows what pitfalls there may be when the EM algorithm is used in more complicated settings where multiple maxima are present.

3.6.3 Example 3.3: Nonconvergence of a Generalized EM Sequence

We now present an example of a GEM sequence for which $L(\boldsymbol{\Psi}^{(k)})$ converges, but for which $\boldsymbol{\Psi}^{(k)}$ does not converge to a single point. This example was originally presented by Boyles (1983) as a counterexample to a claim incorrectly made in Theorem 2 of DLR that, under their conditions (1) and (2), a GEM sequence $\{\boldsymbol{\Psi}^{(k)}\}$ converges to a point $\boldsymbol{\Psi}^*$ in the closure of $\boldsymbol{\Omega}$.

Suppose that $\phi(\boldsymbol{w}; \boldsymbol{\mu}, \boldsymbol{\Sigma})$ denotes the bivariate normal density with mean $\boldsymbol{\mu} = (\mu_1, \mu_2)^T$ and known covariance matrix $\boldsymbol{\Sigma}$ taken equal to the (2×2) identity matrix $\boldsymbol{I}_2$. In this example of Boyles (1983), there are no missing data, so that

$$\boldsymbol{y} = \boldsymbol{x} = (w_1, w_2)^T,$$

where $\boldsymbol{x}$ consists of the single observation $\boldsymbol{w} = (w_1, w_2)^T$. The vector $\boldsymbol{\Psi}$ of unknown parameters is $\boldsymbol{\mu}$, and

$$\begin{aligned} Q(\boldsymbol{\Psi}; \boldsymbol{\Psi}^{(k)}) &= \log L(\boldsymbol{\mu}) \\ &= \log \phi(\boldsymbol{w}; \boldsymbol{\mu}, \boldsymbol{I}_2) \\ &= (2\pi)^{-1} \exp\{-\tfrac{1}{2}(\boldsymbol{w} - \boldsymbol{\mu})^T (\boldsymbol{w} - \boldsymbol{\mu})\}. \end{aligned}$$

The GEM sequence is defined to be

$$\mu_1^{(k)} = w_1 + r^{(k)} \cos \theta^{(k)}$$

and

$$\mu_2^{(k)} = w_2 + r^{(k)} \sin \theta^{(k)},$$

where

$$r^{(k)} = 1 + (k+1)^{-1}$$

and

$$\theta^{(k)} = \sum_{i=1}^{k} (i+1)^{-1}$$

for $k = 1, 2, \ldots,$ and where r and θ are the polar coordinates centered at the single observation $\boldsymbol{w}$. The initial value for r and θ are $r^{(0)} = 2$ and $\theta^{(0)} = 0$.

We have that

$$\begin{aligned} \log L(\boldsymbol{\Psi}^{(k+1)}) - \log L(\boldsymbol{\Psi}^{(k)}) &= \tfrac{1}{2}(r^{(k)^2} - r^{(k+1)^2}) \\ &= \tfrac{1}{2}\left\{ r^{(k)^2} - (2 - \frac{1}{r^{(k)^2}})^2 \right\}, \end{aligned} \qquad (3.39)$$

since

$$r^{(k+1)} = 2 - \frac{1}{r^{(k)}}.$$

Now

$$0 < 2 - u^{-1} \leq u \qquad (3.40)$$

for $u \geq 1$. As $r^{(k)} \geq 1$ for each k, we have from (3.39) and (3.40) that

$$L(\boldsymbol{\Psi}^{(k+1)}) - L(\boldsymbol{\Psi}^{(k)}) \geq 0.$$

Hence $\{\boldsymbol{\Psi}^{(k)}\}$ is an instance of a GEM algorithm.

Since $r^{(k)} \to 1$, as $k \to \infty$, the sequence of likelihood values $\{L(\boldsymbol{\Psi}^{(k)})\}$ converges to the value

$$(2\pi)^{-1}e^{-\frac{1}{2}}.$$

But the sequence of iterates $\{\boldsymbol{\Psi}^{(k)}\}$ converges to the circle of unit radius and center $\boldsymbol{w}$. It does not converge to a single point, although it does satisfy the conditions of Theorem 2 of DLR for the convergence of a GEM sequence to a single point. These conditions are that $\{L(\boldsymbol{\Psi}^{(k)}\}$ is a bounded sequence, which is obviously satisfied here, and that

$$Q(\boldsymbol{\Psi}^{(k+1)}; \boldsymbol{\Psi}^{(k)}) - Q(\boldsymbol{\Psi}^{(k)}; \boldsymbol{\Psi}^{(k)}) \geq \lambda \| \boldsymbol{\Psi}^{k+1)} - \boldsymbol{\Psi}^{(k)} \|^2 \tag{3.41}$$

for some scalar $\lambda > 0$ and all k. Now the left-hand side of (3.41) equals

$$\tfrac{1}{2}(r^{(k)^2} - r^{(k+1)^2})$$

while, concerning the right-hand side of (3.41),

$$\| \boldsymbol{\Psi}^{(k+1)} - \boldsymbol{\Psi}^{(k)} \|^2 = r^{(k)^2} + r^{(k+1)^2} - 2r^{(k)}r^{(k+1} \cos((k+2)^{-1}).$$

Taking $\lambda = \frac{1}{2}$, the inequality (3.41) holds if

$$r^{(k+1)}\{r^{(k)} \cos((k+2)^{-1}) - r^{(k+1)}\} \geq 0,$$

which is equivalent to

$$\cos((k+2)^{-1}) \geq 1 - (k+2)^{-2},$$

and the latter is valid for all $k \geq 0$.

Boyles (1983) notes that although these conditions do not imply the convergence of $\{\boldsymbol{\Psi}^{(k)}\}$, they do imply that

$$\| \boldsymbol{\Psi}^{(k+1)} - \boldsymbol{\Psi}^{(k)} \| \to 0, \text{ as } k \to \infty.$$

But as we have seen in this section, this latter condition is not sufficient to ensure convergence to a single point.

3.7 SCORE STATISTIC

We let

$$\boldsymbol{S}(\boldsymbol{y}; \boldsymbol{\Psi}) = \partial \log L(\boldsymbol{\Psi}) / \partial \boldsymbol{\Psi}$$

be the gradient vector of the log likelihood function $L(\boldsymbol{\Psi})$; that is, the score statistic based on the observed (incomplete) data $\boldsymbol{y}$. The gradient vector of the complete-data log likelihood function is given by

$$\boldsymbol{S}_c(\boldsymbol{x}; \boldsymbol{\Psi}) = \partial \log L_c(\boldsymbol{\Psi}) / \partial \boldsymbol{\Psi}.$$

The incomplete-data score statistic $S(y;\ \Psi)$ can be expressed as the conditional expectation of the complete-data score statistic; that is,

$$S(y;\ \Psi) = E_{\Psi}\{S_c(X;\ \Psi) \mid y\}. \tag{3.42}$$

To see this, we note that

$$\begin{aligned} S(y;\ \Psi) &= \partial \log L(\Psi)/\partial \Psi \\ &= \partial \log g(y;\ \Psi)/\partial \Psi \\ &= g'(y;\ \Psi)/g(y;\ \Psi) \\ &= \left\{ \int_{\mathcal{X}(y)} g_c'(x;\ \Psi)\, dx \right\} \Big/ g(y;\ \Psi), \end{aligned} \tag{3.43}$$

where the prime denotes differentiation with respect to Ψ. On multiplying and dividing the integrand by $g_c(x;\ \Psi)$ in (3.43), we have that

$$\begin{aligned} S(y;\ \Psi) &= \int_{\mathcal{X}(y)} \{\partial \log g_c(x;\ \Psi)/\partial \Psi\}\{g_c(x;\ \Psi)/g(y;\ \Psi)\}\, dx \\ &= \int_{\mathcal{X}(y)} \{\partial \log L_c(\Psi)/\partial \Psi\} k(x \mid y;\ \Psi)\, dx \\ &= E_{\Psi}\{\partial \log L_c(\Psi)/\partial \Psi \mid y\} \\ &= E_{\Psi}\{S_c(X;\ \Psi) \mid y\}. \end{aligned} \tag{3.44}$$

Another way of writing this result is that

$$S(y;\ \Psi) = [\partial Q(\Psi_o;\ \Psi)/\partial \Psi_o]_{\Psi_o = \Psi}, \tag{3.45}$$

which is the self-consistency property of the EM algorithm, as noted in Section 3.2. The result (3.45) follows from (3.44) on interchanging the operations of expectation and differentiation. It is assumed in the above that regularity conditions hold for this interchange.

Note that the result (3.45) was derived by an alternative argument in Section 3.4; see equation (3.15).

3.8 MISSING INFORMATION

3.8.1 Missing Information Principle

We let

$$I(\Psi;\ y) = -\partial^2 \log L(\Psi)/\partial \Psi \partial \Psi^T$$

be the matrix of the negative of the second-order partial derivatives of the (incomplete-data) log likelihood function with respect to the elements of Ψ. Under regularity conditions, the expected (Fisher) information matrix $\mathcal{I}(\Psi)$ is given by

$$\mathcal{I}(\boldsymbol{\Psi}) = E_{\boldsymbol{\Psi}}\{\boldsymbol{S}(\boldsymbol{Y};\, \boldsymbol{\Psi})\boldsymbol{S}^T(\boldsymbol{Y};\, \boldsymbol{\Psi})\}$$
$$= -E_{\boldsymbol{\Psi}}\{\boldsymbol{I}(\boldsymbol{\Psi};\, \boldsymbol{Y})\}.$$

With respect to the complete-data log likelihood, we let

$$\boldsymbol{I}_c(\boldsymbol{\Psi};\, \boldsymbol{x}) = -\partial^2 \log L_c(\boldsymbol{\Psi})/\partial\boldsymbol{\Psi}\partial\boldsymbol{\Psi}^T.$$

The expected (complete-data) information matrix is given then by

$$\mathcal{I}_c(\boldsymbol{\Psi}) = -E_{\boldsymbol{\Psi}}\{\boldsymbol{I}_c(\boldsymbol{\Psi};\, \boldsymbol{X})\}.$$

We have from (3.3) that

$$\log L(\boldsymbol{\Psi}) = \log L_c(\boldsymbol{\Psi}) - \log k(\boldsymbol{x} \mid \boldsymbol{y};\, \boldsymbol{\Psi}). \tag{3.46}$$

On differentiating the negative of both sides of (3.46) twice with respect to $\boldsymbol{\Psi}$, we have that

$$\boldsymbol{I}(\boldsymbol{\Psi};\, \boldsymbol{y}) = \boldsymbol{I}_c(\boldsymbol{\Psi};\, \boldsymbol{x}) + \partial^2 \log k(\boldsymbol{x} \mid \boldsymbol{y};\, \boldsymbol{\Psi})/\partial\boldsymbol{\Psi}\partial\boldsymbol{\Psi}^T. \tag{3.47}$$

Taking the expectation of both sides of (3.47) over the conditional distribution of $\boldsymbol{z}$ given $\boldsymbol{y}$ yields

$$\boldsymbol{I}(\boldsymbol{\Psi};\, \boldsymbol{y}) = \mathcal{I}_c(\boldsymbol{\Psi};\, \boldsymbol{y}) - \mathcal{I}_m(\boldsymbol{\Psi};\, \boldsymbol{y}), \tag{3.48}$$

where

$$\mathcal{I}_c(\boldsymbol{\Psi};\, \boldsymbol{y}) = E_{\boldsymbol{\Psi}}\{\boldsymbol{I}_c(\boldsymbol{\Psi};\, \boldsymbol{X}) \mid \boldsymbol{y}\} \tag{3.49}$$

is the conditional expectation of the complete-data information matrix $\boldsymbol{I}_c(\boldsymbol{\Psi};\, \boldsymbol{X})$ given $\boldsymbol{y}$, and where

$$\mathcal{I}_m(\boldsymbol{\Psi};\, \boldsymbol{y}) = -E_{\boldsymbol{\Psi}}\{\partial^2 \log k(\boldsymbol{X} \mid \boldsymbol{y};\, \boldsymbol{\Psi})/\partial\boldsymbol{\Psi}\partial\boldsymbol{\Psi}^T \mid \boldsymbol{y}\} \tag{3.50}$$

is the expected information matrix for $\boldsymbol{\Psi}$ based on $\boldsymbol{x}$ (or equivalently, the unobservable data $\boldsymbol{z}$) when conditioned on $\boldsymbol{y}$. This latter information denoted by $\mathcal{I}_m(\boldsymbol{\Psi};\, \boldsymbol{y})$ can be viewed as the "missing information" as a consequence of observing only $\boldsymbol{y}$ and not also $\boldsymbol{z}$. Thus (3.48) has the following interpretation: observed information equals the (conditional expected) complete information minus the missing information. This has been called the Missing Information principle by Orchard and Woodbury (1972).

On averaging both sides of (3.48) over the distribution of $\boldsymbol{Y}$, we have

$$\mathcal{I}(\boldsymbol{\Psi}) = \mathcal{I}_c(\boldsymbol{\Psi}) - E_{\boldsymbol{\Psi}}\{\mathcal{I}_m(\boldsymbol{\Psi};\, \boldsymbol{Y})\}$$

as an analogous expression to (3.48) for the expected information $\mathcal{I}(\boldsymbol{\Psi})$. Orchard and Woodbury (1972) give a slightly different form of this expression.

3.8.2 Example 3.4. (Example 1.3 Continued)

To demonstrate the relationship (3.48) between the incomplete-data, complete-data, and missing information matrices, we return to Example 1.3 on censored exponentially distributed survival times.

From (3.48), we have

$$I(\mu;\, \boldsymbol{y}) = \mathcal{I}_c(\mu;\, \boldsymbol{y}) - \mathcal{I}_m(\mu;\, \boldsymbol{y}), \tag{3.51}$$

where $I(\mu;\, \boldsymbol{y})$ is the information about μ in the observed data $\boldsymbol{y}$, $\mathcal{I}_c(\mu;\, \boldsymbol{y})$ is the conditional expectation of the complete-data information, and $\mathcal{I}_m(\mu;\, \boldsymbol{y})$ is the missing information. For this example, we shall show that this relationship holds by calculating these three information quantities from their definitions.

By differentiation of the incomplete-data log likelihood function (1.46), we have that

$$I(\mu;\, \boldsymbol{y}) = -r\mu^{-2} + 2\mu^{3} \sum_{j=1}^{n} c_j. \tag{3.52}$$

Concerning $\mathcal{I}_c(\mu;\, \boldsymbol{y})$, we have on differentiation of minus the complete-data log likelihood twice with respect to μ, that

$$I_c(\mu;\, \boldsymbol{x}) = (-n\mu^{-2} + 2\mu^{-3} \sum_{j=1}^{n} w_j),$$

and so

$$\begin{aligned}
\mathcal{I}_c(\mu;\, \boldsymbol{y}) &= E_\mu\{I_c(\mu;\, \boldsymbol{x}) \mid \boldsymbol{y}\} \\
&= \left\{ -n\mu^{-2} + 2\mu^{-3} \sum_{j=1}^{r} c_j + 2\mu^{-3} \sum_{j=r+1}^{n} (c_j + \mu) \right\} \\
&= (n - 2r)\mu^{-2} + 2\mu^{-3} \sum_{j=1}^{n} c_j,
\end{aligned} \tag{3.53}$$

using (1.50).

The missing information $\mathcal{I}_m(\mu;\, \boldsymbol{y})$ can be calculated directly from its definition (3.50),

$$\mathcal{I}_m(\mu;\, \boldsymbol{y}) = -E_\mu\{\partial^2 \log k(\boldsymbol{X} \mid \boldsymbol{y};\, \mu)/\partial\mu^2 \mid \boldsymbol{y}\}.$$

Given $\boldsymbol{y}$, the conditional joint distribution of $W_1, \ldots, W_r$ is degenerate with mass one at the point $(c_1, \ldots, c_r)^T$. Hence we can effectively work with the conditional distribution of $\boldsymbol{Z}$ given $\boldsymbol{y}$ in calculating $\mathcal{I}_m(\mu;\, \boldsymbol{y})$ from (3.50). From (1.49), the conditional density of $\boldsymbol{Z}$ given $\boldsymbol{y}$ is

$$k(\boldsymbol{z} \mid \boldsymbol{y};\, \mu) = \mu^{-(n-r)} \exp\left\{ -\tfrac{1}{2} \sum_{j=r+1}^{n} (w_j - c_j)/\mu \right\} I_A(\boldsymbol{z}), \tag{3.54}$$

where

$$A = \{z : w_j > c_j \quad (j = r + 1, \ldots, n)\}.$$

On differentiation of (3.54), we have that

$$\begin{aligned}
\mathcal{I}_m(\mu; y) &= -(n-r)\mu^{-2} + 2\mu^{-3} E_\mu \left\{ \sum_{j=r+1}^{n} (W_j - c_j) \mid y \right\} \\
&= -(n-r)\mu^{-2} + 2\mu^{-3} \sum_{j=r+1}^{n} \mu \\
&= (n-r)\mu^{-2}, \qquad (3.55)
\end{aligned}$$

since from (1.50),

$$E_\mu\{(W_j - c_j) \mid y\} = \mu$$

for $j = r + 1, \ldots, n$. On subtracting (3.55) from (3.53), we confirm the expression (3.52) for $I(\mu; y)$ obtained directly from its definition.

In Section 4.2, it will be seen that $\mathcal{I}_m(\mu; y)$ can be expressed as

$$\mathcal{I}_m(\mu; y) = \text{var}_\mu\{S_c(X; \mu) \mid y\}. \qquad (3.56)$$

On using this definition, we have

$$\begin{aligned}
\mathcal{I}_m(\mu; y) &= \text{var}_\mu\{S_c(X; \mu) \mid y\} \qquad (3.57) \\
&= \text{var}_\mu \left\{ \mu^{-2} \sum_{j=1}^{n} (W_j - c_j) \mid y \right\} \\
&= (n-r)\mu^{-2}, \qquad (3.58)
\end{aligned}$$

since from (1.49), it follows that

$$\begin{aligned}
\text{var}_\mu(W_j \mid y) &= 0 \quad (j = 1, \ldots, r), \\
&= \mu^2 \quad (j = r + 1, \ldots, n).
\end{aligned}$$

It can be seen that the result (3.58) agrees with (3.55) as obtained directly from its definition.

In order to compute the expected information $\mathcal{I}(\mu)$, we have to make some assumption about the underlying censoring mechanism. Suppose that an observation W_j is censored if its realized value w_j exceeds C. That is, an observation is censored if it fails after time C from when the experiment was commenced. Then its expected information $\mathcal{I}(\mu)$ is given by

$$\begin{aligned}\mathcal{I}(\mu) &= E_\mu\{I(\mu; \boldsymbol{y})\} \\ &= E_\mu\left\{-r\mu^{-2} + 2\mu^{-3}\sum_{j=1}^{n} W_j\right\} \\ &= \mu^{-2}E_\mu\left\{-r + 2\mu^{-1}\sum_{j=1}^{n} W_j\right\}. \end{aligned} \tag{3.59}$$

Now under the assumed censoring scheme,

$$\begin{aligned}E_\mu(r) &= n\,\mathrm{pr}\{W_j < C\} \\ &= n(1 - e^{-C/\mu}),\end{aligned} \tag{3.60}$$

and

$$E_\mu(W_j) = \mu(1 - e^{-C/\mu}). \tag{3.61}$$

Thus

$$\mathcal{I}(\mu) = n\mu^{-2}(1 - e^{-C/\mu}). \tag{3.62}$$

As the complete-data expected information $\mathcal{I}_c(\mu)$ is $n\mu^{-2}$, it can be seen that the expected missing information about μ is

$$E_\mu\{\mathcal{I}_m(\mu; \boldsymbol{Y})\} = n\mu^{-2}e^{-C/\mu}, \tag{3.63}$$

which tends to zero, as $C \to \infty$. That is, as C becomes large, most items will fail before this censoring time, and so the amount of missing information due to censoring becomes negligible.

For this example, the complete-data density belongs to the regular exponential family with natural parameter $\theta = \mu^{-1}$ and sufficient statistic

$$t(\boldsymbol{X}) = \sum_{j=1}^{n} W_j.$$

Hence if we work directly in terms of the information about θ, rather than μ, verification of (3.48) is trivial. For from (1.58) and (1.59),

$$\begin{aligned}\mathcal{I}_c(\theta; \boldsymbol{y}) &= \mathrm{var}_\theta\{t(\boldsymbol{X})\} \\ &= n\theta^2,\end{aligned}$$

while from (3.57),

$$\begin{aligned}\mathcal{I}_m(\theta; \boldsymbol{y}) &= \mathrm{var}_\theta\{t(\boldsymbol{X}) \mid \boldsymbol{y}\} \\ &= (n - r)\theta^2,\end{aligned}$$

thus establishing

$$\begin{aligned} I(\theta; \boldsymbol{y}) &= r\theta^2 \\ &= r\mu^{-2}. \end{aligned}$$

3.9 RATE OF CONVERGENCE OF THE EM ALGORITHM

3.9.1 Rate Matrix for Linear Convergence

We have seen in Section 1.5 that the EM algorithm implicitly defines a mapping $\boldsymbol{\Psi} \rightarrow \boldsymbol{M}(\boldsymbol{\Psi})$, from the parameter space of $\boldsymbol{\Psi}$, $\boldsymbol{\Omega}$, to itself such that each iteration $\boldsymbol{\Psi}^{(k)} \rightarrow \boldsymbol{\Psi}^{(k+1)}$ is defined by

$$\boldsymbol{\Psi}^{(k+1)} = \boldsymbol{M}(\boldsymbol{\Psi}^{(k)}) \quad (k = 0, 1, 2, \ldots).$$

If $\boldsymbol{\Psi}^{(k)}$ converges to some point $\boldsymbol{\Psi}^*$ and $\boldsymbol{M}(\boldsymbol{\Psi})$ is continuous, then $\boldsymbol{\Psi}^*$ is a fixed point of the algorithm; that is, $\boldsymbol{\Psi}^*$ must satisfy

$$\boldsymbol{\Psi}^* = \boldsymbol{M}(\boldsymbol{\Psi}^*). \tag{3.64}$$

By a Taylor series expansion of $\boldsymbol{\Psi}^{(k+1)} = \boldsymbol{M}(\boldsymbol{\Psi}^{(k)})$ about the point $\boldsymbol{\Psi}^{(k)} = \boldsymbol{\Psi}^*$, we have in a neighborhood of $\boldsymbol{\Psi}^*$ that

$$\boldsymbol{\Psi}^{(k+1)} - \boldsymbol{\Psi}^* \approx \boldsymbol{J}(\boldsymbol{\Psi}^*)(\boldsymbol{\Psi}^{(k)} - \boldsymbol{\Psi}^*), \tag{3.65}$$

where $\boldsymbol{J}(\boldsymbol{\Psi})$ is the $d \times d$ Jacobian matrix for $\boldsymbol{M}(\boldsymbol{\Psi}) = (M_1(\boldsymbol{\Psi}), \ldots, M_d(\boldsymbol{\Psi}))^T$, having (i, j)th element $J_{ij}(\boldsymbol{\Psi})$ equal to

$$J_{ij}(\boldsymbol{\Psi}) = \partial M_i(\boldsymbol{\Psi}) / \partial \Psi_j,$$

where $\Psi_j = (\boldsymbol{\Psi})_j$. Note that $\boldsymbol{J}(\boldsymbol{\Psi})$ is the transpose of the matrix $\boldsymbol{DM}$ that is used by DLR and Meng and Rubin (1991).

Thus, in a neighborhood of $\boldsymbol{\Psi}^*$, the EM algorithm is essentially a linear iteration with rate matrix $\boldsymbol{J}(\boldsymbol{\Psi}^*)$, since $J(\boldsymbol{\Psi}^*)$ is typically nonzero. For this reason, $\boldsymbol{J}(\boldsymbol{\Psi}^*)$ is often referred to as the matrix rate of convergence, or simply, the rate of convergence.

For vector $\boldsymbol{\Psi}$, a measure of the actual observed convergence rate is the global rate of convergence, which is defined as

$$r = \lim_{k \rightarrow \infty} \| \boldsymbol{\Psi}^{(k+1)} - \boldsymbol{\Psi}^* \| / \| \boldsymbol{\Psi}^{(k)} - \boldsymbol{\Psi}^* \|,$$

where $\|\cdot\|$ is any norm on d-dimensional Euclidean space $\mathbb{R}^d$. It is well known that under certain regularity conditions,

$$r = \lambda_{\max} \equiv \text{the largest eigenvalue of } \boldsymbol{J}(\boldsymbol{\Psi}^*).$$

Notice that a large value of r implies slow convergence. To be consistent with the common notion that the higher the value of the measure, the faster the algorithm converges, Meng (1994) defines $s = 1 - r$ as the global speed of convergence. Thus s is the smallest eigenvalue of

$$\boldsymbol{S} = \boldsymbol{I}_d - \boldsymbol{J}(\boldsymbol{\Psi}^*), \tag{3.66}$$

which may be called the (matrix) speed of convergence; $\boldsymbol{S}$ is often referred to as the iteration matrix in the optimization literature.

3.9.2 Measuring the Linear Rate of Convergence

In practice, r is typically assessed as

$$r = \lim_{k\to\infty} \|\boldsymbol{\Psi}^{(k+1)} - \boldsymbol{\Psi}^{(k)}\| / \|\boldsymbol{\Psi}^{(k)} - \boldsymbol{\Psi}^{(k-1)}\|. \tag{3.67}$$

The rate of convergence can also be measured component by component. The ith componentwise rate of convergence is defined as

$$r_i = \lim_{k\to\infty} |\Psi_i^{(k+1)} - \Psi_i^*| / |\Psi_i^{(k)} - \Psi_i^*|,$$

provided that it exists. We define $r_i \equiv 0$ if $\Psi_i^{(k)} \equiv \Psi_i^{(k_0)}$, for all $k \geq k_0$, k_0 fixed. Analogous to (3.67), r_i can be assessed in practice as

$$r_i = \lim_{k\to\infty} |\Psi_i^{(k+1)} - \Psi_i^{(k)}| / |\Psi_i^{(k)} - \Psi_i^{(k-1)}|.$$

Under broad regularity conditions, it can be shown (for example, Meng and Rubin, 1994) that

$$r = \max_{1\leq i\leq d} r_i,$$

which is consistent with the intuition that the algorithm as a whole converges if and only if every component Ψ_i does. A component whose componentwise rate equals the global rate is then called the slowest component for the obvious reason. A component is the slowest if it is not orthogonal to the eigenvector corresponding to $\lambda_{\max}$, and thus typically there is more than one such component; see Meng (1994).

3.9.3 Rate Matrix in Terms of Information Matrices

Suppose that $\{\boldsymbol{\Psi}^{(k)}\}$ is an EM sequence for which

$$\partial Q(\boldsymbol{\Psi}; \boldsymbol{\Psi}^{(k)}) / \partial \boldsymbol{\Psi} = \boldsymbol{0} \tag{3.68}$$

is satisfied by $\boldsymbol{\Psi} = \boldsymbol{\Psi}^{(k+1)}$, which will be the case with standard complete-data estimation. Then DLR showed that if $\boldsymbol{\Psi}^{(k)}$ converges to a point $\boldsymbol{\Psi}^*$, then

$$\boldsymbol{J}(\boldsymbol{\Psi}^*) = \mathcal{I}_c^{-1}(\boldsymbol{\Psi}^*; \boldsymbol{y})\mathcal{I}_m(\boldsymbol{\Psi}^*; \boldsymbol{y}). \tag{3.69}$$

This result was obtained also by Sundberg (1976).

Thus the rate of convergence of the EM algorithm is given by the largest eigenvalue of the information ratio matrix $\mathcal{I}_c^{-1}(\boldsymbol{\Psi}; \boldsymbol{y})\mathcal{I}_m(\boldsymbol{\Psi}; \boldsymbol{y})$, which measures the proportion of information about $\boldsymbol{\Psi}$ that is missing by not also observing $\boldsymbol{z}$ in addition to $\boldsymbol{y}$. The greater the proportion of missing information, the slower the rate of convergence. The fraction of information loss may vary across different components of $\boldsymbol{\Psi}$, suggesting that certain components of $\boldsymbol{\Psi}$ may approach $\boldsymbol{\Psi}^*$ rapidly using the EM algorithm, while other components may require many iterations.

The rate of convergence of the EM algorithm can be expressed equivalently in terms of the largest eigenvalue of

$$\boldsymbol{I}_d - \mathcal{I}_c^{-1}(\boldsymbol{\Psi}^*; \boldsymbol{y})\boldsymbol{I}(\boldsymbol{\Psi}^*; \boldsymbol{y}),$$

where $\boldsymbol{I}_d$ denotes the $d \times d$ identity matrix. This is because we can express $\boldsymbol{J}(\boldsymbol{\Psi}^*)$ also in the form

$$\boldsymbol{J}(\boldsymbol{\Psi}^*) = \boldsymbol{I}_d - \mathcal{I}_c^{-1}(\boldsymbol{\Psi}^*; \boldsymbol{y})\boldsymbol{I}(\boldsymbol{\Psi}^*; \boldsymbol{y}). \tag{3.70}$$

This result follows from (3.69) on noting from (3.48) that

$$\mathcal{I}_m(\boldsymbol{\Psi}^*; \boldsymbol{y}) = \mathcal{I}_c(\boldsymbol{\Psi}^*; \boldsymbol{y}) - \boldsymbol{I}(\boldsymbol{\Psi}^*; \boldsymbol{y}).$$

From (3.70),

$$\begin{aligned} \partial^2 \log L(\boldsymbol{\Psi}^*)/\partial\boldsymbol{\Psi}\,\partial\boldsymbol{\Psi}^T &= -\boldsymbol{I}(\boldsymbol{\Psi}^*; \boldsymbol{y}) \\ &= -\mathcal{I}_c(\boldsymbol{\Psi}^*; \boldsymbol{y})\{\boldsymbol{I}_d - \boldsymbol{J}(\boldsymbol{\Psi}^*)\}. \end{aligned} \tag{3.71}$$

Exceptions to the convergence of the EM algorithm to a local maximum of $L(\boldsymbol{\Psi})$ occur if $\boldsymbol{J}(\boldsymbol{\Psi}^*)$ has eigenvalues exceeding unity. An eigenvalue of $\boldsymbol{J}(\boldsymbol{\Psi}^*)$ which is unity in a neighborhood of $\boldsymbol{\Psi}^*$ implies a ridge in $L(\boldsymbol{\Psi})$ through $\boldsymbol{\Psi}^*$. Generally, we expect $\partial^2 \log L(\boldsymbol{\Psi}^*)/\partial\boldsymbol{\Psi}\partial\boldsymbol{\Psi}^T$ to be negative semidefinite, if not negative definite, in which case the eigenvalues of $\boldsymbol{J}(\boldsymbol{\Psi}^*)$ all lie in $[0, 1]$ or $[0, 1)$, respectively.

The slow convergence of the EM algorithm has been reported in quite a few applications. Some of these are in Hartley (1958) (in contingency tables with missing data), Fienberg (1972) (in incomplete multiway contingency tables), Sundberg (1976), Haberman (1977), Nelder (1977) (when there is more than one missing observation in designed experiments; in latent structure analysis), and Thompson (1977) (in variance components estimation). Horng (1986, 1987) theoretically demonstrates that this behavior of the EM algorithm in some of these situations is due to the rate of convergence being

1 (called sublinear convergence). His examples include (1) mixtures of two normal distributions with a common variance in known proportions; (2) multivariate normal with missing values and with a singular covariance matrix but with nonsingular submatrices corresponding to the observed variables in each of the cases (which is what is required to execute the E-step); (3) a variance components model when one of the variances is close to zero at the limit point; and (4) some factor analysis situations.

3.9.4 Rate Matrix for Maximum *a Posteriori* Estimation

When finding the MAP estimate, the rate of convergence is given by replacing

$$\begin{aligned} -\mathcal{I}_c(\boldsymbol{\Psi}^*; \boldsymbol{y}) &= -E_{\boldsymbol{\Psi}^*}\{\boldsymbol{I}_c(\boldsymbol{\Psi}^*; \boldsymbol{X}) \mid \boldsymbol{Y})\} \\ &= [\partial^2 Q(\boldsymbol{\Psi}; \boldsymbol{\Psi}^*)/\partial\boldsymbol{\Psi}\partial\boldsymbol{\Psi}^T]_{\boldsymbol{\Psi}=\boldsymbol{\Psi}^*} \end{aligned} \tag{3.72}$$

by

$$[\partial^2 Q(\boldsymbol{\Psi}; \boldsymbol{\Psi}^*)/\partial\boldsymbol{\Psi}\partial\boldsymbol{\Psi}^T]_{\boldsymbol{\Psi}=\boldsymbol{\Psi}^*} + \partial^2 \log p(\boldsymbol{\Psi}^*)/\partial\boldsymbol{\Psi}\partial\boldsymbol{\Psi}^T$$

in (3.69), where $p(\boldsymbol{\Psi})$ is the prior density for $\boldsymbol{\Psi}$.

3.9.5 Derivation of Rate Matrix in Terms of Information Matrices

We now show how the result (3.69), or equivalently (3.70), can be established. By a linear Taylor series expansion of the gradient vector $\boldsymbol{S}(\boldsymbol{y}; \boldsymbol{\Psi})$ of the log likelihood function $\log L(\boldsymbol{\Psi})$ about the point $\boldsymbol{\Psi} = \boldsymbol{\Psi}^{(k)}$, we have that

$$\boldsymbol{S}(\boldsymbol{y}; \boldsymbol{\Psi}) \approx \boldsymbol{S}(\boldsymbol{y}; \boldsymbol{\Psi}^{(k)}) - \boldsymbol{I}(\boldsymbol{\Psi}^{(k)}; \boldsymbol{y})(\boldsymbol{\Psi} - \boldsymbol{\Psi}^{(k)}). \tag{3.73}$$

On putting $\boldsymbol{\Psi} = \boldsymbol{\Psi}^*$ in (3.73), it follows since $\boldsymbol{S}(\boldsymbol{y}; \boldsymbol{\Psi})$ vanishes at the point $\boldsymbol{\Psi}^*$ that

$$\boldsymbol{\Psi}^* \approx \boldsymbol{\Psi}^{(k)} + \boldsymbol{I}^{-1}(\boldsymbol{\Psi}^{(k)}; \boldsymbol{y})\boldsymbol{S}(\boldsymbol{y}; \boldsymbol{\Psi}^{(k)}). \tag{3.74}$$

The next step is to expand $\partial Q(\boldsymbol{\Psi}; \boldsymbol{\Psi}^{(k)})/\partial\boldsymbol{\Psi}$ in a linear Taylor series about the point $\boldsymbol{\Psi} = \boldsymbol{\Psi}^{(k)}$ and to evaluate it at $\boldsymbol{\Psi} = \boldsymbol{\Psi}^{(k+1)}$ under the assumption (3.68) that it is zero at this point. This gives

$$\begin{aligned} \boldsymbol{0} &= [\partial Q(\boldsymbol{\Psi}; \boldsymbol{\Psi}^{(k)})/\partial\boldsymbol{\Psi}]_{\boldsymbol{\Psi}=\boldsymbol{\Psi}^{(k+1)}} \\ &\approx [\partial Q(\boldsymbol{\Psi}; \boldsymbol{\Psi}^{(k)})/\partial\boldsymbol{\Psi}]_{\boldsymbol{\Psi}=\boldsymbol{\Psi}^{(k)}} \\ &\quad + [\partial^2 Q(\boldsymbol{\Psi}; \boldsymbol{\Psi}^{(k)})/\partial\boldsymbol{\Psi}\partial\boldsymbol{\Psi}^T]_{\boldsymbol{\Psi}=\boldsymbol{\Psi}^{(k)}}(\boldsymbol{\Psi}^{(k+1)} - \boldsymbol{\Psi}^{(k)}) \\ &= S(\boldsymbol{y}; \boldsymbol{\Psi}^{(k)}) - \mathcal{I}_c(\boldsymbol{\Psi}^{(k)}; \boldsymbol{y})(\boldsymbol{\Psi}^{(k+1)} - \boldsymbol{\Psi}^{(k)}), \end{aligned} \tag{3.75}$$

since from (3.72),

$$[\partial^2 Q(\boldsymbol{\Psi}; \boldsymbol{\Psi}^{(k)})/\partial\boldsymbol{\Psi}\partial\boldsymbol{\Psi}^T]_{\boldsymbol{\Psi}=\boldsymbol{\Psi}^{(k)}} = -\mathcal{I}_c(\boldsymbol{\Psi}^{(k)}; \boldsymbol{y}).$$

From (3.75)

$$\boldsymbol{S}(\boldsymbol{y};\, \boldsymbol{\Psi}^{(k)}) \approx \mathcal{I}_c(\boldsymbol{\Psi}^{(k)};\, \boldsymbol{y})(\boldsymbol{\Psi}^{(k+1)} - \boldsymbol{\Psi}^{(k)}). \tag{3.76}$$

On using now this approximation for $\boldsymbol{S}(\boldsymbol{y};\, \boldsymbol{\Psi}^{(k)})$ in (3.74), we have that

$$\begin{aligned}\boldsymbol{\Psi}^* - \boldsymbol{\Psi}^{(k)} &\approx \boldsymbol{I}^{-1}(\boldsymbol{\Psi}^{(k)};\, \boldsymbol{y})\mathcal{I}_c(\boldsymbol{\Psi}^{(k)};\, \boldsymbol{y})(\boldsymbol{\Psi}^{(k+1)} - \boldsymbol{\Psi}^{(k)}) \\ &= \boldsymbol{I}^{-1}(\boldsymbol{\Psi}^{(k)};\, \boldsymbol{y})\mathcal{I}_c(\boldsymbol{\Psi}^{(k)};\, \boldsymbol{y})(\boldsymbol{\Psi}^{(k+1)} - \boldsymbol{\Psi}^* + \boldsymbol{\Psi}^* - \boldsymbol{\Psi}^{(k)}).\end{aligned}$$

Thus

$$\begin{aligned}\boldsymbol{\Psi}^{(k+1)} - \boldsymbol{\Psi}^* &\approx \{\boldsymbol{I}_d - \mathcal{I}_c^{-1}(\boldsymbol{\Psi}^{(k)};\, \boldsymbol{y})\boldsymbol{I}(\boldsymbol{\Psi}^{(k)};\, \boldsymbol{y})\}(\boldsymbol{\Psi}^{(k)} - \boldsymbol{\Psi}^*) \\ &\approx \{\boldsymbol{I}_d - \mathcal{I}_c^{-1}(\boldsymbol{\Psi}^*;\, \boldsymbol{y})\boldsymbol{I}(\boldsymbol{\Psi}^*;\, \boldsymbol{y})\}(\boldsymbol{\Psi}^{(k)} - \boldsymbol{\Psi}^*) \\ &= \boldsymbol{J}(\boldsymbol{\Psi}^*)(\boldsymbol{\Psi}^{(k)} - \boldsymbol{\Psi}^*),\end{aligned}$$

thus establishing (3.70).

3.9.6 Example 3.5. (Example 3.4 Continued)

On putting $\mu = \hat{\mu}$ in the expression (3.53) for the conditional expectation of the (complete-data) information about μ, it can be seen that

$$\mathcal{I}_c(\hat{\mu};\, \boldsymbol{y}) = n\hat{\mu}^{-2},$$

while from (3.52), the incomplete-data observed information equals

$$I(\hat{\mu};\, \boldsymbol{y}) = r\hat{\mu}^{-2}.$$

Corresponding to (3.65), we have

$$\mu^{(k+1)} - \hat{\mu} = J(\hat{\mu})(\mu^{(k)} - \hat{\mu}), \tag{3.77}$$

where

$$\begin{aligned}J(\hat{\mu}) &= 1 - \mathcal{I}_c^{-1}(\hat{\mu};\, \boldsymbol{y})I(\hat{\mu};\, \boldsymbol{y}) \\ &= 1 - n^{-1}r.\end{aligned}$$

The result (3.77) was established directly in Section 1.5; see equation (1.53).

CHAPTER 4

Standard Errors and Speeding Up Convergence

4.1 INTRODUCTION

In this chapter, we address two issues that have led to some criticism of the EM algorithm. The first concerns the provision of standard errors, or the full covariance matrix in multiparameter situations, of the MLE obtained via the EM algorithm. One initial criticism of the EM algorithm was that it does not automatically provide an estimate of the covariance matrix of the MLE, as do some other methods, such as Newton-type methods.

Hence we shall consider a number of methods for assessing the covariance matrix of the MLE $\hat{\boldsymbol{\Psi}}$ of the parameter vector $\boldsymbol{\Psi}$, obtained via the EM algorithm. Most of these methods are based on the observed information matrix $\boldsymbol{I}(\hat{\boldsymbol{\Psi}}; \boldsymbol{y})$. It will be seen that there are methods that allow $\boldsymbol{I}(\hat{\boldsymbol{\Psi}}; \boldsymbol{y})$ to be calculated within the EM framework. For independent data, $\boldsymbol{I}(\hat{\boldsymbol{\Psi}}; \boldsymbol{y})$ can be approximated without additional work beyond the calculations used to compute the MLE in the first instance.

The other common criticism that has been leveled at the EM algorithm is that its convergence can be quite slow. We shall therefore consider some methods that have been proposed for accelerating the EM algorithm. However, methods to accelerate the EM algorithm do tend to sacrifice the simplicity it usually enjoys. As remarked by Lange (1995b), it is likely that no acceleration method can match the stability and simplicity of the unadorned EM algorithm.

The methods to be discussed in this chapter for speeding up the convergence of the EM algorithm are applicable for a given specification of the complete data. In the next chapter, we consider some recent developments that approach the problem of speeding up convergence in terms of the choice of the missing data in the specification of the complete-data problem in the EM framework. A working parameter is introduced in the specification of the complete data, which thus indexes a class of EM algorithms. We shall see that in some particular cases it is possible to modify the choice of the complete data so that the resulting EM algorithm can be just as easily implemented yet produce striking gains in the speed of convergence.

4.2 OBSERVED INFORMATION MATRIX

4.2.1 Direct Evaluation

As explained in Section 4.1, it is common in practice to estimate the inverse of the covariance matrix of the MLE $\hat{\boldsymbol{\Psi}}$ by the observed information matrix $\boldsymbol{I}(\hat{\boldsymbol{\Psi}}; \boldsymbol{y})$. Hence we shall consider in the following subsections a number of ways for calculating or approximating $\boldsymbol{I}(\hat{\boldsymbol{\Psi}}; \boldsymbol{y})$. One way to proceed is to directly evaluate $\boldsymbol{I}(\hat{\boldsymbol{\Psi}}; \boldsymbol{y})$ after the computation of the MLE $\hat{\boldsymbol{\Psi}}$. However, analytical evaluation of the second-order derivatives of the incomplete-data log likelihood, $\log L(\boldsymbol{\Psi})$, may be difficult, or at least tedious. Indeed, often it is for reasons of this nature that the EM algorithm is used to compute the MLE in the first instance.

4.2.2 Extraction of Observed Information Matrix in Terms of the Complete-Data Log Likelihood

Louis (1982) shows that the missing information matrix $\mathcal{I}_m(\boldsymbol{\Psi}; \boldsymbol{y})$ can be expressed in the form

$$\mathcal{I}_m(\boldsymbol{\Psi}; \boldsymbol{y}) = \text{cov}_{\boldsymbol{\Psi}}\{\boldsymbol{S}_c(\boldsymbol{X}; \boldsymbol{\Psi}) \mid \boldsymbol{y}\} \tag{4.1}$$

$$= E_{\boldsymbol{\Psi}}\{\boldsymbol{S}_c(\boldsymbol{X}; \boldsymbol{\Psi})\boldsymbol{S}_c^T(\boldsymbol{X}; \boldsymbol{\Psi}) \mid \boldsymbol{y}\} - \boldsymbol{S}(\boldsymbol{y}; \boldsymbol{\Psi})\boldsymbol{S}^T(\boldsymbol{y}; \boldsymbol{\Psi}), \tag{4.2}$$

since

$$\boldsymbol{S}(\boldsymbol{y}; \boldsymbol{\Psi}) = E_{\boldsymbol{\Psi}}\{\boldsymbol{S}_c(\boldsymbol{X}; \boldsymbol{\Psi}) \mid \boldsymbol{y}\}.$$

On substituting (4.1) and then (4.2) into (3.48), we have that

$$\begin{aligned}\boldsymbol{I}(\boldsymbol{\Psi}; \boldsymbol{y}) &= \mathcal{I}_c(\boldsymbol{\Psi}; \boldsymbol{y}) - \mathcal{I}_m(\boldsymbol{\Psi}; \boldsymbol{y}) \\ &= \mathcal{I}_c(\boldsymbol{\Psi}; \boldsymbol{y}) - \text{cov}_{\boldsymbol{\Psi}}\{\boldsymbol{S}_c(\boldsymbol{X}; \boldsymbol{\Psi}) \mid \boldsymbol{y}\} \\ &= \mathcal{I}_c(\boldsymbol{\Psi}; \boldsymbol{y}) - E_{\boldsymbol{\Psi}}\{\boldsymbol{S}_c(\boldsymbol{X}; \boldsymbol{\Psi})\boldsymbol{S}_c^T(\boldsymbol{X}; \boldsymbol{\Psi}) \mid \boldsymbol{y}\} \\ &\quad + \boldsymbol{S}(\boldsymbol{y}; \boldsymbol{\Psi})\boldsymbol{S}^T(\boldsymbol{y}; \boldsymbol{\Psi}).\end{aligned} \tag{4.3}$$

The result (4.3) can be established by directly showing that $\mathcal{I}_m(\boldsymbol{\Psi}; \boldsymbol{y})$ can be put in the form of the right-hand side of (4.1). Louis (1982) establishes (4.3) by working with $\boldsymbol{I}(\boldsymbol{\Psi}; \boldsymbol{y})$ as follows. From (3.43), $\boldsymbol{I}(\boldsymbol{\Psi}; \boldsymbol{y})$ can be expressed as

$$
\begin{aligned}
\boldsymbol{I}(\boldsymbol{\Psi};\boldsymbol{y}) &= -\partial S(\boldsymbol{y};\boldsymbol{\Psi})/\partial\boldsymbol{\Psi} \\
&= -\partial\left[\left\{\int_{\mathcal{X}(\boldsymbol{y})} g_c'(\boldsymbol{x};\,\boldsymbol{\Psi})\,d\boldsymbol{x}\right\}\Big/ g(\boldsymbol{y};\,\boldsymbol{\Psi})\right]\Big/\partial\boldsymbol{\Psi} \\
&= -\left\{\int_{\mathcal{X}(\boldsymbol{y})} \partial^2 g_c(\boldsymbol{x};\,\boldsymbol{\Psi})/\partial\boldsymbol{\Psi}\partial\boldsymbol{\Psi}^T\,d\boldsymbol{x}\right\}\Big/ g(\boldsymbol{y};\boldsymbol{\Psi}) \\
&\quad + \left\{\int_{\mathcal{X}(\boldsymbol{y})} g_c'(\boldsymbol{x};\,\boldsymbol{\Psi})\,d\boldsymbol{x}\right\}\left\{\int_{\mathcal{X}(\boldsymbol{y})} g_c'(\boldsymbol{x};\,\boldsymbol{\Psi})\,d\boldsymbol{x}\right\}^T\Big/ \{g(\boldsymbol{y};\,\boldsymbol{\Psi})\}^2.
\end{aligned}
\tag{4.4}
$$

In (4.4) and the equations below, the prime denotes differentiation with respect to $\boldsymbol{\Psi}$. Also, it is assumed in this section and the sequel that regularity conditions hold for the interchange of the operations of differentiation and integration where necessary.

Proceeding as before with the derivation of (3.44),

$$
\boldsymbol{I}(\boldsymbol{\Psi};\,\boldsymbol{y}) = -\left\{\int_{\mathcal{X}(\boldsymbol{y})} \partial^2 g_c \boldsymbol{x};\,\boldsymbol{\Psi})/\partial\boldsymbol{\Psi}\partial\boldsymbol{\Psi}^T\,d\boldsymbol{x}\right\}\Big/ g(\boldsymbol{y};\,\boldsymbol{\Psi}) + \boldsymbol{S}(\boldsymbol{y};\,\boldsymbol{\Psi})\boldsymbol{S}^T(\boldsymbol{y};\,\boldsymbol{\Psi}),
\tag{4.5}
$$

on using the result (3.43) for the last term on the right-hand side of (4.4).

The first term on the right-hand side of (4.5) can be expressed as

$$
\begin{aligned}
&-\left\{\int_{\mathcal{X}(\boldsymbol{y})} \partial^2 g_c(\boldsymbol{x};\,\boldsymbol{\Psi})/\partial\boldsymbol{\Psi}\partial\boldsymbol{\Psi}^T\,d\boldsymbol{x}\right\}\Big/ g(\boldsymbol{y};\,\boldsymbol{\Psi}) \\
&= -\int_{\mathcal{X}(\boldsymbol{y})} [\{\partial^2 \log g_c(\boldsymbol{x};\,\boldsymbol{\Psi})/\partial\boldsymbol{\Psi}\partial\boldsymbol{\Psi}^T\}\{g_c(\boldsymbol{x};\,\boldsymbol{\Psi})/g(\boldsymbol{y};\,\boldsymbol{\Psi}\}]\,d\boldsymbol{x} \\
&\quad - \int_{\mathcal{X}(\boldsymbol{y})} \{g_c'(\boldsymbol{x};\,\boldsymbol{\Psi})/g_c(\boldsymbol{x};\boldsymbol{\Psi})\}\{g_c'(\boldsymbol{x};\,\boldsymbol{\Psi})/g_c(\boldsymbol{x};\,\boldsymbol{\Psi})\}^T\{g_c(\boldsymbol{x};\,\boldsymbol{\Psi})/g(\boldsymbol{y};\,\boldsymbol{\Psi})\}\,d\boldsymbol{x} \\
&= \int_{\mathcal{X}(\boldsymbol{y})} \boldsymbol{I}_c(\boldsymbol{\Psi};\,\boldsymbol{x})k(\boldsymbol{x}\mid\boldsymbol{y};\,\boldsymbol{\Psi})\,d\boldsymbol{x} \\
&\quad - \int_{\mathcal{X}(\boldsymbol{y})} \boldsymbol{S}_c(\boldsymbol{x};\,\boldsymbol{\Psi})\boldsymbol{S}_c^T(\boldsymbol{x};\,\boldsymbol{\Psi})k(\boldsymbol{x}\mid\boldsymbol{y};\,\boldsymbol{\Psi})\,d\boldsymbol{x} \\
&= E_{\boldsymbol{\Psi}}\{\boldsymbol{I}_c(\boldsymbol{\Psi};\,\boldsymbol{X})\mid\boldsymbol{y}\} - E_{\boldsymbol{\Psi}}\{\boldsymbol{S}_c(\boldsymbol{X};\,\boldsymbol{\Psi})\boldsymbol{S}_c^T(\boldsymbol{X};\,\boldsymbol{\Psi})\mid\boldsymbol{y}\}. \\
&= \mathcal{I}_c(\boldsymbol{\Psi};\,\boldsymbol{y}) - E_{\boldsymbol{\Psi}}\{\boldsymbol{S}_c(\boldsymbol{X};\,\boldsymbol{\Psi})\boldsymbol{S}_c^T(\boldsymbol{X};\,\boldsymbol{\Psi})\mid\boldsymbol{y}\}.
\end{aligned}
\tag{4.6}
$$

Substitution of (4.6) into (4.5) gives the expression (4.3) for $\boldsymbol{I}(\boldsymbol{\Psi};\,\boldsymbol{y})$.

From (4.3), the observed information matrix $\boldsymbol{I}(\hat{\boldsymbol{\Psi}})$ can be computed as

$$I(\hat{\boldsymbol{\Psi}}; \boldsymbol{y}) = \mathcal{I}_c(\hat{\boldsymbol{\Psi}}; \boldsymbol{y}) - \mathcal{I}_m(\hat{\boldsymbol{\Psi}}; \boldsymbol{y}) \tag{4.7}$$

$$= \mathcal{I}_c(\hat{\boldsymbol{\Psi}}; \boldsymbol{y}) - [\mathrm{cov}_{\boldsymbol{\Psi}}\{\boldsymbol{S}_c(\boldsymbol{X}; \boldsymbol{\Psi})\boldsymbol{S}_c^T(\boldsymbol{X}; \boldsymbol{\Psi})\}]_{\boldsymbol{\Psi}=\hat{\boldsymbol{\Psi}}} \tag{4.8}$$

$$= \mathcal{I}_c(\hat{\boldsymbol{\Psi}}; \boldsymbol{y}) - [E_{\boldsymbol{\Psi}}\{\boldsymbol{S}_c(\boldsymbol{X}; \boldsymbol{\Psi})\boldsymbol{S}_c^T(\boldsymbol{X}; \boldsymbol{\Psi})\} \mid \boldsymbol{y}]_{\boldsymbol{\Psi}=\hat{\boldsymbol{\Psi}}}, \tag{4.9}$$

since the last term on the right-hand side of (4.3) is zero as $\hat{\boldsymbol{\Psi}}$ satisfies

$$\boldsymbol{S}(\boldsymbol{y}; \boldsymbol{\Psi}) = \boldsymbol{0}.$$

Hence, the observed information matrix for the original (incomplete-data) problem can be computed in terms of the conditional moments of the gradient and curvature of the complete-data log likelihood function introduced within the EM framework.

It is important to emphasize that the aforementioned estimates of the covariance matrix of the MLE are based on the second derivatives of the (complete-data) log likelihood, and so are guaranteed to be valid inferentially only asymptotically. Consequently, from both frequentist and Bayesian perspectives, the practical propriety of the resulting normal theory inferences is improved when the log likelihood is more nearly normal. To this end, Meng and Rubin (1991) transform some of the parameters in their examples, which are to be presented in Section 4.5.

4.2.3 Regular Case

In the case of the regular exponential family with $\boldsymbol{\Psi}$ as the natural parameter vector, the matrix $\boldsymbol{I}_c(\boldsymbol{\Psi}; \boldsymbol{x})$ is not a function of the data and so

$$\mathcal{I}_c(\boldsymbol{\Psi}; \boldsymbol{y}) = \mathcal{I}_c(\boldsymbol{\Psi}) \tag{4.10}$$

$$= \mathrm{cov}_{\boldsymbol{\Psi}}\{\boldsymbol{S}_c(\boldsymbol{X}; \boldsymbol{\Psi})\}$$

$$= \mathrm{cov}_{\boldsymbol{\Psi}}\{\boldsymbol{t}(\boldsymbol{X})\} \tag{4.11}$$

since from (1.56),

$$\boldsymbol{S}_c(\boldsymbol{X}; \boldsymbol{\Psi}) = \boldsymbol{t}(\boldsymbol{X}).$$

From (4.1), we have that

$$\mathcal{I}_m(\boldsymbol{\Psi}; \boldsymbol{y}) = \mathrm{cov}_{\boldsymbol{\Psi}}\{\boldsymbol{S}_c(\boldsymbol{X}; \boldsymbol{\Psi}) \mid \boldsymbol{y}\}$$

$$= \mathrm{cov}_{\boldsymbol{\Psi}}\{\boldsymbol{t}(\boldsymbol{X}) \mid \boldsymbol{y}\}. \tag{4.12}$$

Hence from (4.7), the observed incomplete-data information matrix $\boldsymbol{I}(\hat{\boldsymbol{\Psi}}; \boldsymbol{y})$

can be written as the difference at $\boldsymbol{\Psi} = \hat{\boldsymbol{\Psi}}$ between the unconditional covariance matrix and the conditional covariance matrix of the complete-data score function $\boldsymbol{S}_c(\boldsymbol{X}; \boldsymbol{\Psi})$, which is the sufficient statistic $\boldsymbol{t}(\boldsymbol{X})$. Thus the observed information matrix $\boldsymbol{I}(\hat{\boldsymbol{\Psi}}; \boldsymbol{y})$ is given by

$$\boldsymbol{I}(\hat{\boldsymbol{\Psi}}; \boldsymbol{y}) = [\text{cov}_{\boldsymbol{\Psi}}\{\boldsymbol{t}(\boldsymbol{X})\} - \text{cov}_{\boldsymbol{\Psi}}\{\boldsymbol{t}(\boldsymbol{X}) \mid \boldsymbol{y}\}]_{\boldsymbol{\Psi}=\hat{\boldsymbol{\Psi}}}.$$

4.2.4 Evaluation of the Conditional Expected Complete-Data Information Matrix

If any of the formulas (4.7) to (4.9) is to be used to calculate the observed information matrix $\boldsymbol{I}(\hat{\boldsymbol{\Psi}}; \boldsymbol{y})$, then we have to calculate $\mathcal{I}_c(\hat{\boldsymbol{\Psi}}; \boldsymbol{y})$, the conditional expectation of the complete-data information matrix evaluated at the MLE. Also, it will be seen shortly in Section 4.5 that we need to calculate $\mathcal{I}_c(\hat{\boldsymbol{\Psi}}; \boldsymbol{y})$ when using the Supplemented EM algorithm to compute the observed information matrix $\boldsymbol{I}(\hat{\boldsymbol{\Psi}}; \boldsymbol{y})$.

The calculation of $\mathcal{I}_c(\hat{\boldsymbol{\Psi}}; \boldsymbol{y})$ is readily facilitated by standard complete-data computations if the complete-data density $g_c(\boldsymbol{x}; \boldsymbol{\Psi})$ belongs to the regular exponential family, since then

$$\mathcal{I}_c(\hat{\boldsymbol{\Psi}}; \boldsymbol{y}) = \mathcal{I}_c(\hat{\boldsymbol{\Psi}}). \tag{4.13}$$

This result is obvious from (4.10) if $\boldsymbol{\Psi}$ is the natural parameter. Suppose now that $\boldsymbol{\theta} = \boldsymbol{c}(\boldsymbol{\Psi})$ is the natural parameter. Then to show that (4.13) still holds, we have on using the chain rule to differentiate minus $\log L_c(\boldsymbol{\Psi})$ twice with respect to $\boldsymbol{\Psi}$ that

$$\boldsymbol{I}_c(\boldsymbol{\Psi}; \boldsymbol{x}) = \boldsymbol{A}^T(-\partial^2 \log L_c(\boldsymbol{\Psi})/\partial\boldsymbol{\theta}\partial\boldsymbol{\theta}^T)\boldsymbol{A} + \boldsymbol{B}, \tag{4.14}$$

where

$$\begin{aligned}(\boldsymbol{A})_{ij} &= \partial\theta_i/\partial\Psi_j, \\ (\boldsymbol{B})_{ij} &= \boldsymbol{d}_{ij}^T \partial \log L_c(\boldsymbol{\Psi})/\partial\boldsymbol{\theta},\end{aligned}$$

and

$$\begin{aligned}\boldsymbol{d}_{ij} &= (\partial^2\theta_1/\partial\Psi_i\partial\Psi_j, \ldots, \partial^2\theta_d/\partial\Psi_i\partial\Psi_j)^T \\ &= \partial^2\boldsymbol{\theta}/\partial\Psi_i\partial\Psi_j\end{aligned}$$

for $i, j = 1, \ldots, d$. As $-\partial^2 \log L_c(\boldsymbol{\Psi})/\partial\boldsymbol{\theta}\partial\boldsymbol{\theta}^T$ does not depend on the data, it equals the expected complete-data information matrix for $\boldsymbol{\theta}$, and so

$$\boldsymbol{A}^T(-\partial^2 \log L_c(\boldsymbol{\Psi})/\partial\boldsymbol{\theta}\partial\boldsymbol{\theta}^T)\boldsymbol{A}$$

is equal to $\mathcal{I}_c(\boldsymbol{\Psi})$, the expected complete-data information matrix for $\boldsymbol{\Psi}$. It therefore follows on taking the conditional expectation of (4.14) given $\boldsymbol{y}$ that

$$\mathcal{I}_c(\boldsymbol{\Psi}; \boldsymbol{y}) = \mathcal{I}_c(\boldsymbol{\Psi}) + E_{\boldsymbol{\Psi}}(\boldsymbol{B} \mid \boldsymbol{y}), \tag{4.15}$$

where

$$\begin{aligned} E_{\boldsymbol{\Psi}}(\boldsymbol{B} \mid \boldsymbol{y}) &= \boldsymbol{d}_{ij}^T E_{\boldsymbol{\Psi}}\{\partial \log L_c(\boldsymbol{\Psi})/\partial \boldsymbol{\theta} \mid \boldsymbol{y}\} \\ &= \boldsymbol{d}_{ij}^T \partial \log L(\boldsymbol{\Psi})/\partial \boldsymbol{\theta}. \end{aligned} \tag{4.16}$$

Now

$$\partial \log L(\boldsymbol{\Psi})/\partial \boldsymbol{\theta} = \boldsymbol{0}$$

at $\boldsymbol{\theta}^* = \boldsymbol{c}(\boldsymbol{\Psi}^*)$ for any stationary point $\boldsymbol{\Psi}^*$ of $\mathrm{L}(\boldsymbol{\Psi})$. Thus then from (4.15) and (4.16),

$$\mathcal{I}_c(\boldsymbol{\Psi}^*; \boldsymbol{y}) = \mathcal{I}_c(\boldsymbol{\Psi}^*),$$

for any stationary point $\boldsymbol{\Psi}^*$ of $\mathrm{L}(\boldsymbol{\Psi})$, including the MLE $\hat{\boldsymbol{\Psi}}$ of $\boldsymbol{\Psi}$. When $g_c(\boldsymbol{x}; \boldsymbol{\Psi})$ is from an irregular exponential family, the calculation of $\mathcal{I}_c(\boldsymbol{\Psi}; \boldsymbol{y})$ is still straightforward, as $\log L_c(\boldsymbol{\Psi})$ is linear in $\boldsymbol{t}(\boldsymbol{x})$. Hence $\mathcal{I}_c(\boldsymbol{\Psi}; \boldsymbol{y})$ is formed from $\boldsymbol{I}_c(\boldsymbol{\Psi}; \boldsymbol{x})$ simply by replacing $\boldsymbol{t}(\boldsymbol{x})$ by its conditional expectation.

4.2.5 Examples

Example 4.1. (Example 1.1 Continued). For this example, the observed information $I(\hat{\Psi}; \boldsymbol{y})$ can be obtained directly by evaluating the right-hand side of (1.16) at $\Psi = \hat{\Psi} = 0.6268214980$ to give

$$I(\hat{\Psi}; \boldsymbol{y}) = 377.516907. \tag{4.17}$$

Thus the asymptotic standard error of $\hat{\Psi}$ is

$$1/\sqrt{377.5} = 0.051. \tag{4.18}$$

Concerning the expression (4.8) for the asymptotic variance of $\hat{\Psi}$, we now consider the calculation of $\mathcal{I}_c(\hat{\Psi}; \boldsymbol{y})$. On differentiation of (1.22) with respect to Ψ,

$$S_c(\boldsymbol{x}; \Psi) = \frac{y_{12} + y_4}{\Psi} - \frac{y_2 + y_3}{1 - \Psi} \tag{4.19}$$

and

$$I_c(\Psi; \boldsymbol{x}) = \frac{y_{12} + y_4}{\Psi^2} + \frac{y_2 + y_3}{(1 - \Psi)^2}. \tag{4.20}$$

Taking the conditional expectation of (4.20) given $\boldsymbol{y}$ gives

$$\mathcal{I}_c(\Psi; \boldsymbol{y}) = \frac{E_\Psi(Y_{12} \mid \boldsymbol{y}) + y_4}{\Psi^2} + \frac{y_2 + y_3}{(1 - \Psi)^2}, \tag{4.21}$$

where

$$E_{\Psi}(Y_{12} \mid \boldsymbol{y}) = \tfrac{1}{4}y_1\Psi/(\tfrac{1}{2} + \tfrac{1}{4}\Psi). \tag{4.22}$$

This last result (4.22) follows from the work in Section 1.4.2, where it was noted that the conditional distribution of Y_{12} given $\boldsymbol{y}$ is binomial with sample size $y_1 = 125$ and probability parameter

$$\tfrac{1}{4}\Psi/(\tfrac{1}{2} + \tfrac{1}{4}\Psi).$$

On evaluation at $\Psi = \hat{\Psi} = 0.6268214980$, we obtain

$$\mathcal{I}_c(\hat{\Psi}; \boldsymbol{y}) = 435.318. \tag{4.23}$$

From (4.21), we can express $\mathcal{I}_c(\hat{\Psi}; \boldsymbol{y})$ also as

$$\mathcal{I}_c(\hat{\Psi}; \boldsymbol{y}) = \frac{E_{\hat{\Psi}}(Y_{12} \mid \boldsymbol{y}) + y_2 + y_3 + y_4}{\hat{\Psi}(1 - \hat{\Psi})}. \tag{4.24}$$

This agrees with the fact that $L_c(\Psi)$ has a binomial form with probability parameter Ψ and sample size $Y_{12} + y_2 + y_3 + y_4$, with conditional expectation

$$E_{\Psi}(Y_{12} \mid \boldsymbol{y}) + y_2 + y_3 + y_4.$$

Thus $\mathcal{I}_c(\Psi; \boldsymbol{y})$ is given by the conditional expectation of the inverse of the binomial variance,

$$(Y_{12} + y_2 + y_3 + y_4)/\{\Psi(1 - \Psi)\}.$$

It is also of interest to note here that

$$\mathcal{I}_c(\hat{\Psi}; \boldsymbol{y}) \neq \mathcal{I}_c(\hat{\Psi}).$$

For on taking the expectation of (4.20), we have that

$$\mathcal{I}_c(\Psi) = \tfrac{1}{2}n/\{\Psi(1 - \Psi)\},$$

and so

$$\mathcal{I}_c(\hat{\Psi}) = \tfrac{1}{2}n/\{\hat{\Psi}(1 - \hat{\Psi})\},$$

which does not equal (4.24). This can be explained by the fact that the complete-data density $g_c(\boldsymbol{x}; \Psi)$ is a member of an irregular exponential family.

Moving now to the computation of $\mathcal{I}_m(\hat{\Psi}; \boldsymbol{y})$, we have from (4.1) and (4.19) that

$$\begin{aligned}
\mathcal{I}_m(\Psi; \boldsymbol{y}) &= \text{var}_\Psi\{S_c(\boldsymbol{X}; \Psi) \mid \boldsymbol{y}\} \\
&= \text{var}_\Psi\left\{\left(\frac{Y_{12} + y_4}{\Psi} - \frac{y_2 + y_3}{1 - \Psi}\right) \Big| \boldsymbol{y}\right\} \\
&= \Psi^{-2}\text{var}_\Psi(Y_{12} \mid \boldsymbol{y}) \\
&= \Psi^{-2} y_1\{(\Psi/8)/(\tfrac{1}{2} + \tfrac{1}{4}\Psi)^2\},
\end{aligned} \tag{4.25}$$

which on evaluation at $\Psi = \hat{\Psi}$ gives

$$\mathcal{I}_m(\hat{\Psi}; \boldsymbol{y}) = 57.801. \tag{4.26}$$

From (4.23) and (4.26), we obtain that the observed information is given by

$$\begin{aligned}
I(\hat{\Psi}; \boldsymbol{y}) &= \mathcal{I}_c(\hat{\Psi}; \boldsymbol{y}) - \mathcal{I}_m(\hat{\Psi}; \boldsymbol{y}) \\
&= 435.318 - 57.801 \\
&= 377.517,
\end{aligned}$$

which yields 0.051 as the standard error of $\hat{\Psi}$. This agrees with (4.18) obtained by direct evaluation of the observed information $I(\hat{\Psi}; \boldsymbol{y})$.

From (3.69), we have that the rate of convergence is given approximately by

$$\begin{aligned}
J(\hat{\Psi}) &= \mathcal{I}_m(\hat{\Psi})/\mathcal{I}_c(\hat{\Psi}) \\
&= 57.801/435.318 \\
&= 0.1328,
\end{aligned}$$

which is in agreement with the assessed rate of 0.1328 in Table 1.2.

Example 4.2. Mixture of Two Univariate Normals with Known Common Variance. One of the examples of Louis (1982) to demonstrate the use of the formula (4.8) or (4.9) for calculating the observed information matrix $\boldsymbol{I}(\hat{\boldsymbol{\Psi}}; \boldsymbol{y})$ was ML estimation for a mixture of two univariate normal densities with known common variance set equal to one. That is, the observed data vector is given by

$$\boldsymbol{y} = (w_1, \ldots, w_n)^T,$$

where $w_1, \ldots, w_n$ denote the observed values of a random sample from the density

$$f(w; \boldsymbol{\Psi}) = \pi_1\phi(w; \mu_1, 1) + \pi_2\phi(w; \mu_2, 1),$$

where

$$\boldsymbol{\Psi} = (\pi_1, \mu_1, \mu_2)^T.$$

In illustrating the use of the result (4.8) for this particular example, we shall just consider the diagonal elements of $\mathcal{I}_c(\hat{\boldsymbol{\Psi}}; \boldsymbol{y})$, as it is easy to check that its off-diagonal elements are all zero.

Concerning the calculation of $\mathcal{I}_c(\hat{\boldsymbol{\Psi}}; \boldsymbol{y})$ first, we have that the diagonal elements of $\boldsymbol{I}_c(\boldsymbol{\Psi}; \boldsymbol{x})$ are given by

$$I_{c;11}(\boldsymbol{\Psi}; \boldsymbol{x}) = \sum_{j=1}^{n} (z_{1j}/\pi_1^2 + z_{2j}/\pi_2^2), \tag{4.27}$$

$$I_{c;22}(\boldsymbol{\Psi}; \boldsymbol{x}) = \sum_{j=1}^{n} z_{1j}, \tag{4.28}$$

and

$$I_{c;33}(\boldsymbol{\Psi}; \boldsymbol{x}) = \sum_{j=1}^{n} z_{2j}. \tag{4.29}$$

In the above equations, $I_{c;ii}(\boldsymbol{\Psi}; \boldsymbol{x})$ denotes the ith diagonal element of $\boldsymbol{I}_c(\boldsymbol{\Psi}; \boldsymbol{x})$.

On taking the conditional expectation of (4.27) to (4.29) given $\boldsymbol{y}$, we find that

$$\mathcal{I}_{c;11}(\boldsymbol{\Psi}; \boldsymbol{y}) = \sum_{j=1}^{n} \{\tau_1(w_j; \boldsymbol{\Psi})/\pi_1^2 + \tau_2(w_j; \boldsymbol{\Psi})/\pi_2^2\} \tag{4.30}$$

$$\mathcal{I}_{c;22}(\boldsymbol{\Psi}; \boldsymbol{y}) = \sum_{j=1}^{n} \tau_1(w_j; \boldsymbol{\Psi}), \tag{4.31}$$

and

$$\mathcal{I}_{c;33}(\boldsymbol{\Psi}; \boldsymbol{y}) = \sum_{j=1}^{n} \tau_2(w_j; \boldsymbol{\Psi}), \tag{4.32}$$

where $\tau_1(w_j, \boldsymbol{\Psi})$ and $\tau_2(w_j, \boldsymbol{\Psi})$ are as in Section 2.7. Evaluation of the right-hand sides of (4.30) to (4.32) at the point $\boldsymbol{\Psi} = \hat{\boldsymbol{\Psi}}$ gives

$$\mathcal{I}_c(\hat{\boldsymbol{\Psi}}; \boldsymbol{y}) = \text{diag}\{n/(\hat{\pi}_1 \hat{\pi}_2),\ n\hat{\pi}_1,\ n\hat{\pi}_2\}. \tag{4.33}$$

In this particular example where the complete-data density belongs to the regular exponential family, it is simpler to use the relationship

$$\mathcal{I}_c(\hat{\boldsymbol{\Psi}}; \boldsymbol{y}) = \mathcal{I}_c(\hat{\boldsymbol{\Psi}})$$

to calculate $\mathcal{I}_c(\hat{\boldsymbol{\Psi}}; \boldsymbol{y})$, since it is well known that

$$\mathcal{I}_c(\boldsymbol{\Psi}) = \text{diag}\{n/(\pi_1 \pi_2),\ n\pi_1,\ n\pi_2\}.$$

This last result can be verified by taking the unconditional expectation of (4.27) to (4.29).

Considering the second term on the right-hand side of (4.8), we have that the elements of the complete-data gradient vector $\boldsymbol{S}_c(\boldsymbol{x}; \boldsymbol{\Psi})$ are given by

$$S_{c;1}(\boldsymbol{x};\ \boldsymbol{\Psi}) = \sum_{j=1}^{n}(z_{1j}/\pi_1 - z_{2j}/\pi_2)$$

$$= \left(\sum_{j=1}^{n} z_{1j} - n\pi_1\right) \Big/ (\pi_1\pi_2)$$

and

$$S_{c;i+1}(\boldsymbol{x};\ \boldsymbol{\Psi}) = \sum_{j=1}^{n}(w_j - \mu_{i+1})z_{ij} \quad (i = 1, 2).$$

We now proceed to the calculation of $\mathcal{I}_m(\hat{\boldsymbol{\Psi}};\ \boldsymbol{y})$ via the formula

$$\mathcal{I}_m(\boldsymbol{\Psi};\ \boldsymbol{y}) = \text{cov}_{\boldsymbol{\Psi}}\{\boldsymbol{S}_c(\boldsymbol{X};\ \boldsymbol{\Psi}) \mid \boldsymbol{y}\}.$$

The first diagonal element of $\mathcal{I}_m(\boldsymbol{\Psi};\ \boldsymbol{y})$ is given by

$$\mathcal{I}_{m;11}(\boldsymbol{\Psi};\ \boldsymbol{y}) = \text{var}_{\boldsymbol{\Psi}}\left\{\left(\sum_{j=1}^{n} Z_{1j} - n\pi_1\right) \Big/ (\pi_1\pi_2) \mid \boldsymbol{y}\right\}$$

$$= \pi_1^{-2}\pi_2^{-2}\text{var}_{\boldsymbol{\Psi}}\left\{\sum_{j=1}^{n} Z_{ij} \mid \boldsymbol{y}\right\}$$

$$= \pi_1^{-2}\pi_2^{-2}\sum_{j=1}^{n} \tau_1(w_j;\ \boldsymbol{\Psi})\tau_2(w_j;\ \boldsymbol{\Psi}).$$

Proceeding in a similar manner, it can be shown that

$$\mathcal{I}_{m;i+1,i+1}(\boldsymbol{\Psi};\ \boldsymbol{y}) = \text{var}_{\boldsymbol{\Psi}}\left\{\sum_{j=1}^{n}(w_j - \mu_i)Z_{ij} \mid \boldsymbol{y}\right\}$$

$$= \tau_1(w_j;\ \boldsymbol{\Psi})\tau_2(w_j;\ \boldsymbol{\Psi})\sum_{j=1}^{n}(w_j - \mu_i)^2 \tag{4.34}$$

$$\mathcal{I}_{m;1,i+1}(\boldsymbol{\Psi};\ \boldsymbol{y}) = \text{cov}_{\boldsymbol{\Psi}}\left\{\sum_{j=1}^{n}\frac{Z_{1j} - n\pi_1}{\pi_1\pi_2},\ \sum_{j=1}^{n}(w_j - \mu_i)Z_{ij}\right\}$$

$$= (-1)^{i+1}\pi_1^{-1}\pi_2^{-1}\sum_{j=1}^{n} \tau_1(w_j;\ \boldsymbol{\Psi})\tau_2(w_j;\ \boldsymbol{\Psi})(w_j - \mu_i) \tag{4.35}$$

for $i = 1, 2$ and

$$\begin{aligned}\mathcal{I}_{m;23}(\boldsymbol{\Psi}; \boldsymbol{y}) &= \operatorname{cov}_{\boldsymbol{\Psi}}\left\{\sum_{j=1}^{n}(w_j - \mu_1)Z_{1j}, \sum_{j=1}^{n}(w_j - \mu_2)Z_{2j}\right\} \\ &= -\sum_{j=1}^{n} \tau_1(w_j; \boldsymbol{\Psi})\tau_2(w_j; \boldsymbol{\Psi})(w_j - \mu_1)(w_j - \mu_2).\end{aligned}$$

The observed information matrix $\boldsymbol{I}(\hat{\boldsymbol{\Psi}}; \boldsymbol{y})$ can be obtained on evaluating $\mathcal{I}_m(\boldsymbol{\Psi}; \boldsymbol{y})$ at $\boldsymbol{\Psi} = \hat{\boldsymbol{\Psi}}$ and subtracting it from $\mathcal{I}_c(\hat{\boldsymbol{\Psi}}; \boldsymbol{y})$.

Of course in this simplified example, $\boldsymbol{I}(\hat{\boldsymbol{\Psi}}; \boldsymbol{y})$ could have been obtained by direct differentiation of the (incomplete-data) log likelihood. To illustrate this for the second diagonal term, we have that

$$\begin{aligned}I_{22}(\Psi; \boldsymbol{y}) &= -\partial^2 \log L(\boldsymbol{\Psi})/\partial\mu_1^2 \\ &= n\pi_1 - \tau_1(w_j; \boldsymbol{\Psi})\tau_2(w_j; \boldsymbol{\Psi})\sum_{j=1}^{n}(w_j - \mu_1)^2. \qquad (4.36)\end{aligned}$$

From (4.33) and (4.34), it can be seen that

$$I_{22}(\hat{\boldsymbol{\Psi}}; \boldsymbol{y}) = \mathcal{I}_{c;22}(\hat{\boldsymbol{\Psi}}; \boldsymbol{y}) - \mathcal{I}_{m;22}(\hat{\boldsymbol{\Psi}}; \boldsymbol{y}).$$

4.3 APPROXIMATIONS TO THE OBSERVED INFORMATION MATRIX: i.i.d. CASE

It can be seen from the expression (4.9) for the observed information matrix $\boldsymbol{I}(\hat{\boldsymbol{\Psi}}; \boldsymbol{y})$ that it requires, in addition to the code for the E- and M-steps, the calculation of the conditional (on the observed data $\boldsymbol{y}$) expectation of the complete-data information matrix $\boldsymbol{I}(\boldsymbol{\Psi}; \boldsymbol{x})$ and of the complete-data score statistic vector $S_c(\boldsymbol{X}; \boldsymbol{\Psi})$ times its transpose. For many problems, this can be algebraically tedious or even intractable. Hence, we now consider some practical methods for approximating the observed information matrix.

In the case of independent and identically distributed (i.i.d.) data, an approximation to the observed information matrix is readily available without any additional analyses having to be performed. In the i.i.d. case, the observed data $\boldsymbol{y}$ may be assumed to consist of observations $\boldsymbol{w}_1, \ldots, \boldsymbol{w}_n$ on n independent and identically distributed random variables with common p.d.f., say, $f(\boldsymbol{w}; \boldsymbol{\Psi})$. The log likelihood $\log L(\boldsymbol{\Psi})$ can be expressed then in the form

$$\log L(\boldsymbol{\Psi}) = \sum_{j=1}^{n} \log L_j(\boldsymbol{\Psi}),$$

where

$$L_j(\boldsymbol{\Psi}) = f(\boldsymbol{w}_j; \boldsymbol{\Psi})$$

is the likelihood function for $\boldsymbol{\Psi}$ formed from the single observation $\boldsymbol{w}_j$ ($j = 1, \ldots, n$).

We can now write the score vector $\boldsymbol{S}(\boldsymbol{y};\, \boldsymbol{\Psi})$ as

$$\boldsymbol{S}(\boldsymbol{y};\, \boldsymbol{\Psi}) = \sum_{j=1}^{n} \boldsymbol{s}(\boldsymbol{w}_j;\, \boldsymbol{\Psi}),$$

where

$$\boldsymbol{s}(\boldsymbol{w}_j;\, \boldsymbol{\Psi}) = \partial \log L_j(\boldsymbol{\Psi}) / \partial \boldsymbol{\Psi}.$$

The expected information matrix $\mathcal{I}(\boldsymbol{\Psi})$ can be written as

$$\mathcal{I}(\boldsymbol{\Psi}) = n\boldsymbol{i}(\boldsymbol{\Psi}), \tag{4.37}$$

where

$$\begin{aligned} \boldsymbol{i}(\boldsymbol{\Psi}) &= E_{\boldsymbol{\Psi}}\{\boldsymbol{s}(\boldsymbol{W};\, \boldsymbol{\Psi})\boldsymbol{s}^T(\boldsymbol{W};\, \boldsymbol{\Psi})\} \\ &= \text{cov}_{\boldsymbol{\Psi}}\{\boldsymbol{s}(\boldsymbol{W};\, \boldsymbol{\Psi})\} \end{aligned} \tag{4.38}$$

is the information contained in a single observation. Corresponding to (4.38), the empirical information matrix (in a single observation) can be defined to be

$$\begin{aligned} \bar{\boldsymbol{i}}(\boldsymbol{\Psi}) &= n^{-1} \sum_{j=1}^{n} \boldsymbol{s}(\boldsymbol{w}_j;\, \boldsymbol{\Psi})\boldsymbol{s}^T(\boldsymbol{w}_j;\, \boldsymbol{\Psi}) - \bar{\boldsymbol{s}}\bar{\boldsymbol{s}}^T \\ &= n^{-1} \sum_{j=1}^{n} \boldsymbol{s}(\boldsymbol{w}_j;\, \boldsymbol{\Psi})\boldsymbol{s}^T(\boldsymbol{w}_j;\, \boldsymbol{\Psi}) - n^{-2}\boldsymbol{S}(\boldsymbol{y};\, \boldsymbol{\Psi})\boldsymbol{S}^T(\boldsymbol{y};\, \boldsymbol{\Psi}), \end{aligned} \tag{4.39}$$

where

$$\bar{\boldsymbol{s}} = n^{-1} \sum_{j=1}^{n} \boldsymbol{s}(\boldsymbol{w}_j;\, \boldsymbol{\Psi}).$$

Corresponding to this empirical form (4.39) for $\boldsymbol{i}(\boldsymbol{\Psi})$, $\mathcal{I}(\boldsymbol{\Psi})$ is estimated by

$$\begin{aligned} \boldsymbol{I}_e(\boldsymbol{\Psi};\, \boldsymbol{y}) &= n\bar{\boldsymbol{i}}(\boldsymbol{\Psi}) \\ &= \sum_{j=1}^{n} \boldsymbol{s}(\boldsymbol{w}_j;\, \boldsymbol{\Psi})\boldsymbol{s}^T(\boldsymbol{w}_j;\, \boldsymbol{\Psi}) - n^{-1}\boldsymbol{S}(\boldsymbol{y};\, \boldsymbol{\Psi})\boldsymbol{S}^T(\boldsymbol{y};\, \boldsymbol{\Psi}). \end{aligned} \tag{4.40}$$

On evaluation at $\boldsymbol{\Psi} = \hat{\boldsymbol{\Psi}}$, $\boldsymbol{I}_e(\hat{\boldsymbol{\Psi}};\, \boldsymbol{y})$ reduces to

$$\boldsymbol{I}_e(\hat{\boldsymbol{\Psi}};\, \boldsymbol{y}) = \sum_{j=1}^{n} \boldsymbol{s}(\boldsymbol{w}_j;\, \hat{\boldsymbol{\Psi}})\boldsymbol{s}^T(\boldsymbol{w}_j;\, \hat{\boldsymbol{\Psi}}), \tag{4.41}$$

since $\boldsymbol{S}(\boldsymbol{y};\, \hat{\boldsymbol{\Psi}}) = \boldsymbol{0}$.

Meilijson (1989) terms $\boldsymbol{I}_e(\hat{\boldsymbol{\Psi}}; \boldsymbol{y})$ as the empirical observed information matrix. It is used commonly in practice to approximate the observed information matrix $\boldsymbol{I}(\hat{\boldsymbol{\Psi}}; \boldsymbol{y})$; see, for example, Berndt, Hall, Hall, and Hausman (1974), Redner and Walker (1984), and McLachlan and Basford (1988, Chapter 2).

It follows that $\boldsymbol{I}_e(\hat{\boldsymbol{\Psi}}; \boldsymbol{y})/n$ is a consistent estimator of $\boldsymbol{i}(\boldsymbol{\Psi})$. The use of (4.41) can be justified also in the following sense. Now

$$\begin{aligned}
\boldsymbol{I}(\boldsymbol{\Psi}; \boldsymbol{y}) &= -\partial^2 \log L(\boldsymbol{\Psi})/\partial\boldsymbol{\Psi}\partial\boldsymbol{\Psi}^T \\
&= -\sum_{j=1}^{n} \partial^2 \log L_j(\boldsymbol{\Psi})/\partial\boldsymbol{\Psi}\partial\boldsymbol{\Psi}^T \\
&= \sum_{j=1}^{n} \{\partial \log L_j(\boldsymbol{\Psi})/\partial\boldsymbol{\Psi}\}\{\partial \log L_j(\boldsymbol{\Psi})/\partial\boldsymbol{\Psi}\}^T \\
&\quad - \sum_{j=1}^{n} \{\partial^2 L_j(\boldsymbol{\Psi})/\partial\boldsymbol{\Psi}\partial\boldsymbol{\Psi}^T\}/L_j(\boldsymbol{\Psi}).
\end{aligned} \tag{4.42}$$

The second term on the right-hand side of (4.42) has zero expectation. Hence

$$\begin{aligned}
\boldsymbol{I}(\hat{\boldsymbol{\Psi}}; \boldsymbol{y}) &\approx \sum_{j=1}^{n} \{\partial \log L_j(\hat{\boldsymbol{\Psi}})/\partial\boldsymbol{\Psi}\}\{\partial \log L_j(\hat{\boldsymbol{\Psi}})/\partial\boldsymbol{\Psi}\}^T \\
&= \sum_{j=1}^{n} \boldsymbol{s}(\boldsymbol{w}_j; \hat{\boldsymbol{\Psi}})\boldsymbol{s}^T(\boldsymbol{w}_j; \hat{\boldsymbol{\Psi}}) \\
&= \boldsymbol{I}_e(\hat{\boldsymbol{\Psi}}; \boldsymbol{y}),
\end{aligned} \tag{4.43}$$

where the accuracy of this approximation depends on how close $\hat{\boldsymbol{\Psi}}$ is to $\boldsymbol{\Psi}$.

In the particular case where the complete-data density $g_c(\boldsymbol{x}; \boldsymbol{\Psi})$ is of multinomial form, the second term on the right-hand side of (4.42) is zero, and so (4.43) holds exactly.

It follows from the result (3.42) that

$$\boldsymbol{s}_j(\boldsymbol{w}_j; \boldsymbol{\Psi}) = E_{\boldsymbol{\Psi}}\{\partial \log L_{cj}(\boldsymbol{\Psi})/\partial\boldsymbol{\Psi} \mid \boldsymbol{y}\},$$

where $L_{cj}(\boldsymbol{\Psi})$ is the complete-data likelihood formed from the single observation $\boldsymbol{w}_j$ $(j = 1, \ldots, n)$. Thus the approximation $\boldsymbol{I}_e(\hat{\boldsymbol{\Psi}}; \boldsymbol{y})$ to the observed information matrix $\boldsymbol{I}(\hat{\boldsymbol{\Psi}}; \boldsymbol{y})$ can be expressed in terms of the conditional expectation of the gradient vector of the complete-data log likelihood function evaluated at the MLE $\hat{\boldsymbol{\Psi}}$. It thus avoids the computation of second-order partial derivatives of the complete-data log likelihood.

4.4 OBSERVED INFORMATION MATRIX FOR GROUPED DATA

4.4.1 Approximation Based on Empirical Information

We consider here an approximation to the observed information matrix in the case where the available data are truncated and grouped into r intervals, as formulated in Section 2.8. We let

$$\begin{aligned} \boldsymbol{s}_j(\boldsymbol{\Psi}^{(k)}) &= [\partial Q_j(\boldsymbol{\Psi};\ \boldsymbol{\Psi}^{(k)})/\partial \boldsymbol{\Psi}]_{\boldsymbol{\Psi}=\boldsymbol{\Psi}^{(k)}} \\ &= \partial \log P_j(\boldsymbol{\Psi})/\partial \boldsymbol{\Psi} \end{aligned} \tag{4.44}$$

from (2.67), where $Q_j(\boldsymbol{\Psi};\ \boldsymbol{\Psi}^{(k)})$ is defined by (2.63). It can be seen from Section 2.8 that the quantities $Q_j(\boldsymbol{\Psi};\ \boldsymbol{\Psi}^{(k)})$ are already available from the E-step of the EM algorithm. Hence it is convenient to approximate the observed information matrix in terms of these quantities, or effectively, the $\boldsymbol{s}_j(\boldsymbol{\Psi}^{(k)})$. The latter can be viewed as the incomplete-data score statistic for the jth interval $\mathcal{W}_j$, and so corresponds to $s(\boldsymbol{w}_j;\ \boldsymbol{\Psi}^{(k)})$, the score statistic for the single observation $\boldsymbol{w}_j$ in the case of ungrouped data.

Corresponding to the expression (4.40) for the so-called empirical information matrix for i.i.d. observations available in ungrouped form, we can approximate the observed information matrix for grouped i.i.d. data by

$$\boldsymbol{I}_{e,g}(\boldsymbol{\Psi}^{(k)};\ \boldsymbol{y}) = \sum_{j=1}^{r} n_j \boldsymbol{s}_j(\boldsymbol{\Psi}^{(k)}) \boldsymbol{s}_j^T(\boldsymbol{\Psi}^{(k)}) - n\bar{\boldsymbol{s}}(\boldsymbol{\Psi}^{(k)})\bar{\boldsymbol{s}}^T(\boldsymbol{\Psi}^{(k)}), \tag{4.45}$$

where

$$\bar{\boldsymbol{s}}(\boldsymbol{\Psi}^{(k)}) = \sum_{j=1}^{r} n_j \boldsymbol{s}_j(\boldsymbol{\Psi}^{(k)})/n. \tag{4.46}$$

We can write (4.45) as

$$\boldsymbol{I}_{e,g}(\boldsymbol{\Psi}^{(k)};\ \boldsymbol{y}) = \sum_{j=1}^{r} n_j \{\boldsymbol{s}_j(\boldsymbol{\Psi}^{(k)}) - \bar{\boldsymbol{s}}(\boldsymbol{\Psi}^{(k)})\}\{\boldsymbol{s}_j(\boldsymbol{\Psi}^{(k)}) - \bar{\boldsymbol{s}}(\boldsymbol{\Psi}^{(k)})\}^T, \tag{4.47}$$

demonstrating that it is always positive definite. Note that in the presence of truncation, $\bar{\boldsymbol{s}}(\boldsymbol{\Psi})$ is not zero at the MLE $\hat{\boldsymbol{\Psi}}$. From (2.72), it can be seen that in the case of truncation the score statistic at any point $\boldsymbol{\Psi}_o$ can be expressed as

$$\boldsymbol{S}(\boldsymbol{y};\ \boldsymbol{\Psi}) = \sum_{j=1}^{r} \left\{ n_j - \frac{nP_j(\boldsymbol{\Psi})}{P(\boldsymbol{\Psi})} \right\} s_j(\boldsymbol{\Psi}). \tag{4.48}$$

As $\boldsymbol{S}(\boldsymbol{y};\ \hat{\boldsymbol{\Psi}}) = \boldsymbol{0}$, it implies that

$$\bar{\boldsymbol{s}}(\hat{\boldsymbol{\Psi}}) = \sum_{j=1}^{r} \{P_j(\hat{\boldsymbol{\Psi}})/P(\hat{\boldsymbol{\Psi}})\} s_j(\hat{\boldsymbol{\Psi}}). \tag{4.49}$$

The use of (4.47) as an approximation to the observed matrix is developed in Jones and McLachlan (1992). It can be confirmed that $\boldsymbol{I}_{e,g}(\hat{\boldsymbol{\Psi}}; \boldsymbol{y})/n$ is a consistent estimator of the expected information matrix $\boldsymbol{i}(\boldsymbol{\Psi})$ contained in a single observation. The inverse of $\boldsymbol{I}_{e,g}(\hat{\boldsymbol{\Psi}}; \boldsymbol{y})$ provides an approximation to the covariance matrix of the MLE $\hat{\boldsymbol{\Psi}}$, while $\boldsymbol{I}_{e,g}(\boldsymbol{\Psi}^{(k)}; \boldsymbol{y})$ provides an approximation to $\boldsymbol{I}(\boldsymbol{\Psi}^{(k)}; \boldsymbol{y})$ for use, say, in the Newton-Raphson method of computing the MLE.

The use of $\boldsymbol{I}_{e,g}(\hat{\boldsymbol{\Psi}}; \boldsymbol{y})$ can be justified in the same way that the use of the empirical observation matrix was in Section 4.3 for ungrouped i.i.d. data. For example, corresponding to the relationship (4.47) between the observed information matrix and its empirical analog, we have that

$$\boldsymbol{I}(\hat{\boldsymbol{\Psi}}; \boldsymbol{y}) = \boldsymbol{I}_{e,g}(\hat{\boldsymbol{\Psi}}; \boldsymbol{y}) + R(\hat{\boldsymbol{\Psi}}), \tag{4.50}$$

where

$$\begin{aligned} \boldsymbol{R}(\boldsymbol{\Psi}) = & -\sum_{j=1}^{r} \{n_j / P_j(\boldsymbol{\Psi})\} \partial^2 P_j(\boldsymbol{\Psi}) / \partial\boldsymbol{\Psi} \partial\boldsymbol{\Psi}^T \\ & + \{n / P(\boldsymbol{\Psi})\} \partial^2 P(\boldsymbol{\Psi}) / \partial\boldsymbol{\Psi} \partial\boldsymbol{\Psi}^T \end{aligned}$$

has zero expectation.

To verify (4.50), we have from differentiation of the (incomplete-data) log likelihood that

$$\begin{aligned} \boldsymbol{I}(\boldsymbol{\Psi}; \boldsymbol{y}) = & \sum_{j=1}^{r} n_j \{\partial \log P_j(\boldsymbol{\Psi}) / \partial\boldsymbol{\Psi}\} \{\partial \log P_j(\boldsymbol{\Psi}) / \partial\boldsymbol{\Psi}\}^T \\ & - n \{\partial \log P(\boldsymbol{\Psi}) / \partial\boldsymbol{\Psi}\} \{\partial \log P(\boldsymbol{\Psi}) / \partial\boldsymbol{\Psi}\}^T + \boldsymbol{R}(\boldsymbol{\Psi}). \end{aligned} \tag{4.51}$$

Now from (2.63) and (4.44),

$$\partial \log P_j(\boldsymbol{\Psi}) / \partial\boldsymbol{\Psi} = s_j(\boldsymbol{\Psi}) \tag{4.52}$$

and

$$\partial \log P(\boldsymbol{\Psi}) / \partial\boldsymbol{\Psi} = \sum_{j=1}^{r} \{P_j(\boldsymbol{\Psi}) / P(\boldsymbol{\Psi})\} s_j(\boldsymbol{\Psi}).$$

Since $\boldsymbol{S}(\boldsymbol{y}; \hat{\boldsymbol{\Psi}}) = \boldsymbol{0}$, we have from (2.72) that

$$\begin{aligned} \partial \log P(\hat{\boldsymbol{\Psi}}) / \partial\boldsymbol{\Psi} &= \sum_{j=1}^{r} \{P_j(\hat{\boldsymbol{\Psi}}) / P(\hat{\boldsymbol{\Psi}})\} s_j(\hat{\boldsymbol{\Psi}}) \\ &= \sum_{j=1}^{r} (n_j / n) s_j(\hat{\boldsymbol{\Psi}}) \\ &= \bar{\boldsymbol{s}}(\hat{\boldsymbol{\Psi}}). \end{aligned} \tag{4.53}$$

On using (4.52) and (4.53) in (4.51), we obtain the result (4.50).

4.4.2 Example 4.3: Grouped Data from an Exponential Distribution

We suppose that the random variable W has an exponential distribution with mean μ, as specified by (1.45). The sample space $\mathcal{W}$ of W is partitioned into v mutually exclusive intervals $\mathcal{W}_j$ of equal length d, where

$$\mathcal{W}_j = [w_{j-1}, w_j) \quad (j = 1, \ldots, v)$$

and $w_0 = 0$ and $w_v = \infty$.

The problem is to estimate μ on the basis of the observed frequencies

$$\boldsymbol{y} = (n_1, \ldots, n_v)^T,$$

where n_j denotes the number of observations on W falling in the jth interval $\mathcal{W}_j (j = 1, \ldots, v)$, and where

$$n = \sum_{j=1}^{v} n_j.$$

In Section 2.8, we considered ML estimation from grouped and truncated data in the general case. The present example is simpler in that there is no truncation. Proceeding as in Section 2.8, we declare the complete-data vector $\boldsymbol{x}$ to be

$$\boldsymbol{x}_j = (\boldsymbol{y}^T, \boldsymbol{w}_1^T, \ldots, \boldsymbol{w}_v^T)^T,$$

where

$$\boldsymbol{w}_j = (w_{j1}, \ldots, w_{j,n_j})^T$$

contains the n_j unobservable individual observations on W that fell in the jth interval $\mathcal{W}_j (j = 1, \ldots, v)$.

The complete-data log likelihood is given by

$$\begin{aligned} \log L_c(\mu) &= \sum_{j=1}^{v} \sum_{l=1}^{n_j} \log f(w_{jl}; \mu), \\ &= -n \log \mu - \sum_{j=1}^{v} \sum_{l=1}^{n_j} w_{jl}/\mu \end{aligned} \tag{4.54}$$

on substituting for $f(w_{jk}; \mu)$ from (1.45). Then

$$Q(\mu; \mu^{(k)}) = \sum_{j=1}^{v} n_j^{(k)} Q_j(\mu; \mu^{(k)}),$$

where

$$Q_j(\mu; \mu^{(k)}) = E_{\mu^{(k)}}\{\log f(W; \mu) \mid W \in \mathcal{W}_j\}.$$

In the present case of an exponential distribution,

$$\begin{aligned} Q_j(\mu;\, \mu^{(k)}) &= E_{\mu^{(k)}}\{(-\log\mu - \mu^{-1}W)|W \in \mathcal{W}_j\} \\ &= -\log\mu - \mu^{-1}w_j^{(k)}, \end{aligned} \tag{4.55}$$

where

$$w_j^{(k)} = E_{\mu^{(k)}}(W \mid W \in \mathcal{W}_j) \tag{4.56}$$

for $j = 1, \ldots, v$.

The conditional expectation in (4.56) can be calculated to give

$$w_j^{(k)} = (c_j^{(k)}\mu^{(k)})^{-1} \int_{w_{j-1}}^{w_j} w \exp(-w/\mu^{(k)})\, dw, \tag{4.57}$$

where

$$\begin{aligned} c_j^{(k)} &= \int_{w_{j-1}}^{w_j} \mu^{(k)^{-1}} \exp(-w/\mu^{(k)})\, dw \\ &= \exp(-w_{j-1}/\mu^{(k)})\{1 - \exp(-d/\mu^{(k)})\}. \end{aligned} \tag{4.58}$$

On performing the integration in (4.57), it is easily seen that

$$w_j^{(k)} = \mu^{(k)} + a_j^{(k)}, \tag{4.59}$$

where

$$a_j^{(k)} = w_{j-1} - d/\{\exp(d/\mu^{(k)}) - 1\}. \tag{4.60}$$

On maximizing $Q(\mu;\, \mu^{(k)})$ with respect to μ, it follows that $\mu^{(k+1)}$ is given by

$$\begin{aligned} \mu^{(k+1)} &= \sum_{j=1}^{v}(n_j/n)w_j^{(k)} \\ &= \mu^{(k)} + \overline{a}^{(k)}, \end{aligned} \tag{4.61}$$

where

$$\overline{a}^{(k)} = \sum_{j=1}^{v}(n_j/n)a_j^{(k)}.$$

This completes the implementation of the $(k+1)$th iteration of the EM algorithm. We now consider the application of a modified Newton–Raphson method to this problem. From (2.65), the (incomplete-data) score statistic at $\mu = \mu^{(k)}$ can be computed as

$$\begin{aligned} S(\boldsymbol{y};\, \mu^{(k)}) &= [\partial Q(\mu;\, \mu^{(k)})/\partial\mu]_{\mu=\mu^{(k)}} \\ &= \sum_{j=1}^{v} n_j[\partial Q_j(\mu;\, \mu^{(k)})/\partial\mu]_{\mu=\mu^{(k)}} \\ &= \sum_{j=1}^{v} n_j s_j(\mu^{(k)}), \end{aligned} \tag{4.62}$$

where

$$s_j(\mu^{(k)}) = [\partial Q_j(\mu;\, \mu^{(k)})/\partial\mu]_{\mu=\mu^{(k)}}.$$

On differentiating (4.55) and noting (4.59), we have that

$$\begin{aligned} s_j(\mu^{(k)}) &= [-\mu^{-1} + \mu^{-2} w_j^{(k)}]_{\mu=\mu^{(k)}} \\ &= a_j^{(k)}/\mu^{(k)^2} \qquad (j = 1, \ldots, v). \end{aligned} \tag{4.63}$$

On using (4.63) in (4.62), we have

$$\begin{aligned} S(y;\, \mu^{(k)}) &= \sum_{j=1}^{v} n_j a_j^{(k)}/\mu^{(k)^2} \\ &= n\overline{a}^{(k)}/\mu^{(k)^2}. \end{aligned} \tag{4.64}$$

Corresponding to the use of the so-called empirical covariance matrix to approximate the observed matrix in the ungrouped case of i.i.d. data, Meilijson (1989) considers the Newton–Raphson method with the information $I(\mu^{(k)};\, y)$ after the kth iteration approximated by

$$I_{e,g}(\mu^{(k)};\, y) = \sum_{j=1}^{v} n_j \{s_j(\mu^{(k)})\}^2 - n^{-1}\{S(y;\, \mu^{(k)})\}^2. \tag{4.65}$$

The Newton–Raphson method with the modification (4.65) leads to $\mu^{(k+1)}$ being given by

$$\begin{aligned} \mu^{(k+1)} &= \mu^{(k)} + S(y;\, \mu^{(k)})/I_{e,g}(\mu^{(k)};\, y) \\ &= \mu^{(k)} + h^{(k)}, \end{aligned} \tag{4.66}$$

where

$$h^{(k)} = \frac{\overline{a}^{(k)} \mu^{(k)^2}}{\sum_{j=1}^{v} (n_j/n) a_j^{(k)^2} - \overline{a}^{(k)^2}}. \tag{4.67}$$

Meilijson (1989) contrasts the convergence of $\mu^{(k)}$ as given by (4.61) under the EM algorithm with $\mu^{(k)}$ as given by (4.66) with the modified Newton–Raphson method on various data sets. He demonstrates the superior global convergence when started from any positive value of μ, while the modified Newton–Raphson method diverges when started too close to zero and oscillates between two values on significantly bimodal data. The fit of the model is found to influence the convergence of the modified Newton–Raphson method, and the intervals that of the EM algorithm. When the model fit is good, the former method should be fast converging. When the intervals are sparse, the observed frequencies barely predict the complete-data statistic, and so the convergence of the EM algorithm is slow.

Since

$$S(\boldsymbol{y};\, \hat{\mu}) = 0,$$

we have from (4.63) and (4.65) that

$$I_{e,g}(\hat{\mu};\, \boldsymbol{y}) = n\hat{\mu}^{-4} \sum_{j=1}^{v} (n_j/n)\{a_j(\hat{\mu})\}^2, \tag{4.68}$$

where

$$a_j(\hat{\mu}) = w_{j-1} - d/\{\exp(d/\hat{\mu}) - 1\}.$$

The variance of the MLE $\hat{\mu}$ can be approximated by

$$\text{var}_{\mu}(\hat{\mu}) \approx 1/I_{e,g}(\hat{\mu};\, \boldsymbol{y}).$$

4.5 SUPPLEMENTED EM ALGORITHM

4.5.1 Definition

The methods presented in the previous section are of course applicable only in the specialized case of data arising from i.i.d. observations. For the general case, Meng and Rubin (1989, 1991) define a procedure that obtains a numerically stable estimate of the asymptotic covariance matrix of the EM-computed estimate, using only the code for computing the complete-data covariance matrix, the code for the EM algorithm itself, and the code for standard matrix operations. In particular, neither likelihoods, nor partial derivatives of log likelihoods need to be evaluated.

The basic idea is to use the fact that the rate of convergence is governed by the fraction of the missing information to find the increased variability due to missing information to add to the assessed complete-data covariance matrix. Meng and Rubin (1991) refer to the EM algorithm with their modification for the provision of the asymptotic covariance matrix as the Supplemented EM algorithm.

In his discussion of the DLR paper, Smith (1977) notes the possibility of obtaining the asymptotic variance in single-parameter cases by using the rate of convergence of the EM algorithm. He gives the expression

$$v = v_c/(1 - r), \tag{4.69}$$

where v and v_c denote the asymptotic variance of the maximum likelihood estimator based on the observed (incomplete) and complete data, respectively.

The expression (4.69) can be written in the form

$$v = v_c + \Delta v, \tag{4.70}$$

where

$$\Delta v = \{r/(1-r)\}v_c$$

is the increase in the variance due to not observing the data in $\boldsymbol{z}$. Meng and Rubin (1991) extend this result to the multivariate case of $d > 1$ parameters.

Let $\boldsymbol{V}$ denote the asymptotic covariance matrix of the MLE $\hat{\boldsymbol{\Psi}}$. Then analogous to (4.70), Meng and Rubin (1991) show that

$$\boldsymbol{I}^{-1}(\hat{\boldsymbol{\Psi}}; \boldsymbol{y}) = \mathcal{I}_c^{-1}(\hat{\boldsymbol{\Psi}}; \boldsymbol{y}) + \Delta \boldsymbol{V}, \tag{4.71}$$

where

$$\Delta \boldsymbol{V} = \{\boldsymbol{I}_d - \boldsymbol{J}(\hat{\boldsymbol{\Psi}})\}^{-1}\boldsymbol{J}(\hat{\boldsymbol{\Psi}})\mathcal{I}_c^{-1}(\hat{\boldsymbol{\Psi}}; \boldsymbol{y}).$$

Hence the diagonal elements of $\Delta \boldsymbol{V}$ give the increases in the asymptotic variances of the components of $\hat{\boldsymbol{\Psi}}$ due to missing data.

To derive the result (4.71), note that from (3.48),

$$\begin{aligned}\boldsymbol{I}(\boldsymbol{\Psi}; \boldsymbol{y}) &= \mathcal{I}_c(\boldsymbol{\Psi}; \boldsymbol{y}) - \mathcal{I}_m(\boldsymbol{\Psi}; \boldsymbol{y}) \\ &= \mathcal{I}_c(\boldsymbol{\Psi}; \boldsymbol{y})\{\boldsymbol{I}_d - \mathcal{I}_c^{-1}(\boldsymbol{\Psi}; \boldsymbol{y})\mathcal{I}_m(\boldsymbol{\Psi}; \boldsymbol{y})\}.\end{aligned}$$

From (3.69),

$$\boldsymbol{J}(\hat{\boldsymbol{\Psi}}) = \mathcal{I}_c^{-1}(\hat{\boldsymbol{\Psi}}; \boldsymbol{y})\mathcal{I}_m(\hat{\boldsymbol{\Psi}}; \boldsymbol{y})$$

for an EM sequence satisfying (3.68). Hence the observed information matrix $\boldsymbol{I}(\hat{\boldsymbol{\Psi}}; \boldsymbol{y})$ can be expressed as

$$\boldsymbol{I}(\hat{\boldsymbol{\Psi}}; \boldsymbol{y}) = \mathcal{I}_c(\hat{\boldsymbol{\Psi}}; \boldsymbol{y})\{\boldsymbol{I}_d - \boldsymbol{J}(\hat{\boldsymbol{\Psi}})\},$$

which on inversion, yields

$$\begin{aligned}\boldsymbol{I}^{-1}(\hat{\boldsymbol{\Psi}}; \boldsymbol{y}) &= \{\boldsymbol{I}_d - \boldsymbol{J}(\hat{\boldsymbol{\Psi}})\}^{-1}\mathcal{I}_c^{-1}(\hat{\boldsymbol{\Psi}}; \boldsymbol{y}) \\ &= [\boldsymbol{I}_d + \{\boldsymbol{I}_d - \boldsymbol{J}(\hat{\boldsymbol{\Psi}})\}^{-1}\boldsymbol{J}(\hat{\boldsymbol{\Psi}})]\mathcal{I}_c^{-1}(\hat{\boldsymbol{\Psi}}; \boldsymbol{y}) \\ &= \mathcal{I}_c^{-1}(\hat{\boldsymbol{\Psi}}; \boldsymbol{y}) + \{\boldsymbol{I}_d - \boldsymbol{J}(\hat{\boldsymbol{\Psi}})\}^{-1}\boldsymbol{J}(\hat{\boldsymbol{\Psi}})\mathcal{I}_c^{-1}(\hat{\boldsymbol{\Psi}}; \boldsymbol{y}),\end{aligned} \tag{4.72}$$

thus establishing (4.71).

It can be seen that in order to use (4.71) to compute the observed information matrix, the conditional expected complete-data information matrix $\mathcal{I}_c(\hat{\boldsymbol{\Psi}}; \boldsymbol{y})$ and the Jacobian matrix $\boldsymbol{J}(\hat{\boldsymbol{\Psi}})$ need to be calculated. For a wide class of problems where the complete-data density is from the regular exponential family, the evaluation of $\mathcal{I}_c(\hat{\boldsymbol{\Psi}}; \boldsymbol{y})$ is readily facilitated by standard complete-data computations. Concerning the calculation of $\boldsymbol{J}(\hat{\boldsymbol{\Psi}})$, Meng and Rubin (1991) demonstrate how $\boldsymbol{J}(\hat{\boldsymbol{\Psi}})$ can be readily obtained by using only EM code. The procedure amounts to numerically differentiating the EM map $\boldsymbol{M}(\boldsymbol{\Psi})$, as we now describe.

4.5.2 Calculation of $\mathbf{J}(\hat{\boldsymbol{\Psi}})$ via Numerical Differentiation

We consider here the calculation of $\boldsymbol{J}(\hat{\boldsymbol{\Psi}})$. The component-wise convergence rates of the EM sequence $\{\boldsymbol{\Psi}^{(k)}\}$,

$$\lim_{k\to\infty} (\Psi_i^{(k+1)} - \Psi_i^{(k)})/(\Psi_i^{(k)} - \Psi_i^{(k-1)}) \quad (i = 1, \ldots, d),$$

provide only a few eigenvalues (in most cases, simply the largest eigenvalue) of $\boldsymbol{J}(\hat{\boldsymbol{\Psi}})$, and not the matrix itself. However, as explained by Meng and Rubin (1991), each element of $\boldsymbol{J}(\hat{\boldsymbol{\Psi}})$ is the componentwise rate of convergence of a "forced EM". Let r_{ij} be the (i, j)th element of $\boldsymbol{J}(\hat{\boldsymbol{\Psi}})$ and define $\boldsymbol{\Psi}_{(j)}^{(k)}$ to be

$$\boldsymbol{\Psi}_{(j)}^{(k)} = (\hat{\Psi}_1, \ldots, \hat{\Psi}_{j-1}, \Psi_j^{(k)}, \hat{\Psi}_{j+1}, \ldots, \hat{\Psi}_d)^T, \tag{4.73}$$

where $\Psi_j^{(k)}$ is the value of Ψ_j on the kth iteration in a subsequent application of the EM algorithm, as to be explained below.

By the definition of r_{ij},

$$\begin{aligned} r_{ij} &= \partial M_i(\hat{\boldsymbol{\Psi}})/\partial \Psi_j \\ &= \lim_{\Psi_j \to \hat{\Psi}_j} \frac{M_i(\hat{\Psi}_1, \ldots, \hat{\Psi}_{j-1}, \Psi_j, \hat{\Psi}_{j+1}, \ldots, \hat{\Psi}_d) - M_i(\hat{\boldsymbol{\Psi}})}{\Psi_j - \hat{\Psi}_j} \\ &= \lim_{k\to\infty} \frac{M_i(\boldsymbol{\Psi}_{(j)}^{(k)}) - \hat{\Psi}_i}{\Psi_j^{(k)} - \hat{\Psi}_j} \\ &= \lim_{k\to\infty} r_{ij}^{(k)}, \end{aligned} \tag{4.74}$$

where

$$r_{ij}^{(k)} = \frac{M_i(\boldsymbol{\Psi}_{(j)}^{(k)}) - \hat{\Psi}_i}{\Psi_j^{(k)} - \hat{\Psi}_j}. \tag{4.75}$$

Because $\boldsymbol{M}(\boldsymbol{\Psi})$ is implicitly defined by the output of the E- and M-steps, all quantities in (4.75) can be obtained using only the code for the EM algorithm. Meng and Rubin (1991) suggest the following algorithm for computing the ${r_{ij}}^{(k)}$ after $\hat{\boldsymbol{\Psi}}$ has been found. Firstly, run a sequence of iterations of the EM algorithm, starting from a point that is not equal to $\hat{\boldsymbol{\Psi}}$ in any component. After the kth iteration, compute $\boldsymbol{\Psi}_{(j)}^{(k)}$ from (4.73) and, treating it as the current estimate of $\boldsymbol{\Psi}$, run one iteration of the EM algorithm to obtain $M_i(\boldsymbol{\Psi}_{(j)}^{(k)})$ and, subsequently from (4.75), the ratio $r_{ij}^{(k)}$ for each i $(i = 1, \ldots, d)$. This is done for each j $(j = 1, \ldots, d)$.

The r_{ij} are obtained when the sequence $r_{ij}^{(k^*)}, r_{ij}^{(k^*+1)}, \ldots$ is stable for some k^*. This process may result in using different values of k^* for different r_{ij} elements.

As for numerical accuracy, it is almost always safe to use the original EM starting values as the initial values for the Supplemented EM algorithm in computing $\boldsymbol{J}(\hat{\boldsymbol{\Psi}})$. This choice does not require any additional work, but may result in some unnecessary iterations because the original starting values may be far from the MLE. Meng and Rubin (1991) suggest using a suitable iterate of the original EM sequence (for example, the second) or two complete-data standard deviations from the MLE. The stopping criterion for the Supplemented EM sequence should be less stringent than that for the original EM sequence, because the method for computing $\boldsymbol{J}(\hat{\boldsymbol{\Psi}})$ is essentially numerical differentiation of a function.

4.5.3 Stability

The observed information matrix $\boldsymbol{I}(\hat{\boldsymbol{\Psi}}; \boldsymbol{y})$ can be approximated by using methods such as those described by Carlin (1987) and Meilijson (1989), where the second-order partial derivatives of the log likelihood function or at least its gradient are calculated by numerical differentiation. However, as pointed out by Meng and Rubin (1991), these methods besides requiring the evaluation of this likelihood, are subject to the inaccuracies and difficulties of any numerical differentiation procedure with large matrices. On the other hand, the approach via the Supplemented EM algorithm, is typically more stable than pure numerical differentiation procedures, because the matrix differentiation is being added to an analytically obtained matrix, which is usually the dominant term. Thus the Supplemented EM algorithm is typically more stable than pure numerical differentiation procedures that are used to compute the whole covariance matrix. More specifically, with the Supplemented EM algorithm, there is an automatic self-adjustment that Meng and Rubin (1991) explain as follows. When the fraction of missing data is high, the convergence of the EM algorithm is slow. As a consequence, it provides an excellent sequence of iterates from which the linear rate of convergence of the EM algorithm can be assessed, and thus the results produced by the Supplemented EM algorithm are typically quite accurate. On the other hand, when the fraction of missing data is low and consequently the increases in the variances of the fitted parameters are relatively small, the term $\mathcal{I}_c^{-1}(\hat{\boldsymbol{\Psi}}; \boldsymbol{y})$ in (4.71), which is usually very accurately calculated, dominates the increases in the variances due to the missing data. Thus the Supplemented EM algorithm still provides a satisfactory assessment of the observed information matrix $\boldsymbol{I}(\hat{\boldsymbol{\Psi}}; \boldsymbol{y})$, even though it is not as accurate in assessing the increases in the variances as when convergence of the EM algorithm is slow.

4.5.4 Monitoring Convergence

The matrix

$$\mathcal{I}_c^{-1}(\hat{\boldsymbol{\Psi}}; \boldsymbol{y}) + \Delta \boldsymbol{V}$$

is not numerically constrained to be symmetric, even though it is mathematically symmetric. The asymmetry can arise because of inaccuracies in computing either $\boldsymbol{J}(\hat{\boldsymbol{\Psi}})$ or $\mathcal{I}_c(\hat{\boldsymbol{\Psi}}; \boldsymbol{y})$. For exponential families, $\mathcal{I}_c(\hat{\boldsymbol{\Psi}}; \boldsymbol{y})$ is typically very accurately computed, whereas for nonexponential families, its accuracy depends on large-sample approximations based on linearization methods. In contrast, the accuracy of $\boldsymbol{J}(\hat{\boldsymbol{\Psi}})$ is determined by the accuracy of the EM algorithm itself, which typically is excellent when both the E- and M-steps are simple calculations and is adequate in most cases.

If the matrix $\boldsymbol{I}_d - \boldsymbol{J}(\hat{\boldsymbol{\Psi}})$ is (numerically) symmetric, but not positive semi-definite, then it indicates that the EM algorithm has not converged to a (local) maximum, but rather to a saddle point. In the latter case, the EM algorithm should be rerun, starting near the last iterate but perturbed in the direction of the eigenvector corresponding to the most negative eigenvalue of $\boldsymbol{I}(\hat{\boldsymbol{\Psi}}; \boldsymbol{y})$. In this sense, the Supplemented EM algorithm can also be used to monitor the convergence of the EM algorithm to a local maximum, which cannot be detected by monitoring the increase in the likelihood.

The matrix $\boldsymbol{I}_d - \boldsymbol{J}(\hat{\boldsymbol{\Psi}})$ can be nearly singular when the convergence of the EM algorithm is extremely slow; that is, when the largest eigenvalue of $\boldsymbol{J}(\hat{\boldsymbol{\Psi}})$ is very close to one. Statistically this implies that $L(\boldsymbol{\Psi})$ is very flat along some directions and thus that $\boldsymbol{I}(\hat{\boldsymbol{\Psi}}; \boldsymbol{y})$ is nearly singular. This is a feature of the likelihood function and the data and not a problem created by the Supplemented EM algorithm. Even in these situations, the Supplemented EM algorithm can be very helpful in identifying directions with little information by proceeding as follows.

First obtain the matrix

$$\boldsymbol{P} = \mathcal{I}_c(\hat{\boldsymbol{\Psi}}; \boldsymbol{y})\{\boldsymbol{I}_d - \boldsymbol{J}(\hat{\boldsymbol{\Psi}})\}. \tag{4.76}$$

As discussed above, any lack of (numerical) symmetry in $\boldsymbol{P}$ indicates the existence of programming errors or lack of convergence. Assuming (numerical) symmetry of $\boldsymbol{P}$, standard matrix operations can be used to find the spectral decomposition of $\boldsymbol{P}$,

$$\boldsymbol{P} = \boldsymbol{\Gamma}^T \operatorname{diag}(\lambda_1, \ldots, \lambda_d)\boldsymbol{\Gamma}, \tag{4.77}$$

where $\lambda_1 \geq \lambda_2 \ldots \geq \lambda_d$ are the eigenvalues of $\boldsymbol{P}$ and where the rows of $\boldsymbol{\Gamma}$ are the orthonormalized eigenvectors corresponding to $\lambda_1, \ldots, \lambda_d$. If λ_d is identified as being (numerically) negative, it indicates that the EM algorithm has not converged to a local maximum, but to a saddle point, and that it should be continued in the direction corresponding to λ_d. Otherwise, the eigenvalues can be used to indicate those components of $\boldsymbol{\Gamma\Psi}$ about which the observed data contains little or no information.

4.5.5 Example 4.4: Univariate Contaminated Normal Data

This example is considered by Meng and Rubin (1991) and prior to that by Little and Rubin (1987, Section 10.5). They assumed that the observed data are given by

$$y = (w_1, \ldots, w_n)^T,$$

where $w_1, \ldots, w_n$ denote the realizations of a random sample of size n from the univariate contaminated normal model,

$$f(w; \boldsymbol{\Psi}) = \pi_1 \phi(w; \mu, \sigma^2) + \pi_2 \phi(w; \mu, c\sigma^2), \tag{4.78}$$

and $\pi_1 (0 < \pi_1 < 1)$ and $c (c > 0)$ are both known; $\pi_2 = 1 - \pi_1$. The model (4.78) represents a mixture in known proportions of two univariate normal densities with a common mean and with variances in a known ratio. It is therefore a special case of the general univariate normal mixture model considered in Section 2.7.1. On specializing the equations there for the model (4.78), we have that on the $(k+1)$th iteration of the EM algorithm, $\boldsymbol{\Psi}^{(k+1)} = (\mu^{(k+1)}, \sigma^{(k+1)^2})^T$ satisfies

$$\mu^{(k+1)} = \frac{\sum_{j=1}^{n} \{z_{1j}^{(k)} + c^{-1} z_{2j}^{(k)}\} w_j}{\sum_{j=1}^{n} \{z_{1j}^{(k)} + c^{-1} z_{2j}^{(k)}\}}$$

and

$$\sigma^{(k+1)^2} = \sum_{j=1}^{n} \{z_{1j}^{(k)} (w_j - \mu^{(k+1)})^2 + z_{2j}^{(k)} c^{-1} (w_j - \mu^{(k+1)})^2\} / n,$$

where

$$z_{ij}^{(k)} = \tau_i(w_j; \boldsymbol{\Psi}^{(k)})$$

and

$$\tau_1(w_j; \boldsymbol{\Psi}) = \frac{\pi_1 \phi(w_j; \mu, \sigma^2)}{\pi_1 \phi(w_j; \mu, \sigma^2) + \pi_2 \phi(w_j; \mu, c\sigma^2)}$$

and

$$\tau_2(w_j; \boldsymbol{\Psi}) = 1 - \tau_1(w_j; \boldsymbol{\Psi}) \quad (j = 1, \ldots, n).$$

To illustrate the Supplemented EM algorithm, Meng and Rubin (1991) perform a simulation with

$$\mu = 0, \quad \sigma^2 = 1, \quad c = 0.5, \quad \pi_1 = 0.9, \quad \text{and} \quad n = 100.$$

Meng and Rubin (1991) actually work with the parameter vector

$$(\mu, \log \sigma^2)^T \tag{4.79}$$

in order to improve the normal approximation to the log likelihood function. In the remainder of this section, we therefore let $\boldsymbol{\Psi}$ be the vector (4.79).

Table 4.1. Results of EM Algorithm for Contaminated Normal Example.

k	$\mu^{(k)}$	$d_1^{(k)}$	$d_1^{(k+1)}/d_1^{(k)}$	$\log \sigma^{(k)^2}$	$d_2^{(k)}$	$d_2^{(k+1)}/d_2^{(k)}$
0	0.19866916	−0.00061984	0.01587	0.22943020	0.00923662	0.03502
1	0.19925685	−0.00003215	0.04890	0.22051706	0.00032348	0.03496
2	0.19928743	−0.00000157	0.04708	0.22020489	0.00001131	0.03497
3	0.19928893	−0.00000007	0.04590	0.22019397	0.00000040	0.03498
4	0.19928900	−0.00000000	0.04509	0.22019359	0.00000001	0.03499
5	0.19928900	−0.00000000	0.04443	0.22019358	0.00000000	0.03497
6	0.19928900	−0.00000000	0.04223	0.22019358	0.00000000	0.03386

Source: Adapted from Meng and Rubin (1991), with permission from the Journal of the American Statistical Association.

Table 4.1 gives the EM output with initial values

$$\mu^{(0)} = \overline{w}$$

and

$$\sigma^{(0)^2} = s^2/(\pi_1 + c\pi_2),$$

where $\overline{w}$ and s^2 denote the sample mean and (bias-corrected) sample variance of the observed data $w_1, \ldots, w_n$.

In this table,

$$d_1^{(k+1)} = \mu^{(k+1)} - \hat{\mu}$$

and

$$d_2^{(k+1)} = \log \sigma^{(k+1)^2} - \log \hat{\sigma}^2.$$

The first four rows for Table 4.2 give the corresponding output for $r_{ij}^{(k)}$ $(i, j = 1, 2)$ for $k = 0, 1, 2,$ and 3, obtained by the Supplemented EM algorithm starting from $\boldsymbol{\Psi}^{(0)}$ and taking $\hat{\boldsymbol{\Psi}} = \boldsymbol{\Psi}^{(6)}$. The last row gives the

Table 4.2. Results of SEM Iterations for *J* Matrix in Contaminated Normal Example.

Iteration (k)	$r_{11}^{(k)}$	$r_{12}^{(k)}$	$r_{21}^{(k)}$	$r_{22}^{(k)}$
0	0.04251717	−0.00063697	−0.0012649	0.03494620
1	0.04251677	−0.00063697	−0.0012375	0.03492154
2	0.04251674	−0.00063666	−0.0012359	0.03492066
3	0.04251658	−0.00063663	−0.0012333	0.03492059
True	0.04251675	−0.00063666	−0.00112360	0.03492063

Source: Adapted from Meng and Rubin (1991), with permission from the Journal of the American Statistical Association.

true values of r_{ij} $(i, j = 1, 2)$ obtained by direct computation using analytical expressions.

As noted by Meng and Rubin (1991), it can be seen from Table 4.2 that using only six iterations initially to form $\hat{\boldsymbol{\Psi}}$, we can approximate the r_{ij} very well by $r_{ij}^{(k)}$ for "moderate" k.

From Table 4.2, we can take

$$\boldsymbol{J}(\hat{\boldsymbol{\Psi}}) = \begin{pmatrix} 0.04252 & -0.00064 \\ -0.00112 & 0.03492 \end{pmatrix}. \tag{4.80}$$

Concerning the computation of $\mathcal{I}_c(\hat{\boldsymbol{\Psi}}; \boldsymbol{y})$, we have that the first and second diagonal elements of $\boldsymbol{I}_c(\boldsymbol{\Psi}; \boldsymbol{x})$ are given by

$$-\,\partial^2 \log L_c(\boldsymbol{\Psi})/\partial\mu^2 = \sigma^{-2}\sum_{j=1}^{n}(z_{1j} + c^{-1}z_{2j}) \tag{4.81}$$

and

$$-\,\partial^2 \log L_c(\boldsymbol{\Psi})/\partial\Psi_2^2 = \tfrac{1}{2}n, \tag{4.82}$$

respectively, where $\Psi_2 = \log\sigma^2$. The z_{ij} are the zero-one component indicator variables defined to be one or zero according as the jth observation arises or does not arise from the ith component of the mixture model (4.78). The common value of the two off-diagonal elements of $\boldsymbol{I}_c(\boldsymbol{\Psi}; \boldsymbol{x})$ is not given here as it obviously will have zero conditional expectation.

On taking the conditional expectation of (4.81) given the observed data $\boldsymbol{y}$, we have that

$$\begin{aligned} E_{\boldsymbol{\Psi}}\{-\partial^2 \log L/\partial\mu^2 \mid \boldsymbol{y}\} &= \sigma^{-2}\sum_{j=1}^{n}\{\tau_1(w_j;\, \boldsymbol{\Psi}) + c^{-1}\tau_2(w_j;\, \boldsymbol{\Psi})\} \\ &= \sigma^{-2}\xi(\boldsymbol{\Psi}), \end{aligned}$$

say. It follows that

$$\mathcal{I}_c(\hat{\boldsymbol{\Psi}}; \boldsymbol{y}) = \operatorname{diag}(\hat{\sigma}^{-2}\xi(\hat{\boldsymbol{\Psi}}),\ \tfrac{1}{2}n). \tag{4.83}$$

We have seen in Section 4.2 that if the complete-data density belongs to the regular exponential family, then

$$\mathcal{I}_c(\hat{\boldsymbol{\Psi}}; \boldsymbol{y}) = \mathcal{I}_c(\hat{\boldsymbol{\Psi}}). \tag{4.84}$$

However, for this example, the complete-data density is from an irregular exponential family, and so (4.84) does not necessarily hold. Indeed, it can be confirmed that (4.84) does not hold, since

$$\mathcal{I}_c(\boldsymbol{\Psi}) = \operatorname{diag}\{n(\pi_1 + c^{-1}\pi_2),\ \tfrac{1}{2}n\}.$$

Of course, if π_1 were unknown, the complete-data density belongs to the regular exponential family, and it can be easily confirmed then that (4.84) holds.

From (4.83), we have that

$$\begin{aligned} \mathcal{I}_c^{-1}(\hat{\boldsymbol{\Psi}}; \boldsymbol{y}) &= n^{-1} \begin{pmatrix} \hat{\sigma}^2/\xi(\hat{\boldsymbol{\Psi}}) & 0 \\ 0 & 2 \end{pmatrix} \\ &= \begin{pmatrix} 0.01133 & 0 \\ 0 & 2 \end{pmatrix}, \end{aligned} \tag{4.85}$$

since $n\xi(\hat{\boldsymbol{\Psi}}) = 109.97696$. Thus using (4.80) and (4.85) in (4.72), we have

$$\boldsymbol{I}^{-1}(\hat{\boldsymbol{\Psi}}; \boldsymbol{y}) = \begin{pmatrix} 0.01184 & -0.00001 \\ -0.00001 & 0.02072 \end{pmatrix}.$$

The symmetry of the resulting matrix indicates numerical accuracy, as discussed earlier in this section.

It can be seen from the fourth and seventh columns of Table 4.1 that the iterates $\Psi_1 = \mu$ and $\Psi_2 = \log \sigma^2$ converge at different rates, corresponding to two different eigenvalues of $\boldsymbol{J}(\hat{\boldsymbol{\Psi}})$. This special feature occurs because the MLEs of Ψ_1 and Ψ_2 are asymptotically independent for both complete- and incomplete-data problems.

4.5.6 Example 4.5: Bivariate Normal Data with Missing Values

We now consider the situation where there is no missing information on some components of $\boldsymbol{\Psi}$. Partition $\boldsymbol{\Psi}$ as

$$\boldsymbol{\Psi} = (\boldsymbol{\Psi}_1^T, \boldsymbol{\Psi}_2^T)^T, \tag{4.86}$$

where there is no missing information on the d_1-dimensional subvector $\boldsymbol{\Psi}_1$ of $\boldsymbol{\Psi}$.

Corresponding to the partition (4.86) of $\boldsymbol{\Psi}$, we now partition $\boldsymbol{J}(\boldsymbol{\Psi})$ and $\mathcal{I}_c^{-1}(\boldsymbol{\Psi}; \boldsymbol{y})$ as

$$\boldsymbol{J}(\boldsymbol{\Psi}) = \begin{pmatrix} \boldsymbol{J}_{11}(\boldsymbol{\Psi}) & \boldsymbol{J}_{12}(\boldsymbol{\Psi}) \\ \boldsymbol{J}_{21}(\boldsymbol{\Psi}) & \boldsymbol{J}_{22}(\boldsymbol{\Psi}) \end{pmatrix}$$

and

$$\mathcal{I}_c^{-1}(\boldsymbol{\Psi}; \boldsymbol{y}) = \begin{pmatrix} \boldsymbol{G}_{11}(\boldsymbol{\Psi}; \boldsymbol{y}) & \boldsymbol{G}_{12}(\boldsymbol{\Psi}; \boldsymbol{y}) \\ \boldsymbol{G}_{21}(\boldsymbol{\Psi}; \boldsymbol{y}) & \boldsymbol{G}_{22}(\boldsymbol{\Psi}; \boldsymbol{y}) \end{pmatrix}$$

The EM sequence $\{\boldsymbol{\Psi}_1^{(k)}\}$ for $\boldsymbol{\Psi}_1$ will converge in one iteration for $\boldsymbol{\Psi}_1$ regardless of the starting value, with the result that the corresponding components of $\boldsymbol{M}(\boldsymbol{\Psi})$ will be a constant with zero derivative. That is, $\boldsymbol{J}_{11}(\boldsymbol{\Psi})$ and $\boldsymbol{J}_{21}(\boldsymbol{\Psi})$ are null matrices.

Meng and Rubin (1991) show that the observed information matrix for $\hat{\boldsymbol{\Psi}} = (\hat{\boldsymbol{\Psi}}_1^T, \hat{\boldsymbol{\Psi}}_2^T)^T$ is given by

$$\boldsymbol{I}(\hat{\boldsymbol{\Psi}}; \boldsymbol{y}) = \mathcal{I}_c^{-1}(\hat{\boldsymbol{\Psi}}; \boldsymbol{y}) + \Delta \boldsymbol{V},$$

where

$$\Delta \boldsymbol{V} = \begin{pmatrix} \boldsymbol{0} & \boldsymbol{0} \\ \boldsymbol{0} & \Delta \boldsymbol{V}_{22} \end{pmatrix}$$

and where

$$\Delta \boldsymbol{V}_{22} = \{\boldsymbol{I}_d - \boldsymbol{J}_{22}(\hat{\boldsymbol{\Psi}})\}^{-1} \boldsymbol{J}_{22}(\hat{\boldsymbol{\Psi}}) \boldsymbol{A}(\hat{\boldsymbol{\Psi}}; \boldsymbol{y}) \tag{4.87}$$

and

$$\boldsymbol{A}(\hat{\boldsymbol{\Psi}}; \boldsymbol{y}) = \boldsymbol{G}_{22}(\hat{\boldsymbol{\Psi}}; \boldsymbol{y}) - \boldsymbol{G}_{21}(\hat{\boldsymbol{\Psi}}; \boldsymbol{y}) \boldsymbol{G}_{11}^{-1}(\hat{\boldsymbol{\Psi}}; \boldsymbol{y}) \boldsymbol{G}_{12}(\hat{\boldsymbol{\Psi}}; \boldsymbol{y}).$$

The submatrix $\boldsymbol{J}_{22}(\hat{\boldsymbol{\Psi}})$ of $\boldsymbol{J}(\hat{\boldsymbol{\Psi}})$ in (4.87) can be assessed by the same method used to assess $\boldsymbol{J}(\hat{\boldsymbol{\Psi}})$, as explained in the last section.

To illustrate the application of the Supplemented EM algorithm with no missing information on some of the components of $\boldsymbol{\Psi}$, Meng and Rubin (1991) consider ML estimation of

$$\boldsymbol{\Psi} = (\mu_1, \log \sigma_1^2, \mu_2, \log \sigma_2^2, \zeta)^T$$

in the bivariate normal distribution from $n = 18$ bivariate observations as listed below. The value of the second variable for the last six cases is missing (indicated by ?).

Variate 1:	8	6	11	22	14	17	18	24	19
Variate 2:	59	58	56	53	50	45	43	42	39

Variate 1:	23	26	40	4	4	5	6	8	10
Variate 2:	38	30	27	?	?	?	?	?	?

We have arranged the components in $\boldsymbol{\Psi}$ so that

$$\boldsymbol{\Psi} = (\boldsymbol{\Psi}_1^T, \boldsymbol{\Psi}_2^T)^T,$$

where there is no missing information on the subvector

$$\boldsymbol{\Psi}_1 = (\mu_1, \log \sigma_1^2)^T.$$

Here

$$\zeta = \tfrac{1}{2} \log\{(1 + \rho)/(1 - \rho)\}$$

is Fisher's Z transformation of the correlation ρ.

As the first variate is fully observed, the MLE of $\boldsymbol{\Psi}_1$ is simply the mean and log of the sample variance of the observations on it.

Using the Supplemented EM algorithm as described in Section 4.5.1, the submatrix $\boldsymbol{J}_{22}(\hat{\boldsymbol{\Psi}})$ of $\boldsymbol{J}(\hat{\boldsymbol{\Psi}})$ corresponding to the subvector

$$\boldsymbol{\Psi}_2 = (\mu_2, \log \sigma_2^2, \zeta)^T$$

is obtained by Meng and Rubin (1991) to be

$$\boldsymbol{J}_{22}(\hat{\boldsymbol{\Psi}}) = \begin{pmatrix} 0.33333 & 1.44444 & -0.64222 \\ 0.05037 & 0.29894 & 0.01529 \\ -0.02814 & 0.01921 & 0.32479 \end{pmatrix}.$$

Since the complete-data density is from a regular exponential family, $\mathcal{I}_c(\hat{\boldsymbol{\Psi}}; \boldsymbol{y})$ is equal to $\mathcal{I}_c(\hat{\boldsymbol{\Psi}})$. From standard asymptotic results for the MLE of the vector of parameters of the bivariate normal distribution, we have that

$$\mathcal{I}_c(\boldsymbol{\Psi}) = \begin{pmatrix} \boldsymbol{G}_{11}(\boldsymbol{\Psi}) & \boldsymbol{G}_{12}(\boldsymbol{\Psi}) \\ \boldsymbol{G}_{21}(\boldsymbol{\Psi}) & \boldsymbol{G}_{22}(\boldsymbol{\Psi}) \end{pmatrix}$$

where

$$\boldsymbol{G}_{11}(\boldsymbol{\Psi}) = \text{diag}\{n\sigma_1^{-2}(1-\rho^2)^{-1},\ (n/4)(2-\rho^2)(1-\rho^2)^{-1}\}^T,$$

$$\boldsymbol{G}_{21}(\boldsymbol{\Psi}) = \begin{pmatrix} -n\sigma_1^{-1}\sigma_2^{-1}\rho(1-\rho^2)^{-1} & 0 \\ 0 & -(n/4)\rho^2(1-\rho^2)^{-1} \\ 0 & -\frac{1}{2}n\rho \end{pmatrix},$$

and

$$G_{22}(\boldsymbol{\Psi}) = \begin{pmatrix} n\sigma_2^{-2}(1-\rho^2)^{-1} & 0 & 0 \\ 0 & (n/4)(2-\rho^2)(1-\rho^2)^{-1} & -\frac{1}{2}n\rho \\ 0 & -\frac{1}{2}n\rho & n(1+\rho^2) \end{pmatrix},$$

and $G_{12}(\boldsymbol{\Psi}) = G_{21}^T(\boldsymbol{\Psi})$.

On evaluation at $\boldsymbol{\Psi} = \hat{\boldsymbol{\Psi}}$, Meng and Rubin (1991) find that

$$\boldsymbol{G}_{11}(\hat{\boldsymbol{\Psi}}) = \text{diag}(4.9741, 0.1111),$$

$$\boldsymbol{G}_{12}(\hat{\boldsymbol{\Psi}}) = \begin{pmatrix} -5.0387 & 0 & 0 \\ 0 & 0.0890 & -0.0497 \end{pmatrix},$$

$$\boldsymbol{G}_{22}(\hat{\boldsymbol{\Psi}}) = \begin{pmatrix} 6.3719 & 0 & 0 \\ 0 & 0.1111 & -0.0497 \\ 0 & -0.0497 & 0.0556 \end{pmatrix}.$$

Using the formula (4.87), Meng and Rubin (1991) find that

$$\Delta V_{22} = \begin{pmatrix} 1.0858 & 0.1671 & -0.0933 \\ 0.1671 & 0.0286 & -0.0098 \\ -0.0933 & -0.0098 & 0.0194 \end{pmatrix}.$$

The observed information matrix $\boldsymbol{I}(\hat{\boldsymbol{\Psi}}; \boldsymbol{y})$ is obtained by adding $\Delta \boldsymbol{V}$ to $\mathcal{I}_c^{-1}(\hat{\boldsymbol{\Psi}}; \boldsymbol{y})$. For example, the standard error of $\hat{\mu}_2$ is given by

$$(6.3719 + 1.0858)^{1/2} \cong 2.73.$$

4.6 BOOTSTRAP APPROACH TO STANDARD ERROR APPROXIMATION

The bootstrap was introduced by Efron (1979), who has investigated it further in a series of articles; see Efron and Tibshirani (1993) and the references therein. More recently, Efron (1994) considers the application of the bootstrap to missing-data problems. Over the past fifteen years, the bootstrap has become one of the most popular recent developments in statistics. Hence there now exists an extensive literature on it.

The bootstrap is a powerful technique that permits the variability in a random quantity to be assessed using just the data at hand. An estimate $\hat{F}$ of the underlying distribution is formed from the observed sample. Conditional on the latter, the sampling distribution of the random quantity of interest with F replaced by $\hat{F}$, defines its so-called bootstrap distribution, which provides an approximation to its true distribution. It is assumed that $\hat{F}$ has been so formed that the stochastic structure of the model has been preserved. Usually, it is impossible to express the bootstrap distribution in simple form, and it must be approximated by Monte Carlo methods whereby pseudo-random samples (bootstrap samples) are drawn from $\hat{F}$. In recent times there has been a number of papers written on improving the efficiency of the bootstrap computations with the latter approach. The nonparametric bootstrap uses $\hat{F} = F_n$, the empirical distribution function of $\boldsymbol{Y}$ formed from $\boldsymbol{y}$. If a parametric form is adopted for the distribution function of $\boldsymbol{Y}$, where $\boldsymbol{\Psi}$ denotes the vector of unknown parameters, then the parametric bootstrap uses an estimate $\hat{\boldsymbol{\Psi}}$ formed from $\boldsymbol{y}$ in place of $\boldsymbol{\Psi}$. That is, if we write F as $F_{\boldsymbol{\Psi}}$ to signify its dependence on $\boldsymbol{\Psi}$, then the bootstrap data are generated from $\hat{F} = F_{\hat{\boldsymbol{\Psi}}}$.

Standard error estimation of $\hat{\boldsymbol{\Psi}}$ may be implemented according to the bootstrap as follows.

Step 1. A new set of data, $\boldsymbol{y}^*$, called the bootstrap sample, is generated according to $\hat{F}$, an estimate of the distribution function of $\boldsymbol{Y}$ formed from the original observed data $\boldsymbol{y}$. That is, in the case where $\boldsymbol{y}$ contains the observed values of a random sample of size n on a random vector $\boldsymbol{W}$, $\boldsymbol{y}^*$ consists of the observed values of the random sample

$$\boldsymbol{W}_1^*, \ldots, \boldsymbol{W}_n^* \overset{\text{i.i.d.}}{\sim} \hat{F}_W, \tag{4.88}$$

where $\hat{F}_W$ is held fixed at its observed value.

Step 2. The EM algorithm is applied to the bootstrap observed data $\boldsymbol{y}^*$ to compute the MLE for this data set $\hat{\boldsymbol{\Psi}}^*$.

Step 3. The bootstrap covariance matrix of $\hat{\boldsymbol{\Psi}}^*$ is given by

$$\text{cov}^*(\hat{\boldsymbol{\Psi}}^*) = E^*[\{(\hat{\boldsymbol{\Psi}} - E^*(\hat{\boldsymbol{\Psi}})\}\{(\hat{\boldsymbol{\Psi}}^* - E^*(\hat{\boldsymbol{\Psi}}^*)\}^T], \tag{4.89}$$

where E^* denotes expectation over the distribution of $\boldsymbol{Y}^*$ specified by $\hat{F}$.

The bootstrap covariance matrix can be approximated by Monte Carlo methods. Steps (1) and (2) are repeated independently a number of times (say, B) to give B independent realizations of $\hat{\boldsymbol{\Psi}}^*$, denoted by $\hat{\boldsymbol{\Psi}}_1^*, \ldots, \hat{\boldsymbol{\Psi}}_B^*$. Then (4.89) can be approximated by the sample covariance matrix of these B bootstrap replications to give

$$\text{cov}^*(\hat{\boldsymbol{\Psi}}^*) \approx \sum_{b=1}^{B} (\hat{\boldsymbol{\Psi}}_b^* - \overline{\hat{\boldsymbol{\Psi}}}^*)(\hat{\boldsymbol{\Psi}}_b^* - \overline{\hat{\boldsymbol{\Psi}}}^*)^T/(B-1), \tag{4.90}$$

where

$$\overline{\hat{\boldsymbol{\Psi}}}^* = \sum_{b=1}^{B} \hat{\boldsymbol{\Psi}}^*/B. \tag{4.91}$$

The standard error of the ith element of $\hat{\boldsymbol{\Psi}}$ can be estimated by the positive square root of the ith diagonal element of (4.90). It has be shown that 50 to 100 bootstrap replications are generally sufficient for standard error estimation (Efron and Tibshirani, 1993).

In Step 1 of the above algorithm, the nonparametric version of the bootstrap would take $\hat{F}$ to be the empirical distribution function. Given that we are concerned here with ML estimation in the context of a parametric model, we would tend to use the parametric version of the bootstrap instead of the nonparametric version. Situations where we may still wish to use the latter include problems where the observed data are censored or are missing in the conventional sense. In these cases the use of the nonparametric bootstrap avoids having to postulate a suitable model for the underlying mechanism that controls the censorship or the absence of the data.

4.7 ACCELERATION OF THE EM ALGORITHM VIA AITKEN'S METHOD

4.7.1 Aitken's Acceleration Method

The most commonly used method for EM acceleration is the multivariate version of Aitken's acceleration method. Suppose that $\boldsymbol{\Psi}^{(k)} \rightarrow \boldsymbol{\Psi}^*$, as $k \rightarrow \infty$. Then we can express $\boldsymbol{\Psi}^*$ as

$$\boldsymbol{\Psi}^* = \boldsymbol{\Psi}^{(k)} + \sum_{h=1}^{\infty}(\boldsymbol{\Psi}^{(h+k)} - \boldsymbol{\Psi}^{(h+k-1)}). \tag{4.92}$$

Now

$$\begin{aligned}\boldsymbol{\Psi}^{(h+k)} - \boldsymbol{\Psi}^{(h+k-1)} &= \boldsymbol{M}(\boldsymbol{\Psi}^{(h+k-1)}) - \boldsymbol{M}(\boldsymbol{\Psi}^{(h+k-2)}) &(4.93)\\ &\approx \boldsymbol{J}(\boldsymbol{\Psi}^{(h+k-2)})(\boldsymbol{\Psi}^{(h+k-1)} - \boldsymbol{\Psi}^{(h+k-2)}) &(4.94)\\ &\approx \boldsymbol{J}(\boldsymbol{\Psi}^*)(\boldsymbol{\Psi}^{(h+k-1)} - \boldsymbol{\Psi}^{(h+k-2)}), &(4.95)\end{aligned}$$

since

$$\boldsymbol{J}(\boldsymbol{\Psi}^{(h+k)}) = \boldsymbol{J}(\boldsymbol{\Psi}^*)$$

for k sufficiently large. The approximation (4.94) to (4.93) is obtained by a linear Taylor series expansion of $\boldsymbol{M}(\boldsymbol{\Psi}^{(h+k-1)})$ about the point $\boldsymbol{\Psi}^{(h+k-2)}$, as in (3.65). Repeated application of (4.95) in (4.92) gives

$$\begin{aligned}\boldsymbol{\Psi}^* &\approx \boldsymbol{\Psi}^{(k)} + \sum_{h=0}^{\infty}\{\boldsymbol{J}(\boldsymbol{\Psi}^*)\}^h(\boldsymbol{\Psi}^{(k+1)} - \boldsymbol{\Psi}^{(k)})\\ &= \boldsymbol{\Psi}^{(k)} + \{\boldsymbol{I}_d - \boldsymbol{J}(\boldsymbol{\Psi}^*)\}^{-1}(\boldsymbol{\Psi}^{(k+1)} - \boldsymbol{\Psi}^{(k)}), \qquad (4.96)\end{aligned}$$

as the power series

$$\sum_{h=0}^{\infty}\{\boldsymbol{J}(\boldsymbol{\Psi}^*)\}^h$$

converges to $\{\boldsymbol{I}_d - \boldsymbol{J}(\boldsymbol{\Psi}^*)\}^{-1}$ if $\boldsymbol{J}(\boldsymbol{\Psi}^*)$ has all its eigenvalues between 0 and 1.

4.7.2 Louis' Method

The multivariate version (4.96) of Aitken's acceleration method suggests trying the sequence of iterates $\{\boldsymbol{\Psi}_A^{(k)}\}$, where $\boldsymbol{\Psi}_A^{(k+1)}$ is defined by

$$\boldsymbol{\Psi}_A^{(k+1)} = \boldsymbol{\Psi}_A^{(k)} + \{\boldsymbol{I}_d - \boldsymbol{J}(\boldsymbol{\Psi}_A^{(k)})\}^{-1}(\boldsymbol{\Psi}_{EMA}^{(k+1)} - \boldsymbol{\Psi}_A^{(k)}), \tag{4.97}$$

where $\boldsymbol{\Psi}_{EMA}^{(k+1)}$ is the EM iterate produced using $\boldsymbol{\Psi}_A^{(k)}$ as the current fit for $\boldsymbol{\Psi}$.

Hence this method proceeds on the $(k+1)$th iteration by first producing $\boldsymbol{\Psi}_{EMA}^{(k+1)}$ using an EM iteration with $\boldsymbol{\Psi}_A^{(k)}$ as the current fit for $\boldsymbol{\Psi}$. One then uses the EM iterate $\boldsymbol{\Psi}_{EMA}^{(k+1)}$ in Aitken's acceleration procedure (4.96) to yield the final iterate $\boldsymbol{\Psi}_A^{(k+1)}$ on the $(k+1)$th iteration. This is the method proposed by Louis (1982) for speeding up the convergence of the EM algorithm.

Louis (1982) suggests making use of the relationship (3.70) to estimate $\boldsymbol{J}(\boldsymbol{\Psi}_A^{(k)})$ in (4.97). Using (3.70) in (4.97) gives

$$\boldsymbol{\Psi}_A^{(k+1)} = \boldsymbol{\Psi}_A^{(k)} + \boldsymbol{I}^{-1}(\boldsymbol{\Psi}_A^{(k)}; \boldsymbol{y})\mathcal{I}_c(\boldsymbol{\Psi}_A^{(k)}; \boldsymbol{y})(\boldsymbol{\Psi}_{EMA}^{(k+1)} - \boldsymbol{\Psi}_A^{(k)}). \tag{4.98}$$

As cautioned by Louis (1982), the relationship (3.70) is an approximation useful only local to the MLE, and so should not be used until some EM iterations have been performed. As noted by Meilijson (1989), the use of (4.98) is approximately equivalent to using the Newton-Raphson algorithm to find a zero of the (incomplete-data) score statistic $\boldsymbol{S}(\boldsymbol{y}; \boldsymbol{\Psi})$. To see this, suppose that the EM iterate $\boldsymbol{\Psi}_{EMA}^{(k+1)}$ satisfies the condition

$$[\partial Q(\boldsymbol{\Psi}; \boldsymbol{\Psi}^{(k)})/\partial \boldsymbol{\Psi}]_{\boldsymbol{\Psi}=\boldsymbol{\Psi}_{EMA}^{(k+1)}} = \boldsymbol{0}.$$

Then from (3.76), we have that

$$\boldsymbol{S}(\boldsymbol{y}; \boldsymbol{\Psi}_A^{(k)}) \approx \mathcal{I}_c(\boldsymbol{\Psi}_A^{(k)}; \boldsymbol{y})(\boldsymbol{\Psi}_{EMA}^{(k+1)} - \boldsymbol{\Psi}_A^{(k)}). \tag{4.99}$$

On substituting this result in (4.98), we obtain

$$\boldsymbol{\Psi}_A^{(k+1)} \approx \boldsymbol{\Psi}_A^{(k)} + \boldsymbol{I}^{-1}(\boldsymbol{\Psi}_A^{(k)}; \boldsymbol{y})\boldsymbol{S}(\boldsymbol{y}; \boldsymbol{\Psi}_A^{(k)}). \tag{4.100}$$

It can be seen from (1.6) that the right-hand side of (4.100) is the iterate produced on the $(k+1)$th iteration of the Newton–Raphson procedure applied to finding a zero of $\boldsymbol{S}(\boldsymbol{y}; \boldsymbol{\Psi})$. Hence the use of Aitken's procedure as applied by Louis (1982) is essentially equivalent to the Newton–Raphson procedure applied to $\boldsymbol{S}(\boldsymbol{y}; \boldsymbol{\Psi})$.

More recently, Meilijson (1989) and Jamshidian and Jennrich (1993) note that the accelerated sequence (4.97) as proposed by Louis (1982) is precisely the same as that obtained by applying the Newton–Raphson method to find a zero of the difference

$$\boldsymbol{\delta}(\boldsymbol{\Psi}) = \boldsymbol{M}(\boldsymbol{\Psi}) - \boldsymbol{\Psi},$$

where $\boldsymbol{M}$ is the map defined by the EM sequence. To see this, we can write (4.97) as

$$\boldsymbol{\Psi}_A^{(k+1)} = \boldsymbol{\Psi}_A^{(k)} + \{\boldsymbol{I}_d - \boldsymbol{J}(\boldsymbol{\Psi}_A^{(k)})\}^{-1}\{\boldsymbol{M}(\boldsymbol{\Psi}_A^{(k)}) - \boldsymbol{\Psi}_A^{(k)}\}, \tag{4.101}$$

since

$$\boldsymbol{\Psi}_{EMA}^{(k+1)} = \boldsymbol{M}(\boldsymbol{\Psi}_A^{(k)}).$$

As the gradient of $\boldsymbol{\delta}(\boldsymbol{\Psi})$ is $-\{\boldsymbol{I}_d - \boldsymbol{J}(\boldsymbol{\Psi})\}$, (4.101) corresponds exactly to applying the Newton–Raphson procedure to find a zero of $\boldsymbol{\delta}(\boldsymbol{\Psi})$.

On further ways to approximate $\{\boldsymbol{I}_d - \boldsymbol{J}(\boldsymbol{\Psi}_A^{(k)})\}$ for use in (4.101), Meilijson (1989) suggests using symmetric quasi-Newton updates. However, as cautioned by Jamshidian and Jennrich (1993), these will not work, as $\{\boldsymbol{I}_d - \boldsymbol{J}(\boldsymbol{\Psi})\}$ is in general not symmetric.

4.7.3 Example 4.6: Multinomial Data

As a simple illustration of the use of Aitken's acceleration procedure as proposed by Louis (1982), we consider an example from his paper in which he applies his formula (4.100) to the multinomial data in Example 1.1. It was applied after two consecutive EM iterations. In this case,

$$\begin{aligned}\Psi_A^{(2)} &= \Psi^{(2)} \\ &= 0.626338,\end{aligned}$$

and

$$\begin{aligned}I^{-1}(\Psi_A^{(2)}; \boldsymbol{y})\mathcal{I}_c(\Psi_A^{(2)}; \boldsymbol{y}) &= (434.79/376.95) \\ &= 1.153442.\end{aligned}$$

Thus

$$\begin{aligned}\Psi_A^{(3)} &= 0.626338 + 1.153442(\Psi_{EMA}^{(3)} - 0.626338) \\ &= 0.6268216,\end{aligned}$$

since $\Psi_{EMA}^{(3)} = 0.626757$. It can be seen that $\Psi_A^{(3)}$ is closer to $\hat{\Psi} = 0.6268215$ than $\Psi^{(4)}$ (0.626812), the fourth iterate with the original (unaccelerated) EM algorithm.

4.7.4 Example 4.7: Geometric Mixture

We report here an example from Meilijson (1989), who considers the application of the EM algorithm to ML estimation of the parameter vector

$$\boldsymbol{\Psi} = (\pi_1, p_1, p_2)^T$$

in the two-component geometric mixture

$$f(w; \boldsymbol{\Psi}) = \sum_{i=1}^{2} \pi_i f(w; p_i),$$

where

$$f(w;\, p_i) = p_i(1 - p_i)^{w-1}, \quad w = 1, 2, \ldots \; (0 \leq p_i \leq 1),$$

for $i = 1, 2$.

Proceeding as in Section 1.4 for the general case of a finite mixture model, we declare the complete-data vector $\boldsymbol{x}$ as

$$\boldsymbol{x} = (\boldsymbol{y}^T, \boldsymbol{z}^T)^T,$$

where $\boldsymbol{y} = (w_1, \ldots, w_n)^T$ contains the observed data and where the missing data vector $\boldsymbol{z}$ is taken to be

$$\boldsymbol{z} = (\boldsymbol{z}_1^T, \ldots, \boldsymbol{z}_n^T)^T.$$

As in Section 1.4, the $z_{ij} = (\boldsymbol{z}_j)_i$ is taken to be one or zero, according as to whether the jth observation arises or does not arise from the ith component of the mixture ($i = 1, 2;\ j = 1, \ldots, n$).

The complete-data log likelihood is given by

$$\log L_c(\boldsymbol{\Psi}) = \sum_{i=1}^{2} \sum_{j=1}^{n} z_{ij}\{\log \pi_i + \log p_i + (w_j - 1)\log(1 - p_i)\} \tag{4.102}$$

$$= \sum_{i=1}^{2} \left\{ n_i(\log \pi_i + \log p_i) + \left(\sum_{j=1}^{n} z_{ij} w_j - n_i \right) \log(1 - p_i) \right\}, \tag{4.103}$$

where

$$n_i = \sum_{j=1}^{n} z_{ij}.$$

Corresponding to the E-step at the $(k+1)$th iteration of the EM algorithm, the Q-function is given by

$$Q(\boldsymbol{\Psi};\, \boldsymbol{\Psi}^{(k)}) = \sum_{i=1}^{2} \left\{ n_i^{(k)}(\log \pi_i + \log p_i) + \left(\sum_{j=1}^{n} z_{ij}^{(k)} w_j - n_i^{(k)} \right) \log(1 - p_i) \right\}, \tag{4.104}$$

where

$$n_i^{(k)} = \sum_{j=1}^{n} z_{ij}^{(k)}$$

and

$$z_{ij}^{(k)} = \frac{\pi_i^{(k)} p_i^{(k)} (1 - p_i^{(k)})^{w_j - 1}}{\sum_{h=1}^{2} \pi_h^{(k)} p_h^{(k)} (1 - p_h^{(k)})^{w_j - 1}}$$

is the current conditional expectation of Z_{ij} given $\boldsymbol{y}$ (that is, the current posterior probability that the jth observation arises from the ith component of the mixture ($i = 1, 2$)).

As in the general case (1.37) of a finite mixture model,

$$\pi_i^{(k+1)} = n_i^{(k)}/n,$$

while specific to geometric components in the mixture, we have on differentiation of (4.104) that

$$p_i^{(k+1)} = \left(\sum_{j=1}^{n} z_{ij}^{(k)} w_j / n_i^{(k)} \right)^{-1} \quad (i = 1, 2).$$

We now consider the calculation of the empirical information matrix $\boldsymbol{I}_e(\boldsymbol{\Psi}; \boldsymbol{y})$ as defined by (4.41) after the kth iteration. It can be seen from this definition, that we have to calculate $\boldsymbol{s}(\boldsymbol{w}_j)$, the (incomplete-data) score statistic based on just the jth observation for each j ($j = 1, \dots, n$). It can be expressed in terms of the complete-data single-observation score statistics as

$$\begin{aligned} \boldsymbol{s}(w_j; \boldsymbol{\Psi}) &= \partial \log L_j(\boldsymbol{\Psi})/\partial \boldsymbol{\Psi} \\ &= E_{\boldsymbol{\Psi}^{(k)}}\{\partial \log L_{cj}(\boldsymbol{\Psi})/\partial \boldsymbol{\Psi} \mid \boldsymbol{y}\} \\ &= \partial Q_j(\boldsymbol{\Psi}; \boldsymbol{\Psi}^{(k)})/\partial \boldsymbol{\Psi}, \end{aligned}$$

where

$$Q_j(\boldsymbol{\Psi}; \boldsymbol{\Psi}^{(k)}) = E_{\boldsymbol{\Psi}^{(k)}}\{\log L_{cj}(\boldsymbol{\Psi}) \mid \boldsymbol{y}\}.$$

From (4.103),

$$\log L_{cj}(\boldsymbol{\Psi}) = \sum_{i=1}^{2} z_{ij}\{\log \pi_i + \log p_i + (w_j - 1)\log(1 - p_i)\},$$

and so

$$Q_j(\boldsymbol{\Psi}; \boldsymbol{\Psi}^{(k)}) = \sum_{i=1}^{2} z_{ij}^{(k)}\{\log \pi_i + \log p_i + (w_j - 1)\log(1 - p_i)\}. \quad (4.105)$$

We let $s_i(w_j; \boldsymbol{\Psi})$ denote the ith element of $\boldsymbol{s}(\boldsymbol{w}_j; \boldsymbol{\Psi})$ for $i = 1, 2$, and 3, corresponding to the parameters π_1, p_1, and p_2. On differentiation of (4.105) with respect to $\boldsymbol{\Psi}$, we have that

$$s_1(w_j; \boldsymbol{\Psi}) = (z_{1j}^{(k)} - \pi_1)/(\pi_1 \pi_2) \quad (4.106)$$

and

$$s_{i+1}(w_j; \boldsymbol{\Psi}) = z_{ij}^{(k)}(1 - w_j p_i)/\{p_i(1 - p_i)\} \quad (i = 1, 2). \quad (4.107)$$

The empirical covariance matrix $\boldsymbol{I}_e(\boldsymbol{\Psi}; \boldsymbol{y})$ is therefore given by (4.41) with $\boldsymbol{s}(w_j; \boldsymbol{\Psi})$ specified by (4.106) and (4.107).

Meilijson (1989) tries various methods to fit the geometric mixture model by maximum likelihood to a data set involving approximately 5000 observations with values of W ranging from 1 to 14. The MLE of $\boldsymbol{\Psi}$ is

$$\hat{\boldsymbol{\Psi}} = (0.40647, 0.35466, 0.90334)^T.$$

The initial value of $\boldsymbol{\Psi}$ used for the EM algorithm and its progress on the 5th, 10th, and 100th iterations are displayed in Table 4.3.

Meilijson (1989) finds that the Newton–Raphson and Louis methods, which are practically the same, converged after five iterations (monotonically and very smoothly), starting from the EM iterate $\boldsymbol{\Psi}^{(5)}$. The Newton–Raphson method modified by using the empirical information matrix $\boldsymbol{I}_e(\boldsymbol{\Psi}^{(k)}; \boldsymbol{y})$ in place of the observed information matrix $\boldsymbol{I}(\boldsymbol{\Psi}^{(k)}; \boldsymbol{y})$ on each iteration k, converged after eight iterations from $\boldsymbol{\Psi}^{(5)}$, the second of which was an unstable overshoot, as can be seen from Table 4.4.

The data set was modified by Meilijson (1989) to fit the geometric mixture model less perfectly. For this modified set, the MLE was

$$\hat{\boldsymbol{\Psi}} = (0.64, 0.35, 0.86)^T.$$

The Newton–Raphson method with the modification of the empirical covariance matrix converged in six iterations from

$$\boldsymbol{\Psi} = (0.4, 0.8, 0.9)^T,$$

but diverged from

$$\boldsymbol{\Psi} = (0.2, 0.5, 0.7)^T.$$

However, it did converge in eight iterations from the latter point when negative probabilities were interpreted as 0.1 and those exceeding unity as 0.99. The Newton–Raphson and Louis methods were found to be fast and generally smooth, but with a smaller radius of convergence than under the perfect fit.

Table 4.3. Results of the EM Algorithm for the Geometric Mixture Model.

Iteration (k)	$\pi_1^{(k)}$	$p_1^{(k)}$	$p_2^{(k)}$
0	0.2	0.1	0.2
5	0.146	0.248	0.704
10	0.201	0.264	0.767
100	0.40465	0.35395	0.90220

Table 4.4. Results of the Newton-Raphson Method Using the Empirical Information Matrix for a Geometric Mixture Model.

Iteration (k)	$\pi_1^{(k)}$	$p_1^{(k)}$	$p_2^{(k)}$
0	0.146	0.248	0.704
1	0.43	0.52	0.87
2	0.60	0.49	0.986
3	0.45	0.42	0.91
4	0.42	0.37	0.91
5	0.406	0.356	0.903
.	.	.	.
.	.	.	.
8	0.40647	0.35466	0.90334

4.7.5 Grouped and Truncated Data (Example 2.8 Continued)

We return now to Example 2.8 in Section 2.8.7 with the numerical illustration thereof, where the two-component log normal mixture model (2.53) was fitted to some grouped and truncated data on the volume of red blood cells of a cow. The EM algorithm was found to take 117 iterations to converge. For this data set, the largest eigenvalue λ_{max} of the rate matrix (3.69) is 0.975. Values of λ_{max} near one indicate a slow rate of convergence. Jones and McLachlan (1992) investigated various ways of speeding up the EM algorithm in its application to this data set. In Table 4.5, we report their results obtained using the EM algorithm and its accelerated version using Louis' method.

Table 4.5. Comparison of Methods to Speed up the EM Algorithm for Red Blood Cell Volume Data.

Parameter	Initial Value	EM	Louis Method
π_1	0.45	0.4521	0.4530
		(0.0521)	(0.0522)
μ_1	4.00	4.0728	4.0734
		(0.0384)	(0.0384)
μ_2	4.60	4.7165	4.7169
		(0.0242)	(0.0242)
σ_1^2	0.08	0.0575	0.0577
		(0.0107)	(0.0107)
σ_2^2	0.05	0.0438	0.0436
		(0.0099)	(0.0099)
No. of iterations		107	11

Source: Adapted from Jones and McLachlan (1992).

It can be seen that the latter is effective here in reducing the number of iterations.

They also tried the modified Newton–Raphson method with the information matrix replaced by the empirical information matrix for grouped data, as given by (4.45). It gave identical results for this data set. This is not unexpected, as we have noted that Louis' method is essentially the same as the Newton–Raphson method. But Louis' method and the modified Newton–Raphson method will not always give the same results, as demonstrated by Jones and McLachlan (1992) in their application of these two methods to another data set of this type, where Louis' method was not as effective as the modified Newton-Raphson method in terms of number of iterations. The standard errors of the estimates, given in parentheses in Table 4.5, are based on using the inverse of the empirical information matrix (4.45) to approximate the covariance matrix of the MLE.

4.8 AN AITKEN ACCELERATION-BASED STOPPING CRITERION

The stopping criterion usually adopted with the EM algorithm is in terms of either the size of the relative change in the parameter estimates or the log likelihood. As Lindstrom and Bates (1988) emphasize, this is a measure of lack of progress but not of actual convergence. Recently, Böhning, Dietz, Schaub, Schlattmann, and Lindsay (1994) exploit Aitken's acceleration procedure in its application to the sequence of log likelihood values to provide a useful estimate of its limiting value. It is applicable in the case where the sequence of log likelihood values $\{l^{(k)}\}$ is linearly convergent to some value l^*, where here for brevity of notation

$$l^{(k)} = \log L(\boldsymbol{\Psi}^{(k)}).$$

Under this assumption,

$$l^{(k+1)} - l^* \approx c\,(l^{(k)} - l^*), \tag{4.108}$$

for all k and some $c\,(0 < c < 1)$. The equation (4.108) can be rearranged to give

$$l^{(k+1)} - l^{(k)} \approx (1 - c)(l^* - l^{(k)}), \tag{4.109}$$

for all k. It can be seen from (4.109) that, if c is very close to one, a small increment in the log likelihood, $l^{(k+1)} - l^{(k)}$, does not necessarily mean that $l^{(k)}$ is very close to l^*.

From (4.109), we have that

$$l^{(k+1)} - l^{(k)} \approx c\,(l^{(k)} - l^{(k-1)}), \tag{4.110}$$

for all k, corresponding to the (multivariate) expression (4.95) for successive increments in the estimates of the parameter vector. Just as Aitken's acceleration procedure was applied to (4.92) to obtain (4.96), we can apply it to (4.110) to obtain the corresponding result for the limit l^* of the sequence of log likelihood values

$$l^* = l^{(k)} + \frac{1}{(1-c)}(l^{(k+1)} - l^{(k)}). \tag{4.111}$$

Since c is unknown, it has to be estimated in (4.111), for example, by the ratio of successive increments,

$$c^{(k)} = (l^{(k+1)} - l^{(k)})/(l^{(k)} - l^{(k-1)}).$$

This leads to the Aitken accelerated estimate of l^*,

$$l_A^{(k+1)} = l^{(k)} + \frac{1}{(1-c^{(k)})}(l^{(k+1)} - l^{(k)}). \tag{4.112}$$

In applications where the primary interest in on the sequence of log likelihood values rather than the sequence of parameter estimates, Böhning et al. (1994) suggest the EM algorithm can be stopped if

$$| \, l_A^{(k+1)} - l_A^{(k)} \, | < \text{tol},$$

where tol is the desired tolerance. An example concerns the resampling approach (McLachlan, 1987) to the problem of assessing the null distribution of the likelihood ratio test statistic for the number of components in a mixture model. The criterion (4.112) is applicable for any log likelihood sequence that is linearly convergent.

4.9 CONJUGATE GRADIENT ACCELERATION OF THE EM ALGORITHM

4.9.1 Conjugate Gradient Method

Jamshidian and Jennrich (1993) propose an alternative method based on conjugate gradients. The key is that the change in $\boldsymbol{\Psi}$ after an EM iteration

$$\boldsymbol{\delta}(\boldsymbol{\Psi}) = \boldsymbol{M}(\boldsymbol{\Psi}) - \boldsymbol{\Psi}$$

can be viewed (approximately at least) as a generalized gradient, making it natural to apply generalized conjugate gradient methods in an attempt to accelerate the EM. Though not identified as such, this was done by Golub and Nash (1982) as their alternative to the Yates (1933) EM algorithm for fitting unbalanced analysis of variance models. The proposed method is relatively simple and can handle problems with a large number of parameters. The latter is an area where the EM algorithm is particularly important and where it is often the only algorithm used. Before we discuss the generalized gradient approach to accelerating the performance of the EM algorithm, we describe the generalized conjugate gradient algorithm used by Jamshidian and Jennrich (1993).

4.9.2 A Generalized Conjugate Gradient Algorithm

We present the algorithm in Jamshidian and Jennrich (1993) for finding the maximum of a function $h(\boldsymbol{\Psi})$, where $\boldsymbol{\Psi}$ ranges over a subset of $\mathbb{R}^d$. Let $\boldsymbol{q}(\boldsymbol{\Psi})$ denote the gradient of $h(\boldsymbol{\Psi})$ and consider the generalized norm

$$||\boldsymbol{\Psi}|| = (\boldsymbol{\Psi}^T \boldsymbol{A} \boldsymbol{\Psi})^{1/2}$$

on $\mathbb{R}^d$, where $\boldsymbol{A}$ is a positive-definite matrix. Also, let $\tilde{\boldsymbol{q}}(\boldsymbol{\Psi})$ be the gradient of $h(\boldsymbol{\Psi})$ with respect to this norm. That is,

$$\tilde{\boldsymbol{q}}(\boldsymbol{\Psi}) = \boldsymbol{A}^{-1}\boldsymbol{q}(\boldsymbol{\Psi}).$$

The vector $\tilde{\boldsymbol{q}}(\boldsymbol{\Psi})$ is called the generalized gradient of $h(\boldsymbol{\Psi})$ defined by $\boldsymbol{A}$.

The generalized conjugate gradient algorithm is implemented as follows. Let $\boldsymbol{\Psi}^{(0)}$ be the initial value of $\boldsymbol{\Psi}$ and let

$$\boldsymbol{d}^{(0)} = \tilde{q}(\boldsymbol{\Psi}^{(0)}).$$

One sequentially computes for $k = 0, 1, 2, ...,$

1. $\alpha^{(k)}$, the value of α that maximizes

$$h(\boldsymbol{\Psi}^{(k)} + \alpha \boldsymbol{d}^{(k)});$$

2.

$$\boldsymbol{\Psi}^{(k+1)} = \boldsymbol{\Psi}^{(k)} + \alpha^{(k)} \boldsymbol{d}^{(k)};$$

3.

$$\beta^{(k)} = \frac{\{\boldsymbol{q}^T(\boldsymbol{\Psi}^{(k+1)}) - \boldsymbol{q}^T(\boldsymbol{\Psi}^{(k)})\}\tilde{\boldsymbol{q}}(\boldsymbol{\Psi}^{(k+1)})}{\{\boldsymbol{q}^T(\boldsymbol{\Psi}^{(k+1)}) - \boldsymbol{q}^T(\boldsymbol{\Psi}^{(k)})\}\boldsymbol{d}^{(k)}};$$

4.

$$\boldsymbol{d}^{(k+1)} = \tilde{\boldsymbol{q}}(\boldsymbol{\Psi}^{(k+1)}) - \beta^{(k)} \boldsymbol{d}^{(k)}.$$

This is called a generalized conjugate gradient algorithm, because it uses generalized gradients to define the search direction $\boldsymbol{d}^{(k)}$ and because, for negative-definite quadratic functions h, the $\boldsymbol{d}^{(k)}$ are orthogonal in the metric defined by the negative of the Hessian of h.

4.9.3 Accelerating the EM Algorithm

From (4.99),

$$\begin{aligned}\boldsymbol{\Psi}^{(k+1)} - \boldsymbol{\Psi}^{(k)} &= \boldsymbol{M}(\boldsymbol{\Psi}^{(k)}) - \boldsymbol{\Psi}^{(k)} \\ &\approx \mathcal{I}_c^{-1}(\boldsymbol{\Psi}^{(k)}; \boldsymbol{y})\boldsymbol{S}(\boldsymbol{y}; \boldsymbol{\Psi}^{(k)}).\end{aligned}$$

Thus the change in $\boldsymbol{\Psi}$ after an EM iteration,

$$\boldsymbol{\delta}(\boldsymbol{\Psi}) = \boldsymbol{M}(\boldsymbol{\Psi}) - \boldsymbol{\Psi},$$

is approximately equal to

$$\boldsymbol{\delta}(\boldsymbol{\Psi}) \approx \mathcal{I}_c^{-1}(\boldsymbol{\Psi}; \boldsymbol{y})\boldsymbol{S}(\boldsymbol{y}; \boldsymbol{\Psi}), \tag{4.113}$$

where $\boldsymbol{S}(\boldsymbol{y}; \boldsymbol{\Psi})$ is the gradient of the log likelihood function $\log L(\boldsymbol{\Psi})$. Typically $\mathcal{I}_c(\boldsymbol{\Psi}; \boldsymbol{y})$ is positive definite. Thus (4.113) shows that when $\boldsymbol{M}(\boldsymbol{\Psi})$ is near to $\boldsymbol{\Psi}$, the difference $\boldsymbol{\delta}(\boldsymbol{\Psi}) = \boldsymbol{M}(\boldsymbol{\Psi}) - \boldsymbol{\Psi}$ is, to a good approximation, a generalized gradient of $\log L(\boldsymbol{\Psi})$.

Jamshidian and Jennrich (1993) propose accelerating the EM algorithm by applying the generalized conjugate gradient algorithm in the previous section with

$$h(\boldsymbol{\Psi}) = \log L(\boldsymbol{\Psi}),$$

and

$$\boldsymbol{q}(\boldsymbol{\Psi}) = \boldsymbol{S}(\boldsymbol{y}; \boldsymbol{\Psi}),$$

and where the generalized gradient $\tilde{\boldsymbol{q}}(\boldsymbol{\Psi})$ is given by $\delta(\boldsymbol{\Psi})$, the change in $\boldsymbol{\Psi}$ after performing each subsequent EM iteration. They call the resulting algorithm the accelerated EM (AEM) algorithm. This algorithm is applied after a few EM iterations. More specifically, Jamshidian and Jennrich (1993) suggest running the EM algorithm until twice the difference between successive values of the log likelihood falls below one. In their examples, this usually occurred after five EM iterations.

Although the AEM algorithm is fairly simple, it is more complex than the EM algorithm itself. In addition to the EM iterations, one must compute the gradient of log $L(\boldsymbol{\Psi})$ (that is, the score statistic $\boldsymbol{S}(\boldsymbol{y}; \boldsymbol{\Psi})$). Frequently, however, the latter is either available from the EM code or is obtainable with minor modification. The biggest complication is the line search required in Step 1 above. An algorithm for this search is given in the Appendix of Jamshidian of Jennrich (1993). Their experience suggests that a simple line search is sufficient. To demonstrate the effectiveness of the AEM algorithm, Jamshidian and Jennrich (1993) applied it to several problems in areas that included the estimation of a covariance matrix from incomplete multivariate normal data, confirmatory factor analysis, and repeated measures analysis. In terms of floating-point operation counts, for all of the comparative examples considered, the AEM algorithm increases the speed of the EM algorithm, in some cases by a factor of 10 or more.

4.10 HYBRID METHODS FOR FINDING THE MAXIMUM LIKELIHOOD ESTIMATE

4.10.1 Introduction

Various authors, including Redner and Walker (1984), propose a hybrid approach to the computation of the MLE that would switch from the EM algorithm after a few iterations to the Newton–Raphson or some quasi-Newton method. The idea is to use the EM algorithm initially to take advantage of its good global convergence properties and to then exploit the rapid local convergence of Newton-type methods by switching to such a method. It can be seen that the method of Louis (1982) is a hybrid algorithm of this nature, for in effect after a few initial EM iterations, it uses the Newton–Raphson method to accelerate convergence after performing each subsequent EM iteration. Of course there is no guarantee that these hybrid algorithms increase the likelihood $L(\boldsymbol{\Psi})$ monotonically. Hybrid algorithms have been considered also by Atkinson (1992), Heckman and Singer (1984), Jones and McLachlan (1992), and Aitkin and Aitkin (1996).

4.10.2 Combined EM and Modified Newton–Raphson Algorithm

We now consider the recent work of Aitkin and Aitkin (1996) on a hybrid method that combines the EM algorithm with a modified Newton–Raphson method whereby the information matrix is replaced by the empirical information matrix. In the context of fitting finite normal mixture models, they construct a hybrid algorithm that starts with five EM iterations before switching to the modified Newton–Raphson method until convergence or until the log likelihood decreases. In the case of the latter, Aitkin and Aitkin (1996) propose halving the step size up to five times. As further step-halvings would generally leave the step size smaller than that of an EM step, if the log likelihood decreases after five step-halves, the algorithm of Aitkin and Aitkin (1996) returns to the previous EM iterate and runs the EM algorithm for a further five iterations, before switching back again to the modified Newton–Raphson method. Their choice of performing five EM iterations initially is based on the work of Redner and Walker (1984), who report that, in their experience, 95 percent of the change in the log likelihood from its initial value to its maximum generally occurs in five iterations.

Aitkin and Aitkin (1996) replicate part of the study by Everitt (1988) of the fitting of a mixture of two normal densities with means μ_1 and μ_2 and variances σ_1^2 and σ_2^2 in proportions π_1 and π_2,

$$f(w; \boldsymbol{\Psi}) = \pi_1 \phi(w; \mu_1, \sigma_1^2) + \pi_2 \phi(w; \mu_2, \sigma_2^2). \tag{4.114}$$

Table 4.6. Parameter Values Used in the Simulation of Two-Component Normal Mixture Models.

Mixture	π_1	μ_1	μ_2	σ_1^2	σ_2^2
I	0.4	0	3	0.5	1
II	0.4	0	3	1	2
III	0.2	0	3	1	2

They simulated ten random samples of size $n = 50$ from this normal mixture in each of three cases with the parameter vector

$$\boldsymbol{\Psi} = (\pi_1, \mu_1, \mu_2, \sigma_1^2, \sigma_2^2)^T$$

specified as in Table 4.6. These cases are increasingly difficult to fit. The normal mixture model (4.114) is fitted to each simulated random sample using the starting values, as specified in Table 4.7. The stopping criterion of Aitkin and Aitkin (1996) is a difference in successive values of the log likelihood of 10^{-5}.

Aitkin and Aitkin (1996) find that their hybrid algorithm required 70 percent of the time required for the EM algorithm to converge, consistently over all starting values of $\boldsymbol{\Psi}$. They find that the EM algorithm is impressively stable. Their hybrid algorithm almost always decreases the log likelihood when the switch to the modified Newton–Raphson is first applied, and sometimes requires a large number of EM controlling steps (after full step-halving) before finally increasing the log likelihood, and then usually converging rapidly to the same maximizer as with the EM algorithm. Local maxima are encountered by both algorithms about equally often.

As is well known, mixture likelihoods for small sample sizes are badly behaved with multiple maxima. Aitkin and Aitkin (1996) liken the maximization of the normal mixture log likelihood in their simulation studies as

Table 4.7. Choice of Starting Values in the Fitting of Normal Mixture Models to Simulated Data.

Mixture	π_1	μ_1	μ_2	σ_1^2	σ_2^2
	0.4	0	3	0.5	1
I	0.4	0	1	0.5	0.5
	0.2	1	2	1	0.5
	0.4	0	3	1	2
II	0.4	0	1	1	1
	0.2	1	2	0.5	1
	0.2	0	3	1	2
III	0.2	0	1	1	1
	0.4	1	2	0.5	1

to the progress of a "traveler following the narrow EM path up a hazardous mountain with chasms on all sides. When in sight of the summit, the modified Newton–Raphson method path leapt to the top, but when followed earlier, it caused repeated falls into the chasms, from which the traveler had to be pulled back onto the EM track."

4.11 A GENERALIZED EM ALGORITHM BASED ON ONE NEWTON–RAPHSON STEP

4.11.1 Derivation of a Condition to be a Generalized EM Sequence

In Section 1.5 in formulating the GEM algorithm, we considered as an example of a GEM algorithm the sequence of iterates $\{\boldsymbol{\Psi}^{(k)}\}$, where $\boldsymbol{\Psi}^{(k+1)}$ is defined to be

$$\boldsymbol{\Psi}^{(k+1)} = \boldsymbol{\Psi}^{(k)} + a^{(k)}\boldsymbol{\delta}^{(k)}, \tag{4.115}$$

where

$$\boldsymbol{\delta}^{(k)} = -[\partial^2 Q(\boldsymbol{\Psi};\ \boldsymbol{\Psi}^{(k)})/\partial\boldsymbol{\Psi}\partial\boldsymbol{\Psi}^T]^{-1}_{\boldsymbol{\Psi}=\boldsymbol{\Psi}^{(k)}}[\partial Q(\boldsymbol{\Psi};\ \boldsymbol{\Psi}^{(k)})/\partial\boldsymbol{\Psi}]_{\boldsymbol{\Psi}=\boldsymbol{\Psi}^{(k)}}, \tag{4.116}$$

and where $0 < a^{(k)} \leq 1$. The constant $a^{(k)}$ is chosen so that

$$Q(\boldsymbol{\Psi}^{(k+1)};\ \boldsymbol{\Psi}^{(k)}) \geq Q(\boldsymbol{\Psi}^{(k)};\ \boldsymbol{\Psi}^{(k)}). \tag{4.117}$$

holds, that is, so that (4.115) defines a GEM algorithm. In the case of $a^{(k)} = 1$, (4.115) defines the first iterate obtained when using the Newton–Raphson procedure to obtain a root of the equation

$$\partial Q(\boldsymbol{\Psi};\ \boldsymbol{\Psi}^{(k)})/\partial\boldsymbol{\Psi} = \mathbf{0}$$

on the M-step where the intent is to maximize the Q-function.

We now derive the result (1.65) that

$$Q(\boldsymbol{\Psi}^{(k+1)};\ \boldsymbol{\Psi}^{(k)}) - Q(\boldsymbol{\Psi}^{(k)};\ \boldsymbol{\Psi}^{(k)}) = a^{(k)}\boldsymbol{S}(\boldsymbol{y};\ \boldsymbol{\Psi}^{(k)})^T\boldsymbol{A}^{(k)}\boldsymbol{S}(\boldsymbol{y};\ \boldsymbol{\Psi}^{(k)}), \tag{4.118}$$

where

$$\boldsymbol{A}^{(k)} = \mathcal{I}_c^{-1}(\boldsymbol{\Psi}^{(k)};\ \boldsymbol{y})\{\boldsymbol{I}_d - \tfrac{1}{2}a^{(k)}\tilde{\mathcal{I}}_c^{(k)}(\boldsymbol{y})\mathcal{I}_c^{-1}(\boldsymbol{\Psi}^{(k)};\ \boldsymbol{y})\} \tag{4.119}$$

and where

$$\begin{aligned}\tilde{\mathcal{I}}_c^{(k)}(\boldsymbol{y}) &= -[\partial^2 Q(\boldsymbol{\Psi};\ \boldsymbol{\Psi}^{(k)})/\partial\boldsymbol{\Psi}\boldsymbol{\Psi}^T]_{\boldsymbol{\Psi}=\tilde{\boldsymbol{\Psi}}^{(k)}} \\ &= E_{\boldsymbol{\Psi}^{(k)}}\{\boldsymbol{I}_c(\tilde{\boldsymbol{\Psi}}^{(k)};\ \boldsymbol{X}) \mid \boldsymbol{y}\},\end{aligned} \tag{4.120}$$

and $\tilde{\boldsymbol{\Psi}}^{(k)}$ is a point on the line segment between $\boldsymbol{\Psi}^{(k)}$ and $\boldsymbol{\Psi}^{(k+1)}$.

From (3.45) and (3.72), we have that

$$[\partial Q(\boldsymbol{\Psi};\, \boldsymbol{\Psi}^{(k)})/\partial \boldsymbol{\Psi}]_{\boldsymbol{\Psi}=\boldsymbol{\Psi}^{(k}} = \boldsymbol{S}(\boldsymbol{y};\, \boldsymbol{\Psi}^{(k)}) \tag{4.121}$$

and

$$[\partial^2 Q(\boldsymbol{\Psi};\, \boldsymbol{\Psi}^{(k)})/\partial \boldsymbol{\Psi} \partial \boldsymbol{\Psi}^T]_{\boldsymbol{\Psi}=\boldsymbol{\Psi}^{(k)}} = -\mathcal{I}_c(\boldsymbol{\Psi}^{(k)};\, \boldsymbol{y}). \tag{4.122}$$

On using (4.121) and (4.122), we can express $\boldsymbol{\Psi}^{(k+1)}$ as

$$\boldsymbol{\Psi}^{(k+1)} = \boldsymbol{\Psi}^{(k)} + a^{(k)} \mathcal{I}_c^{-1}(\boldsymbol{\Psi};\, \boldsymbol{y}) \boldsymbol{S}(\boldsymbol{y};\, \boldsymbol{\Psi}^{(k)}). \tag{4.123}$$

Now on expanding $Q(\boldsymbol{\Psi};\, \boldsymbol{\Psi}^{(k)}) - Q(\boldsymbol{\Psi}^{(k)};\, \boldsymbol{\Psi}^{(k)})$ in a linear Taylor series expansion about the point $\boldsymbol{\Psi} = \boldsymbol{\Psi}^{(k+1)}$ and using the relationships (4.120), (4.121), and (4.122), we obtain

$$\begin{aligned} &Q(\boldsymbol{\Psi}^{(k+1)};\, \boldsymbol{\Psi}^{(k)}) - Q(\boldsymbol{\Psi}^{(k)};\, \boldsymbol{\Psi}^{(k)}) \\ &\quad = (\boldsymbol{\Psi}^{(k+1)} - \boldsymbol{\Psi}^{(k)})^T \boldsymbol{S}(\boldsymbol{y};\, \boldsymbol{\Psi}^{(k)}) + \tfrac{1}{2}(\boldsymbol{\Psi}^{(k+1)} - \boldsymbol{\Psi}^{(k)})^T \tilde{\mathcal{I}}_c^{(k)}(\boldsymbol{y})(\boldsymbol{\Psi}^{(k+1)} - \boldsymbol{\Psi}^{(k)}). \end{aligned} \tag{4.124}$$

On substituting (4.123) for $\boldsymbol{\Psi}^{(k+1)}$ in (4.124), we obtain the result (4.118).

Typically, $\mathcal{I}_c(\boldsymbol{\Psi}^{(k)}; \boldsymbol{y})$ is positive definite, and so from (4.119), (4.115) yields a GEM sequence if the matrix

$$\boldsymbol{I}_d - \tfrac{1}{2} a^{(k)} \tilde{\mathcal{I}}_c^{(k)}(\boldsymbol{y}) \mathcal{I}_c^{-1}(\boldsymbol{\Psi}^{(k)};\, \boldsymbol{y})$$

is positive definite; that is, if $a^{(k)}$ is chosen sufficiently small, as discussed in Section 1.57. In the case that the complete-data density is a member of the regular exponential family with natural parameter $\boldsymbol{\Psi}$, we have from (4.10) that

$$\mathcal{I}_c(\boldsymbol{\Psi}^{(k)};\, \boldsymbol{y}) = \mathcal{I}_c(\boldsymbol{\Psi}^{(k)}),$$

and so $\mathcal{I}_c(\boldsymbol{\Psi}^{(k)};\, \boldsymbol{y})$ is positive definite.

4.11.2 Simulation Experiment

Rai and Matthews (1993) perform a simulation study involving multinomial data in the context of a carcinogenicity experiment to compare the GEM algorithm based on one Newton–Raphson iteration with the EM algorithm using the limiting value of the Newton–Raphson iterations on the M-step. They find that this GEM algorithm can lead to significant computational savings. This is because it was found on average that the GEM algorithm required only a few more E-steps than the EM algorithm and, with only one iteration per M-step, it thus required significantly fewer Newton–Raphson iterations over the total number of M-steps. The results of their simulation are reported in Table 4.8.

Table 4.8. Results of a Simulation Study Comparing EM and a GEM Algorithm for a Multinomial Problem.

	EM Algorithm		GEM Algorithm	
	No. of E-Steps	No. of M-Steps	No. of E-Steps	No. of M-Steps
Average	108	474	112	112
Standard Error	12	49	14	14

Source: Adapted from Rai and Matthews (1993), with permission of the Biometric Society.

4.12 EM GRADIENT ALGORITHM

As noted in Section 1.5.7, Lange (1995a) considers the sequence of iterates defined by (4.115), but with $a^{(k)} = 1$. He calls this algorithm the EM gradient algorithm. He also considers a modified EM gradient algorithm that has $a^{(k)} = a$ in (4.115) as a means of inflating the current EM gradient step by a factor a to speed up convergence. Lange (1995b) subsequently uses the EM gradient algorithm to form the basis of a quasi-Newton approach to accelerate convergence of the EM algorithm. But as pointed out by Lange (1995b), the EM gradient algorithm is an interesting algorithm in its own right. Since the Newton–Raphson method converges quickly, Lange (1995b) notes that the local properties of the EM gradient algorithm are almost identical with those of the EM algorithm.

The matrix

$$[\partial^2 Q(\boldsymbol{\Psi}; \boldsymbol{\Psi}^{(k)})/\partial\boldsymbol{\Psi}\,\partial\boldsymbol{\Psi}^T]_{\boldsymbol{\Psi}=\boldsymbol{\Psi}^{(k)}} = -\mathcal{I}_c(\boldsymbol{\Psi}^{(k)}; \boldsymbol{y}) \tag{4.125}$$

may not be negative definite, and consequently the EM gradient algorithm is not necessarily an ascent algorithm. In practice, the matrix (4.125) is typically negative definite. In the case of the complete-data density being a member of the regular exponential family with natural parameter $\boldsymbol{\Psi}$, we have from (3.72) and (4.13) that

$$\begin{aligned}[\partial^2 Q(\boldsymbol{\Psi}; \boldsymbol{\Psi}^{(k)})/\partial\boldsymbol{\Psi}\boldsymbol{\Psi}^T]_{\boldsymbol{\Psi}=\boldsymbol{\Psi}^{(k)}} &= -\mathcal{I}_c(\boldsymbol{\Psi}^{(k)}; \boldsymbol{y}) \\ &= -\mathcal{I}_c(\boldsymbol{\Psi}^{(k)}),\end{aligned} \tag{4.126}$$

and so negative definiteness is automatic. In this case, the EM gradient algorithm coincides with the ascent algorithm of Titterington (1984).

One advantage of Titterington's (1984) algorithm is that it is necessarily uphill. This means that a fractional step in the current direction will certainly lead to an increase in $L(\boldsymbol{\Psi})$. The EM gradient algorithm also will have this property if $\mathcal{I}_c(\boldsymbol{\Psi}^{(k)}; \boldsymbol{y})$ is positive definite. In some cases, a reparameterization is needed to ensure that $\mathcal{I}_c(\boldsymbol{\Psi}^{(k)}; \boldsymbol{y})$ is positive definite. For example, as noted above, if the complete-data density is a member of the regular exponential family with natural parameter $\boldsymbol{\Psi}$, then

$$\mathcal{I}_c(\boldsymbol{\Psi}; \boldsymbol{y}) = \mathcal{I}_c(\boldsymbol{\Psi}), \tag{4.127}$$

and so $\mathcal{I}_c(\boldsymbol{\Psi}; \boldsymbol{y})$ is positive definite. However, as seen in Section 4.2.4, if the natural parameter is $c(\boldsymbol{\Psi})$, then although (4.127) holds at the MLE, it will not hold in general at $\boldsymbol{\Psi} = \boldsymbol{\Psi}^{(k)}$. Hence we need to transform $\boldsymbol{\Psi}$ to $\boldsymbol{\theta} = c(\boldsymbol{\Psi})$. The reparameterization does not affect the EM algorithm; see Lansky et al. (1992).

Lange (1995a) suggests improving the speed of convergence of the EM gradient algorithm by inflating the current EM gradient step by a factor a to give

$$\boldsymbol{\Psi}^{(k+1)} = \boldsymbol{\Psi}^{(k)} + a\delta^{(k)} \tag{4.128}$$

where, on using (4.121) and (4.122) in(4.116),

$$\delta^{(k)} = \mathcal{I}_c^{-1}(\boldsymbol{\Psi}^{(k)}; \boldsymbol{y})\boldsymbol{S}(\boldsymbol{y}; \boldsymbol{\Psi}^{(k)}).$$

In Section 4.11.2, we noted that Rai and Matthews (1993) had considered a sequence of iterates of the form (4.128), but with $a = a^{(k)}$ and where $a^{(k)}$ was chosen so as to ensure that it implies a GEM sequence and thereby the monotonicity of $L(\boldsymbol{\Psi}^{(k)})$.

Lange (1995b) considers the choice of a in (4.128) directly in terms of $L(\boldsymbol{\Psi})$. He shows that the modified EM gradient sequence has the desirable property of being locally monotonic when $0 < a < 2$. By a second-order Taylor series expansion of $\log L(\boldsymbol{\Psi}) - \log L(\boldsymbol{\Psi}^{(k)})$ about the point $\boldsymbol{\Psi} = \boldsymbol{\Psi}^{(k)}$, we have on evaluation at the point $\boldsymbol{\Psi} = \boldsymbol{\Psi}^{(k+1)}$ that

$$\log L(\boldsymbol{\Psi}^{(k+1)}) - \log L(\boldsymbol{\Psi}^{(k)}) = \tfrac{1}{2}(\boldsymbol{\Psi}^{(k+1)} - \boldsymbol{\Psi}^{(k)})^T \boldsymbol{C}^{(k)}(\boldsymbol{\Psi}^{(k+1)} - \boldsymbol{\Psi}^{(k)}), \tag{4.129}$$

where

$$\boldsymbol{C}^{(k)} = \frac{2}{a}\mathcal{I}_c(\boldsymbol{\Psi}^{(k)}; \boldsymbol{y}) - \boldsymbol{I}(\tilde{\boldsymbol{\Psi}}^{(k)}; \boldsymbol{y}), \tag{4.130}$$

and $\tilde{\boldsymbol{\Psi}}^{(k)}$ is a point on the line segment from $\boldsymbol{\Psi}^{(k)}$ to $\boldsymbol{\Psi}^{(k+1)}$. Now

$$\begin{aligned}\lim_{k\to\infty} \boldsymbol{C}^{(k)} &= \lim_{k\to\infty} \left\{\frac{2}{a}\mathcal{I}_c(\boldsymbol{\Psi}^{(k)}; \boldsymbol{y}) - \boldsymbol{I}(\tilde{\boldsymbol{\Psi}}^{(k)}; \boldsymbol{y})\right\} \\ &= \left(\frac{2}{a} - 1\right)\mathcal{I}_c(\boldsymbol{\Psi}^*; \boldsymbol{y}) - \{\boldsymbol{I}(\boldsymbol{\Psi}^*; \boldsymbol{y}) - \mathcal{I}_c(\boldsymbol{\Psi}^*; \boldsymbol{y})\}. \qquad (4.131)\end{aligned}$$

where it is assumed that the sequence $\{\boldsymbol{\Psi}^{(k)}\}$ converges to some point $\boldsymbol{\Psi}^*$.

Assuming that $\mathcal{I}_c(\boldsymbol{\Psi}; \boldsymbol{y})$ is positive definite, the first term on the right-hand side of (4.131) is positive definite if $0 < a < 2$.

As seen in Section 3.2,

$$H(\boldsymbol{\Psi};\, \boldsymbol{\Psi}^{(k)}) = Q(\boldsymbol{\Psi};\, \boldsymbol{\Psi}^{(k)}) - \log L(\boldsymbol{\Psi})$$

has a maximum at $\boldsymbol{\Psi} = \boldsymbol{\Psi}^{(k)}$, and so

$$[\partial^2 H(\boldsymbol{\Psi};\, \boldsymbol{\Psi}^{(k)})/\partial\boldsymbol{\Psi}\,\partial\boldsymbol{\Psi}^T]_{\boldsymbol{\Psi}=\boldsymbol{\Psi}^{(k)}}$$

is negative semidefinite.

The second term on the right-side of (4.131) is negative semidefinite, since

$$\begin{aligned}
&\boldsymbol{I}(\boldsymbol{\Psi}^*;\, \boldsymbol{y}) - \mathcal{I}_c(\boldsymbol{\Psi}^*;\, \boldsymbol{y}) \\
&\quad = [\partial^2 Q(\boldsymbol{\Psi};\, \boldsymbol{\Psi}^*)/\partial\boldsymbol{\Psi}\,\partial\boldsymbol{\Psi}^T]_{\boldsymbol{\Psi}=\boldsymbol{\Psi}^*} - (\partial^2 \log L(\boldsymbol{\Psi}^*)/\partial\boldsymbol{\Psi}\,\partial\boldsymbol{\Psi}^T) \\
&\quad = [\partial^2 H(\boldsymbol{\Psi};\, \boldsymbol{\Psi}^*)/\partial\boldsymbol{\Psi}\,\partial\boldsymbol{\Psi}^T]_{\boldsymbol{\Psi}=\boldsymbol{\Psi}^*} \\
&\quad = \lim_{k\to\infty} [\partial^2 H(\boldsymbol{\Psi};\, \boldsymbol{\Psi}^{(k)})/\partial\boldsymbol{\Psi}\,\partial\boldsymbol{\Psi}^T]_{\boldsymbol{\Psi}=\boldsymbol{\Psi}^{(k)}},
\end{aligned} \tag{4.132}$$

which is negative semidefinite from above. Hence if $0 < a < 2$, $\lim_{k\to\infty} \boldsymbol{C}^{(k)}$ is a positive definite matrix because it is expressible as the difference between a positive definite matrix and a negative semidefinite matrix.

Since the eigenvalues of a matrix are defined continuously in its entries, it follows that if $0 < a < 1$, the quadratic function (4.129) is positive for k sufficiently large and $\boldsymbol{\Psi}^{(k+1)} \neq \boldsymbol{\Psi}^{(k)}$.

Lange (1995b) also investigates the global convergence of the EM gradient algorithm. Although he concludes that monotonicity appears to be the rule in practice, to establish convergence in theory, monotonicity has to be enforced. Lange (1995b) elects to do this by using a line search at every EM gradient step. The rate of convergence of the EM gradient algorithm is identical with that of the EM algorithm. It is possible for the EM and EM gradient algorithms started from the same point to converge to different points.

4.13 A QUASI-NEWTON ACCELERATION OF THE EM ALGORITHM

4.13.1 The Method

We have seen above that the EM gradient algorithm approximates the M-step of the EM algorithm by using one step of the Newton–Raphson method applied to find a zero of the equation

$$\partial Q(\boldsymbol{\Psi};\, \boldsymbol{\Psi}^{(k)})/\partial\boldsymbol{\Psi} = \boldsymbol{0}. \tag{4.133}$$

That is, it uses

$$\begin{aligned}\boldsymbol{\Psi}^{(k+1)} &= \boldsymbol{\Psi}^{(k)} - [\partial^2 Q(\boldsymbol{\Psi};\, \boldsymbol{\Psi}^{(k)})/\partial\boldsymbol{\Psi}\partial\boldsymbol{\Psi}^T]_{\boldsymbol{\Psi}=\boldsymbol{\Psi}^{(k)}}[\partial Q(\boldsymbol{\Psi};\, \boldsymbol{\Psi}^{(k)})/\partial\boldsymbol{\Psi}]_{\boldsymbol{\Psi}=\boldsymbol{\Psi}^{(k)}} \\ &= \boldsymbol{\Psi}^{(k)} + \mathcal{I}_c(\boldsymbol{\Psi}^{(k)};\, \boldsymbol{y})\boldsymbol{S}(\boldsymbol{y};\, \boldsymbol{\Psi}^{(k)}).\end{aligned} \tag{4.134}$$

If the Newton–Raphson method is applied to find a zero of the incomplete-data likelihood equation directly, we have that

$$\boldsymbol{\Psi}^{(k+1)} = \boldsymbol{\Psi}^{(k)} + \boldsymbol{I}_c(\boldsymbol{\Psi}^{(k)};\, \boldsymbol{y})\boldsymbol{S}(\boldsymbol{y};\, \boldsymbol{\Psi}^{(k)}). \tag{4.135}$$

Thus the use of the EM gradient algorithm can be viewed as using the Newton–Raphson method to find a zero of the likelihood equation, but with the approximation

$$\boldsymbol{I}(\boldsymbol{\Psi}^{(k)};\, \boldsymbol{y}) \approx \mathcal{I}_c(\boldsymbol{\Psi}^{(k)};\, \boldsymbol{y}). \tag{4.136}$$

The quasi-Newton acceleration procedure proposed by Lange (1995b) defines $\boldsymbol{\Psi}^{(k+1)}$ to be

$$\boldsymbol{\Psi}^{(k+1)} = \boldsymbol{\Psi}^{(k)} + \{\mathcal{I}_c(\boldsymbol{\Psi}^{(k)};\, \boldsymbol{y}) + \boldsymbol{B}^{(k)}\}^{-1}\boldsymbol{S}(\boldsymbol{y};\, \boldsymbol{\Psi}^{(k)}). \tag{4.137}$$

We have from (3.48) that

$$\boldsymbol{I}(\boldsymbol{\Psi}^{(k)};\, \boldsymbol{y}) = \mathcal{I}_c(\boldsymbol{\Psi}^{(k)};\, \boldsymbol{y}) - \mathcal{I}_m(\boldsymbol{\Psi}^{(k)};\, \boldsymbol{y}), \tag{4.138}$$

where from (3.5) and (3.50),

$$\begin{aligned}\mathcal{I}_m(\boldsymbol{\Psi}^{(k)};\, \boldsymbol{y}) &= -E_{\boldsymbol{\Psi}^{(k)}}\{[\partial^2 \log k(\boldsymbol{X} \mid \boldsymbol{y};\, \boldsymbol{\Psi})/\partial\boldsymbol{\Psi}\partial\boldsymbol{\Psi}^T]_{\boldsymbol{\Psi}=\boldsymbol{\Psi}^{(k)}} \mid \boldsymbol{y}\} \\ &= -[\partial^2 H(\boldsymbol{\Psi};\, \boldsymbol{\Psi}^{(k)})/\partial\boldsymbol{\Psi}\partial\boldsymbol{\Psi}^T]_{\boldsymbol{\Psi}=\boldsymbol{\Psi}^{(k)}}.\end{aligned} \tag{4.139}$$

Thus the presence of the term $\boldsymbol{B}^{(k)}$ in (4.137) can be viewed as an attempt to approximate the Hessian of $H(\boldsymbol{\Psi};\, \boldsymbol{\Psi}^{(k)})$ at the point $\boldsymbol{\Psi} = \boldsymbol{\Psi}^{(k)}$.

Lange (1995b) takes $\boldsymbol{B}^{(k)}$ to be based on Davidon's (1959) symmetric, rank-one update defined by

$$\boldsymbol{B}^{(k)} = \boldsymbol{B}^{(k-1)} + c^{(k)}\boldsymbol{v}^{(k)}\boldsymbol{v}^{(k)^T} \tag{4.140}$$

and where the constant $c^{(k)}$ and the vector $\boldsymbol{v}^{(k)}$ are specified as

$$c^{(k)} = -1/(\boldsymbol{v}^{(k)^T}\boldsymbol{d}^{(k)}), \tag{4.141}$$

$$\boldsymbol{v}^{(k)} = \boldsymbol{h}^{(k)} + \boldsymbol{B}^{(k-1)}\boldsymbol{d}^{(k)}. \tag{4.142}$$

Here

$$\boldsymbol{d}^{(k)} = \boldsymbol{\Psi}^{(k)} - \boldsymbol{\Psi}^{(k-1)} \tag{4.143}$$

and

$$\boldsymbol{h}^{(k)} = [\partial H(\boldsymbol{\Psi};\, \boldsymbol{\Psi}^{(k)})/\partial \boldsymbol{\Psi}]_{\boldsymbol{\Psi}=\boldsymbol{\Psi}^{(k-1)}}. \tag{4.144}$$

Lange (1995b) suggests taking $\boldsymbol{B}^{(0)} = \boldsymbol{I}_d$, which corresponds to initially performing an EM gradient step. When the inner product on the denominator on the right-hand side of (4.141) is zero or is very small relative to

$$\|\boldsymbol{h}^{(k)} + \boldsymbol{B}^{(k-1)}\boldsymbol{d}^{(k)}\| \,.\, \|\boldsymbol{d}^{(k)}\|,$$

Lange (1995b) suggests omitting the update and taking $\boldsymbol{B}^{(k)} = \boldsymbol{B}^{(k-1)}$.

When the matrix

$$\mathcal{I}_c(\boldsymbol{\Psi}^{(k)};\, \boldsymbol{y}) + \boldsymbol{B}^{(k)}$$

fails to be positive definite, Lange (1995b) suggests replacing it by

$$\mathcal{I}_c(\boldsymbol{\Psi}^{(k)};\, \boldsymbol{y}) + (\tfrac{1}{2})^m \boldsymbol{B}^{(k)}, \tag{4.145}$$

where m is the smallest positive integer such that (4.145) is positive definite.

In view of the identities in $\boldsymbol{\Psi}$ and $\boldsymbol{\Psi}^{(k)}$, namely that

$$\partial H(\boldsymbol{\Psi};\, \boldsymbol{\Psi}^{(k)})/\partial \boldsymbol{\Psi} + \partial \log L(\boldsymbol{\Psi})/\partial \boldsymbol{\Psi} = \partial Q(\boldsymbol{\Psi};\, \boldsymbol{\Psi}^{(k)})/\partial \boldsymbol{\Psi}$$

and that

$$[\partial H(\boldsymbol{\Psi};\, \boldsymbol{\Psi}^{(k)})/\partial \boldsymbol{\Psi}]_{\boldsymbol{\Psi}=\boldsymbol{\Psi}^{(k)}=\boldsymbol{0}},$$

we can express $\boldsymbol{h}^{(k)}$ as

$$\begin{aligned} \boldsymbol{h}^{(k)} &= [\partial Q(\boldsymbol{\Psi};\, \boldsymbol{\Psi}^{(k)})/\partial \boldsymbol{\Psi}]_{\boldsymbol{\Psi}=\boldsymbol{\Psi}^{(k-1)}} - \partial \log L(\boldsymbol{\Psi}^{(k-1)})/\partial \boldsymbol{\Psi} \\ &= [\partial Q(\boldsymbol{\Psi};\, \boldsymbol{\Psi}^{(k)})/\partial \boldsymbol{\Psi}]_{\boldsymbol{\Psi}=\boldsymbol{\Psi}^{(k-1)}} - [\partial Q(\boldsymbol{\Psi};\, \boldsymbol{\Psi}^{(k-1)})/\partial \boldsymbol{\Psi}]_{\boldsymbol{\Psi}=\boldsymbol{\Psi}^{(k-1)}}. \end{aligned}$$

It can be seen that almost all the relevant quantities for the quasi-Newton acceleration of the EM algorithm can be expressed in terms of $Q(\boldsymbol{\Psi};\, \boldsymbol{\Psi}^{(k)})$ and its derivatives. The only relevant quantity not so expressible is the likelihood $L(\boldsymbol{\Psi})$. The latter needs to be computed if the progress of the algorithm is to be monitored. If it is found that (4.137) modified according to (4.145) overshoots at any given iteration, then some form of step-decrementing can be used to give an increase in $L(\boldsymbol{\Psi})$. Lange (1995b) follows Powell's (1978) suggestion and fits a quadratic to the function in r, $\log L(\boldsymbol{\Psi}(r))$, through the values $L(\boldsymbol{\Psi}^{(k)})$ and $L(\boldsymbol{\Psi}^{(k+1)})$ with slope

$$\{\partial \log L(\boldsymbol{\Psi}^{(k)})/\partial \boldsymbol{\Psi}\}^T \boldsymbol{d}^{(k)}$$

at $r = 0$, where

$$\boldsymbol{\Psi}(r) = \boldsymbol{\Psi}^{(k)} + r\,\boldsymbol{d}^{(k)}.$$

If the maximum of the quadratic occurs at $r_{\max}$, then one steps back with

$$r = \max\{r_{\max}, 0.1\}.$$

If this procedure still does not yield an increase in $L(\boldsymbol{\Psi})$, then it can be repeated. Lange (1995b) notes that one or two step decrements invariably give the desired increase in $L(\boldsymbol{\Psi})$. Because the algorithm moves uphill, step decrementing is bound to succeed. Lange (1995b) surmises that a step-halving strategy would be equally effective.

As Lange (1995b) points out, the early stages of the algorithm resemble a close approximation to the EM algorithm, later stages approximate Newton–Raphson, and intermediate stages make a graceful transition between the two extremes.

4.13.2 Example 4.8: Dirichlet Distribution

We now consider the example given by Lange (1995b) on ML estimation of the parameters of the Dirichlet distribution. This distribution is useful in modeling data on proportions (Kingman, 1993). We let $U_1, \ldots, U_m$ be m independent random variables with U_i having a gamma $(\theta_i, 1)$ density function,

$$f(u;\, \alpha, 1) = \{u^{\theta_i - 1}/\Gamma(\theta_i)\} \exp(-u) I_{(0,\infty)}(u)\,; \quad (\theta_i > 0),$$

for $i = 1, \ldots, m$.

Put

$$W_i = U_i \Big/ \sum_{h=1}^{m} U_h \quad (i = 1, \ldots, m). \tag{4.146}$$

Then the random vector

$$\boldsymbol{W} = (W_1, \ldots, W_m)^T$$

has the Dirichlet density

$$\frac{\Gamma(\sum_{i=1}^{m} \theta_i)}{\prod_{i=1}^{m} \Gamma(\theta_i)} \prod_{i=1}^{m} w_i^{\theta_i - 1} \tag{4.147}$$

over the simplex

$$\left\{ \boldsymbol{w} : w_i > 0 \, (i = 1, \ldots, m);\, \sum_{i=1}^{m} w_i = 1 \right\}.$$

It can be seen that (4.147) is a member of the regular exponential family.

Let $\boldsymbol{w}_1, \ldots, \boldsymbol{w}_n$ denote the observed values of a random sample of size n from the Dirichlet distribution (4.147). The problem is to find the MLE of the parameter vector

$$\boldsymbol{\Psi} = (\theta_1, \ldots, \theta_m)^T$$

on the basis of the observed data

$$\boldsymbol{y} = (\boldsymbol{w}_1^T, \ldots, \boldsymbol{w}_n^T)^T.$$

An obvious choice for the complete-data vector is

$$\boldsymbol{x} = (\boldsymbol{u}_1^T, \ldots, \boldsymbol{u}_n^T)^T,$$

where

$$\boldsymbol{u}_j = (u_{1j}, \ldots, u_{mj})^T$$

and where corresponding to (4.147), the u_{ij} are defined by

$$w_{ij} = u_{ij} \Big/ \sum_{h=1}^{m} u_{hj} \quad (i = 1, \ldots, m;\ j = 1, \ldots, n),$$

and $w_{ij} = (\boldsymbol{w}_j)_i$.

The log likelihood function is given by

$$\log L(\boldsymbol{\Psi}) = \sum_{i=1}^{m}\sum_{j=1}^{n}(\theta_i - 1)\log w_{ij} + n \log \Gamma\left(\sum_{i=1}^{m} \theta_i\right) - n \sum_{i=1}^{m} \log \Gamma(\theta_i), \tag{4.148}$$

while the complete-data log likelihood is

$$\log L_c(\boldsymbol{\Psi}) = -\sum_{i=1}^{m}\sum_{j=1}^{n}\{(\theta_i - 1)\log u_{ij} - u_{ij}\} - n \sum_{i=1}^{m} \log \Gamma(\theta_i). \tag{4.149}$$

On the E-step at the $(k+1)$th iteration of the EM algorithm applied to this problem, we have that

$$\begin{aligned} Q(\boldsymbol{\Psi}; \boldsymbol{\Psi}^{(k)}) = & \sum_{i=1}^{m}\sum_{j=1}^{n}\left\{(\theta_i - 1)\sum_{i=1}^{m} E_{\boldsymbol{\Psi}^{(k)}}(\log U_{ij} \mid \boldsymbol{w}_j) - E_{\boldsymbol{\Psi}^{(k)}}(U_{ij} \mid \boldsymbol{w}_j)\right\} \\ & -n \sum_{i=1}^{m} \log \Gamma(\theta_i), \end{aligned} \tag{4.150}$$

It can be seen from (4.150) that, in order to carry out the M-step, we need to calculate the term

$$E_{\boldsymbol{\Psi}^{(k)}}(\log U_{ij} \mid \boldsymbol{w}_j).$$

Now

$$u_{ij} = w_{ij} \sum_{h=1}^{m} u_{hj},$$

and so

$$E_{\boldsymbol{\Psi}^{(k)}}(\log U_{ij} \mid \boldsymbol{w}_j) = \log w_{ij} + E_{\boldsymbol{\Psi}^{(k)}}(\log U_{.j} \mid \boldsymbol{w}_j), \tag{4.151}$$

where

$$U_{.j} = \sum_{h=1}^{m} U_{hj}$$

is distributed independently of $\boldsymbol{w}_j$ according to a gamma $(\sum_{h=1}^{m} \theta_h,\ 1)$ density. Thus in order to compute the second term on the right-side of (4.151), we need the result that if a random variable R has a gamma $(\alpha,\ \beta)$ distribution, then

$$E(\log R) = \psi(\alpha) - \log \beta, \tag{4.152}$$

where

$$\begin{aligned}\psi(s) &= \partial \log \Gamma(s)/\partial s \\ &= \{\partial \Gamma(s)/\partial s\}/\Gamma(s)\end{aligned}$$

is the Digamma function. On using this result, we have from (4.151) that

$$E_{\boldsymbol{\Psi}^{(k)}}(\log U_{ij} \mid \boldsymbol{w}_j) = \log w_{ij} + \psi\left(\sum_{h=1}^{m} \theta_h^{(k)}\right). \tag{4.153}$$

Lange (1995b) notes that the calculation of the term (4.151) can be avoided if one make uses of the identity,

$$\boldsymbol{S}(\boldsymbol{y};\ \boldsymbol{\Psi}^{(k)}) = [\partial Q(\boldsymbol{\Psi};\ \boldsymbol{\Psi}^{(k)})/\partial \boldsymbol{\Psi}]_{\boldsymbol{\Psi}=\boldsymbol{\Psi}^{(k)}}.$$

On evaluating $S_i(\boldsymbol{y};\ \boldsymbol{\Psi}^{(k)})$, the derivative of (4.148) with respect to θ_i at the point $\boldsymbol{\Psi} = \boldsymbol{\Psi}^{(k)}$, we have that

$$\begin{aligned}S_i(\boldsymbol{y};\ \boldsymbol{\Psi}^{(k)}) &= \partial \log L(\boldsymbol{\Psi}^{(k)})/\partial \theta_i \\ &= \sum_{j=1}^{n} \log w_{ij} + n \partial \log \Gamma\left(\sum_{h=1}^{m} \theta_h^{(k)}\right) \Big/ \partial \theta_i - n \partial \log \Gamma(\theta_i^{(k)})/\partial \theta_i \\ &= \sum_{j=1}^{n} \log w_{ij} + n \psi\left(\sum_{h=1}^{m} \theta_h^{(k)}\right) - n \psi(\theta_i^{(k)})\end{aligned} \tag{4.154}$$

for $i = 1, \ldots, m$. On equating $S_i(\boldsymbol{y};\, \boldsymbol{\Psi}^{(k)})$ equal to the derivative of (4.150) with respect to θ_i at the point $\boldsymbol{\Psi} = \boldsymbol{\Psi}^{(k)}$, we obtain

$$\sum_{j=1}^{n} E_{\boldsymbol{\Psi}^{(k)}}(\log U_{ij} \mid \boldsymbol{w}_j) = S_i(\boldsymbol{y};\, \boldsymbol{\Psi}^{(k)}) + n\psi(\theta_i^{(k)})$$

$$= \sum_{j=1}^{n} \log w_{ij} + n\psi\left(\sum_{h=1}^{m} \theta_h^{(k)}\right), \qquad (4.155)$$

which agrees with sum of the right-hand side of (4.153) as obtained above by working directly with the conditional distribution of $\log U_{ij}$ given $\boldsymbol{w}_j$.

Concerning now the M-step, it can be seen that the presence of terms like $\log\Gamma(\theta_i)$ in (4.150) prevent a closed-form solution for $\boldsymbol{\Psi}^{(k+1)}$. Lange (1995b) notes that the EM gradient algorithm is easy to implement for this problem. The components of the score statistic $\boldsymbol{S}(\boldsymbol{y};\, \boldsymbol{\Psi}^{(k)})$ are available from (4.154). The information matrix $\mathcal{I}_c(\boldsymbol{\Psi}^{(k)};\, \boldsymbol{y})$, which equals $\mathcal{I}_c(\boldsymbol{\Psi}^{(k)})$, since the complete-data density belongs to the regular exponential family with natural parameter $\boldsymbol{\Psi}$, is diagonal with ith diagonal element

$$-n\partial^2 \log\Gamma(\theta_i)/\partial\theta_i^2 \quad (i = 1, \ldots, m),$$

which is positive as $\log\Gamma(u)$ is a strictly concave function.

Lange (1995b) applied the EM gradient and the quasi-Newton accelerated EM algorithms to a data set from Mosimann (1962) on the relative frequencies of $m = 3$ serum proteins in $n = 23$ young white Pekin ducklings. Starting from $\boldsymbol{\Psi}^{(0)} = (1, 1, 1)^T$, Lange (1995b) reports that the three algorithms converge smoothly to the point $\hat{\boldsymbol{\Psi}} = (3.22, 20.38, 21.69)^T$, with the log likelihood showing a steady increase along the way. The EM gradient algorithm took

Table 4.9. Performance of Accelerated EM on the Dirichlet Distribution Data of Mosimann (1962).

Iteration	Extra	Exponent	$L(\boldsymbol{\theta})$	θ_1	θ_2	θ_3
1	0	0	15.9424	1.000	1.000	1.000
2	0	0	24.7300	0.2113	1.418	1.457
3	0	0	41.3402	0.3897	2.650	2.760
4	0	1	49.1425	0.6143	3.271	3.445
5	0	1	53.3627	0.8222	3.827	4.045
6	0	0	73.0122	3.368	22.19	23.59
7	0	2	73.0524	3.445	22.05	23.47
8	0	0	73.1250	3.217	20.40	21.70
9	0	0	73.1250	3.217	20.39	21.69
10	0	0	73.1250	3.125	20.38	21.69

Source: Adapted from Lange (1995b).

287 iterations, while the quasi-Newton accelerated EM algorithm took eight iterations. Lange (1995b) notes that the scoring algorithm is an attractive alternative for this problem (Narayanan, 1991), and on application to this data set, it took nine iterations.

Table 4.9 gives some details on the performance of the quasi-Newton acceleration of the EM algorithm for this problem. The column headed "Exponent" refers to the minimum nonnegative integer m required to make

$$\mathcal{I}_c(\boldsymbol{\Psi}^{(k)}; \boldsymbol{y}) + (\tfrac{1}{2})^m \boldsymbol{B}^{(k)}, \tag{4.156}$$

positive definite. The column headed "Extra" refers to the number of step decrements taken in order to produce an increase in $L(\boldsymbol{\Psi})$ at a given iteration. It can be seen that in this problem, step decrementing was not necessary.

CHAPTER 5

Extensions of the EM Algorithm

5.1 INTRODUCTION

In this chapter, we consider some extensions of the EM algorithm. In particular, we focus on the ECM and ECME algorithms. The ECM algorithm as proposed by Meng and Rubin (1993), is a natural extension of the EM algorithm in situations where the maximization process on the M-step is relatively simple when conditional on some function of the parameters under estimation. The ECM algorithm therefore replaces the M-step of the EM algorithm by a number of computationally simpler conditional maximization (CM) steps. As a consequence, it typically converges more slowly than the EM algorithm in terms of number of iterations, but can be faster in total computer time. More importantly, the ECM algorithm preserves the appealing convergence properties of the EM algorithm, such as its monotone convergence.

Liu and Rubin (1994) propose the ECME algorithm, which is an extension of the ECM algorithm. They find it to be nearly always faster than both the EM and ECM algorithms in terms of number of iterations and moreover that it can be faster in total computer time by orders of magnitude. This improvement in speed of convergence is obtained by conditionally maximizing on some or all of the CM-steps the actual (that is, the incomplete-data) log likelihood rather than a current approximation to it as given by the Q-function with the EM and ECM algorithms. Thus in general it is more tedious to code than the ECM algorithm, but the potential gain in faster convergence allows convergence to be more easily assessed. As with the EM and ECM algorithms, the ECME algorithm monotonically increases the likelihood and reliably converges to a local maximizer of the likelihood function. Several illustrative examples of the ECM and ECME algorithms are given in this chapter. A further development to be considered is the AECM algorithm of Meng and van Dyk (1995), which is obtained by combining the ECME algorithm with the Space-Alternating Generalized EM (SAGE) algorithm of Fessler and Hero (1994). It allows the specification of the complete data to vary where necessary over the CM-steps. We consider also in this chapter the other proposal of Meng and van Dyk (1995), which concerns speeding up convergence of the EM algorithm through the choice of the complete data.

5.2 ECM ALGORITHM

5.2.1 Motivation

As noted earlier, one of the major reasons for the popularity of the EM algorithm is that the M-step involves only complete-data ML estimation, which is often computationally simple. But if the complete-data ML estimation is rather complicated, then the EM algorithm is less attractive because the M-step is computationally unattractive. In many cases, however, complete-data ML estimation is relatively simple if maximization is undertaken conditional on some of the parameters (or some functions of the parameters). To this end, Meng and Rubin (1993) introduce a class of generalized EM algorithms, which they call the ECM algorithm for expectation–conditional maximization algorithm. The ECM algorithm takes advantage of the simplicity of complete-data conditional maximization by replacing a complicated M-step of the EM algorithm with several computationally simpler CM-steps. Each of these CM-steps maximizes the conditional expectation of the complete-data log likelihood function found in the preceding E-step subject to constraints on $\boldsymbol{\Psi}$, where the collection of all constraints is such that the maximization is over the full parameter space of $\boldsymbol{\Psi}$.

A CM-step might be in closed form or it might itself require iteration, but because the CM maximizations are over smaller dimensional spaces, often they are simpler, faster, and more stable than the corresponding full maximizations called for on the M-step of the EM algorithm, especially when iteration is required.

5.2.2 Formal Definition

To define formally the ECM algorithm, we suppose that the M-step is replaced by $S > 1$ steps. We let $\boldsymbol{\Psi}^{(k+s/S)}$ denote the value of $\boldsymbol{\Psi}$ on the sth CM-step of the $(k+1)$th iteration, where $\boldsymbol{\Psi}^{(k+s/S)}$ is chosen to maximize

$$Q(\boldsymbol{\Psi}; \boldsymbol{\Psi}^{(k)})$$

subject to the constraint

$$\boldsymbol{g}_s(\boldsymbol{\Psi}) = \boldsymbol{g}_s(\boldsymbol{\Psi}^{(k+(s-1)/S)}). \tag{5.1}$$

Here $C = \{\boldsymbol{g}_s(\boldsymbol{\Psi}), s = 1, \ldots, S\}$ is a set of S preselected (vector) functions. Thus $\boldsymbol{\Psi}^{(k+s/S)}$ satisfies

$$Q(\boldsymbol{\Psi}^{(k+s/S)}; \boldsymbol{\Psi}^{(k)}) \geq Q(\boldsymbol{\Psi}; \boldsymbol{\Psi}^{(k)}) \quad \text{for all } \boldsymbol{\Psi} \in \boldsymbol{\Omega}_s(\boldsymbol{\Psi}^{(k+(s-1)/S)}), \tag{5.2}$$

where

$$\boldsymbol{\Omega}_s(\boldsymbol{\Psi}^{(k+(s-1)/S)}) \equiv \{\boldsymbol{\Psi} \in \boldsymbol{\Omega} : \boldsymbol{g}_s(\boldsymbol{\Psi}) = \boldsymbol{g}_s(\boldsymbol{\Psi}^{(k+(s-1)/S)})\}. \tag{5.3}$$

The value of $\boldsymbol{\Psi}$ on the final CM-step, $\boldsymbol{\Psi}^{(k+S/S)} = \boldsymbol{\Psi}^{(k+1)}$, is taken to be the input on the $(k+2)$th iteration.

From (5.2), we have that

$$\begin{aligned} Q(\boldsymbol{\Psi}^{(k+1)}; \boldsymbol{\Psi}^{(k)}) &\geq Q(\boldsymbol{\Psi}^{(k+(S-1)/S)}; \boldsymbol{\Psi}^{(k)}) \\ &\geq Q(\boldsymbol{\Psi}^{(k+(S-2)/S)}; \boldsymbol{\Psi}^{(k)}) \\ &\;\;\vdots \\ &\geq Q(\boldsymbol{\Psi}^{(k)}; \Psi^{(k)}). \end{aligned} \tag{5.4}$$

This shows that the ECM algorithm is a GEM algorithm and so possesses its desirable convergence properties. As noted before in Section 3.1, the inequality (5.4) is a sufficient condition for

$$L(\boldsymbol{\Psi}^{(k+1)}) \geq L(\boldsymbol{\Psi}^{(k)})$$

to hold.

Under the assumption that $\boldsymbol{g}_s(\boldsymbol{\Psi}), s = 1, \ldots, S$, is differentiable and that the corresponding gradient $\boldsymbol{\nabla g}_s(\boldsymbol{\Psi})$ is of full rank at $\boldsymbol{\Psi}^{(k)}$, for all k, almost all of the convergence properties of the EM established in DLR and Wu (1983) hold. The only extra condition needed is the "space-filling" condition:

$$\bigcap_{s=1}^{S} G_s(\boldsymbol{\Psi}^{(k)}) = \{\boldsymbol{0}\} \quad \text{for all } k, \tag{5.5}$$

where $G_s(\boldsymbol{\Psi})$ is the column space of $\boldsymbol{\nabla g}_s(\boldsymbol{\Psi})$; that is,

$$\boldsymbol{G}_s(\boldsymbol{\Psi}) = \{\boldsymbol{\nabla g}_s(\boldsymbol{\Psi})\boldsymbol{\eta} : \boldsymbol{\eta} \in \mathbb{R}^{d_s}\}$$

and d_s is the dimensionality of the vector function $\boldsymbol{g}_s(\boldsymbol{\Psi})$. By taking the complement of both sides of (5.5), this condition is equivalent to saying that at any $\boldsymbol{\Psi}^{(k)}$, the convex hull of all feasible directions determined by the constraint spaces $\boldsymbol{\Omega}_s(\boldsymbol{\Psi}^{(k+(s-1)/S)}), s = 1, \ldots, S$, is the whole Euclidean space $\mathbb{R}^d$, and thus the resulting maximization is over the whole parameter space $\boldsymbol{\Omega}$ and not a subspace of it. Note that the EM algorithm is a special case of the ECM algorithm with $S = 1$ and $g_1(\boldsymbol{\Psi}) \equiv$ constant (that is, no constraint), whereby (5.5) is automatically satisfied because

$$\boldsymbol{\nabla} g_1(\boldsymbol{\Psi}) \equiv \boldsymbol{0}.$$

In many applications of the ECM algorithm, the S CM-steps correspond to the situation where the parameter vector $\boldsymbol{\Psi}$ is partitioned into S subvectors,

$$\boldsymbol{\Psi} = (\boldsymbol{\Psi}_1^T, \ldots, \boldsymbol{\Psi}_S^T)^T.$$

The sth CM-step then requires the maximization of the Q-function with respect to the sth subvector $\boldsymbol{\Psi}_s$ with the other $(S-1)$ subvectors held fixed at their current values; that is, $\boldsymbol{g}_s(\boldsymbol{\Psi})$ is the vector containing all the subvectors of $\boldsymbol{\Psi}$ except $\boldsymbol{\Psi}_s$ $(s = 1, \ldots, S)$. This is the situation in two of the three illustrative examples that shall be given shortly. In each of these examples, the M-step is not simple, being iterative, and it is replaced by a number of CM-steps which either exist in closed form or require a lower dimensional search.

5.2.3 Convergence Properties

In the case where the data are complete so that the E-step becomes an identity operation, that is,

$$Q(\boldsymbol{\Psi}; \boldsymbol{\Psi}^{(k)}) \equiv L(\boldsymbol{\Psi}),$$

an ECM algorithm becomes a CM algorithm. Meng and Rubin (1993) mention two related issues. First, if the set of constraint functions C is not space-filling, then the CM algorithm will converge to a stationary point of the likelihood in a subspace of $\boldsymbol{\Omega}$, which may or may not be a stationary point of the likelihood in the whole parameter space. Second, since the space-filling condition on C does not involve data, one would expect that if C leads to appropriate convergence of a CM algorithm with complete data, it should also lead to appropriate convergence of an ECM algorithm with incomplete data. This conjecture is established rigorously by Meng and Rubin (1993) when the complete-data density is from an exponential family, where the ECM algorithm is especially useful. The advantage of this is that it enables one to conclude that the ECM algorithm will converge appropriately whenever the CM does so. For instance, one can immediately conclude the appropriate convergence of the ECM algorithm in Example 5.3 to follow without having to verify the space-filling condition, because the monotone convergence of Iterative Proportional Fitting with complete data has been established (Bishop, Fienberg, and Holland, 1975, Chapter 3).

Meng and Rubin (1993) show that if all the conditional maximizations of an ECM algorithm are unique, then all the limit points of any ECM sequence $\{\boldsymbol{\Psi}^{(k)}\}$ are stationary points of $L(\boldsymbol{\Psi})$ if C is space-filling at all $\boldsymbol{\Psi}^{(k)}$. The assumption that all conditional maximizations are unique is very weak in the sense that it is satisfied in many practical problems. But even this condition can be eliminated; see Meng and Rubin (1993, page 275). They also provide the formal work that shows that the ECM algorithm converges to a stationary point under essentially the same conditions that guarantee the convergence of the EM algorithm as considered in Chapter 3. Intuitively, suppose that the ECM algorithm has converged to some limit point $\boldsymbol{\Psi}^*$ and that the required derivatives of the Q-function are all well defined. Then the stationarity of each CM-step implies that the corresponding directional derivatives of Q at $\boldsymbol{\Psi}^*$ are zero, which, under the space-filling condition on the set C of constraints, implies that

$$[\partial Q(\boldsymbol{\Psi}; \boldsymbol{\Psi}^*)/\partial \boldsymbol{\Psi}]_{\boldsymbol{\Psi}=\boldsymbol{\Psi}^*} = \mathbf{0},$$

just as with the M-step of the EM algorithm. Thus, as with the theory of the EM algorithm, if the ECM algorithm converges to $\boldsymbol{\Psi}^*$, then $\boldsymbol{\Psi}^*$ must be a stationary point of the likelihood function $L(\boldsymbol{\Psi})$.

5.2.4 Speed of Convergence

From (3.66) and (3.70), the (matrix) speed of convergence of the EM algorithm in a neighborhood of a limit point $\boldsymbol{\Psi}^*$ is given by

$$\boldsymbol{S}_{\text{EM}} = \boldsymbol{I}_d - \boldsymbol{J}(\boldsymbol{\Psi}^*) \tag{5.6}$$

$$= \boldsymbol{\mathcal{I}}_c^{-1}(\boldsymbol{\Psi}^*; \boldsymbol{y})\boldsymbol{I}(\boldsymbol{\Psi}^*; \boldsymbol{y}), \tag{5.7}$$

and its global speed of convergence s_{EM} by the smallest eigenvalue of (5.7). Under mild conditions, Meng (1994) shows that

$$\text{speed of ECM} = \text{speed of EM} \times \text{speed of CM} ; \tag{5.8}$$

that is,

$$\boldsymbol{S}_{\text{ECM}} = \boldsymbol{S}_{\text{EM}}\boldsymbol{S}_{\text{CM}}, \tag{5.9}$$

where $\boldsymbol{S}_{\text{CM}}$ and $\boldsymbol{S}_{\text{ECM}}$ denote the speed of the CM and ECM algorithms, respectively, corresponding to (5.7). This result is consistent with intuition, since an ECM iteration can be viewed as a composition of two linear iterations, EM and CM.

Let s_{ECM} and s_{CM} denote the global speed of convergence of the ECM and CM algorithms, respectively. As Meng (1994) comments, although it would be naive to expect from (5.9) that

$$s_{\text{ECM}} = s_{\text{EM}}s_{\text{CM}}, \tag{5.10}$$

it seems intuitive to expect that

$$s_{\text{EM}}s_{\text{CM}} \leq s_{\text{ECM}} \leq s_{\text{EM}}. \tag{5.11}$$

Because the increase in the Q-function is less with the ECM algorithm than with the EM algorithm, it would appear reasonable to expect that the former algorithm converges more slowly in terms of global speed, which would imply the upper bound on s_{ECM} in (5.11). But intuition suggests that the slowest convergence of the ECM algorithm occurs when the EM and CM algorithms share a slowest component. In this case, the speed of the ECM algorithm is the product of the speeds of the EM and CM algorithms, leading to the lower bound on s_{ECM} in (5.11). However, neither of the inequalities in (5.11) holds in general, as Meng (1994) uses a simple bivariate normal example to provide counterexamples to both inequalities.

5.2.5 Discussion

The CM algorithm is a special case of the cyclic coordinate ascent method for function maximization in the optimization literature; see, for example, Zangwill (1969, Chapter 5). It can also be viewed as the Gauss-Seidel iteration method applied in an appropriate order to the likelihood equation (for example, Thisted (1988, Chapter 4)). Although these optimization methods are well known for their simplicity and stability, because they typically converge only linearly, they have been less preferred in practice for handling complete-data problems than superlinear methods like quasi-Newton. When used for the M-step of the EM algorithm or a CM-step of the ECM algorithm, however, simple and stable linear converging methods are often more suitable than superlinear converging but less stable algorithms. The reasons are first, that the advantage of superlinear convergence in each M- or CM-step does not transfer to the overall convergence of the EM or ECM algorithms, since the EM and ECM algorithms always converge linearly regardless of the maximization method employed within the maximization step, and secondly, that the stability of the maximization method is critical for preserving the stability of the EM or ECM algorithms since it is used repeatedly within each maximization step in all iterations. Finally, if one performs just one iteration of a superlinear converging algorithm within each M-step of the EM algorithm, then the resulting algorithm is no longer guaranteed to increase the likelihood monotonically.

As emphasized by Meng (1994), although the ECM algorithm may converge slower than the EM algorithm in terms of global speed of convergence, it does not necessarily imply that it takes longer to converge in real time since the actual time in running the M-step and the CM-steps is not taken into account in the speed of convergence. This time can be quite different, especially if the M-step requires iteration.

5.3 MULTICYCLE ECM ALGORITHM

In many cases, the computation of an E-step may be much cheaper than the computation of the CM-steps. Hence one might wish to perform one E-step before each CM-step or a few selected CM-steps. For descriptive simplicity, we focus here on the case with an E-step preceding each CM-step. A cycle is defined to be one E-step followed by one CM-step. Meng and Rubin (1993) called the corresponding algorithm a multicycle ECM.

For instance, consider an ECM algorithm with $S = 2$ CM-steps. Then an E-step would be performed between the two CM-steps. That is, after the first CM-step, an E-step is performed, by which the Q-function is updated from

$$Q(\boldsymbol{\Psi}; \boldsymbol{\Psi}^{(k)}) \tag{5.12}$$

to

$$Q(\boldsymbol{\Psi}; \boldsymbol{\Psi}^{(k+1/2)}). \tag{5.13}$$

The second CM-step of this multicycle ECM algorithm is then undertaken where now $\boldsymbol{\Psi}^{(k+2/S)} = \boldsymbol{\Psi}^{(k+2)}$ is chosen to maximize conditionally (5.13) instead of (5.12).

Since the second argument in the Q-function is changing at each cycle within each iteration, a multicycle ECM may not necessarily be a GEM algorithm; that is, the inequality (5.4) may not hold. However, it is not difficult to show that the likelihood function is not decreased after a multicycle ECM iteration, and hence, after each ECM iteration. To see this, we have that the definition of the sth CM-step implies

$$Q(\boldsymbol{\Psi}^{(k+s/S)}; \boldsymbol{\Psi}^{(k+(s-1)/S)}) \geq Q(\boldsymbol{\Psi}^{(k+(s-1)/S)}; \boldsymbol{\Psi}^{(k+(s-1)/S)}),$$

which we have seen is sufficient to establish that

$$L(\boldsymbol{\Psi}^{(k+s/S)}) \geq L(\boldsymbol{\Psi}^{(k+(s-1)/S)}).$$

Hence the multicycle ECM algorithm monotonically increases the likelihood function $L(\boldsymbol{\Psi})$ after each cycle, and hence, after each iteration. The convergence results of the ECM algorithm apply to a multicycle version of it. An obvious disadvantage of using a multicycle ECM algorithm is the extra computation at each iteration. Intuitively, as a tradeoff, one might expect it to result in larger increases in the log likelihood function per iteration since the Q-function is being updated more often. Meng and Rubin (1993) report that practical implementations do show this potential, but it is not true in general. They note that there are cases where the multicycle ECM algorithm converges more slowly than the ECM algorithm. Details of these examples, which are not typical in practice, are given in Meng (1994).

We shall now give some examples, illustrating the application of the ECM algorithm and its multicycle version. The first few examples show that when the parameters are restricted to particular subspaces, the conditional maximizations either have analytical solutions or require lower dimensional iteration.

5.4 EXAMPLE 5.1: NORMAL MIXTURES WITH EQUAL CORRELATIONS

5.4.1 Normal Components with Equal Correlations

As an example of the application of the ECM algorithm, we consider the fitting of a mixture of two multivariate normal densities with equal correlation matrices to an observed random sample $\boldsymbol{w}_1, \ldots, \boldsymbol{w}_n$. In Section 2.7.2, we considered the application of the EM algorithm to fit a mixture of g normal component densities with unrestricted covariance matrices.

The case of two equal component-correlation matrices can be represented by writing the covariance matrices $\boldsymbol{\Sigma}_1$ and $\boldsymbol{\Sigma}_2$ as

$$\boldsymbol{\Sigma}_1 = \boldsymbol{\Sigma}_0 \tag{5.14}$$

and

$$\boldsymbol{\Sigma}_2 = \boldsymbol{K}\boldsymbol{\Sigma}_0\boldsymbol{K}, \tag{5.15}$$

where $\boldsymbol{\Sigma}_0$ is a positive definite symmetric matrix and

$$\boldsymbol{K} = \mathrm{diag}(\kappa_1, \ldots, \kappa_p).$$

For the application of the ECM algorithm, the vector $\boldsymbol{\Psi}$ of unknown parameters can be partitioned as

$$\boldsymbol{\Psi} = (\boldsymbol{\Psi}_1^T, \boldsymbol{\Psi}_2^T)^T, \tag{5.16}$$

where $\boldsymbol{\Psi}_1$ consists of the mixing proportion π_1, the elements of $\boldsymbol{\mu}_1$ and $\boldsymbol{\mu}_2$, and the distinct elements of $\boldsymbol{\Sigma}_0$, and where $\boldsymbol{\Psi}_2$ contains $\kappa_1, \ldots, \kappa_p$.

5.4.2 Application of ECM Algorithm

The E-step on the $(k+1)$th iteration of this ECM algorithm is the same as for the EM algorithm. In this particular case where the two CM-steps correspond to the maximization of the Q-function with respect to each of the two subvectors in the partition (5.16) of $\boldsymbol{\Psi}$, we can write the values of $\boldsymbol{\Psi}$ obtained after each CM-step as

$$\boldsymbol{\Psi}^{(k+1/2)} = (\boldsymbol{\Psi}_1^{(k+1)^T}, \boldsymbol{\Psi}_2^{(k)^T})^T$$

and

$$\boldsymbol{\Psi}^{(k+1)} = (\boldsymbol{\Psi}_1^{(k+1)^T}, \boldsymbol{\Psi}_2^{(k+1)^T})^T,$$

respectively. The two CM-steps of the ECM algorithm can then be expressed as follows.

CM-Step 1. Calculate $\boldsymbol{\Psi}_1^{(k+1)}$ as the value of $\boldsymbol{\Psi}_1$ that maximizes $Q(\boldsymbol{\Psi};\ \boldsymbol{\Psi}^{(k)})$ with $\boldsymbol{\Psi}_2$ fixed at

$$\boldsymbol{\Psi}_2^{(k)} = (\kappa_1^{(k)}, \ldots, \kappa_p^{(k)})^T.$$

CM-Step 2. Calculate $\boldsymbol{\Psi}_2^{(k+1)}$ as the value of $\boldsymbol{\Psi}_2$ that maximizes $Q(\boldsymbol{\Psi};\ \boldsymbol{\Psi}^{(k)})$ with $\boldsymbol{\Psi}_1$ fixed at $\boldsymbol{\Psi}_1^{(k+1)}$.

The CM-Step 1 can be implemented by proceeding as with the M-step in Section 2.7. There is a slight modification to allow for the fact that the two component-covariance matrices are related by (5.14) and (5.15); that is, have the same correlation structure. It follows that $\boldsymbol{\Psi}_1^{(k+1)}$ consists of $\pi_1^{(k+1)}$, the elements of $\boldsymbol{\mu}_1^{(k+1)}$ and $\boldsymbol{\mu}_2^{(k+1)}$, and the distinct elements of $\boldsymbol{\Sigma}_0^{(k+1)}$, where

$$\pi_i^{(k+1)} = \sum_{j=1}^{n} z_{ij}^{(k)}/n \quad (1 = 1, 2),$$

$$\boldsymbol{\mu}_i^{(k+1)} = \sum_{j=1}^{n} z_{ij}^{(k)} \boldsymbol{w}_j \Big/ \sum_{j=1}^{n} z_{ij}^{(k)} \quad (i = 1, 2),$$

and

$$\boldsymbol{\Sigma}_0^{(k+1)} = \pi_1^{(k+1)} \boldsymbol{V}_1^{(k+1)} + \pi_2^{(k+1)} \boldsymbol{K}^{(k)^{-1}} \boldsymbol{V}_2^{(k+1)} \boldsymbol{K}^{(k)^{-1}},$$

where

$$\boldsymbol{V}_i^{(k+1)} = \sum_{j=1}^{n} z_{ij}^{(k)} (\boldsymbol{w}_j - \boldsymbol{\mu}_i^{(k+1)})(\boldsymbol{w}_j - \boldsymbol{\mu}_i^{(k+1)})^T \Big/ \sum_{j=1}^{n} z_{ij}^{(k)} \quad (i = 1, 2), \tag{5.17}$$

and

$$\boldsymbol{K}^{(k)} = \operatorname{diag}(\kappa_1^{(k)}, \ldots, \kappa_p^{(k)})^T.$$

As in Section 2.7,

$$z_{ij}^{(k)} = \tau_i(\boldsymbol{w}_j;\ \boldsymbol{\Psi}^{(k)}) \quad (i = 1, 2),$$

but here now

$$\tau_1(\boldsymbol{w}_j;\ \boldsymbol{\Psi}^{(k)}) = \frac{\pi_1^{(k)} \phi(\boldsymbol{w}_j;\ \boldsymbol{\mu}_1^{(k)}, \boldsymbol{\Sigma}_0^{(k)})}{\pi_1^{(k)} \phi(\boldsymbol{w}_j;\ \mu_1^{(k)}, \boldsymbol{\Sigma}_0^{(k)}) + \pi_2^{(k)} \phi(\boldsymbol{w}_j;\ \mu_2^{(k)}, \boldsymbol{\Sigma}_2^{(k)})},$$

where

$$\boldsymbol{\Sigma}_2^{(k)} = \boldsymbol{K}^{(k)} \boldsymbol{\Sigma}_0^{(k)} \boldsymbol{K}^{(k)},$$

and

$$\tau_2(\boldsymbol{w}_j;\ \boldsymbol{\Psi}^{(k)}) = 1 - \tau_1(\boldsymbol{w}_j;\ \boldsymbol{\Psi}^{(k)}).$$

Concerning CM-Step 2, it is not difficult to show (McLachlan (1992, Chapter 5)) that $\kappa_v^{(k+1)}$ is a solution of the equation

$$\kappa_v = \sum_{i=1}^{p} \left(\boldsymbol{\Sigma}_0^{(k+1)^{-1}}\right)_{iv} \left(\boldsymbol{V}_2^{(k+1)}\right)_{iv} / \kappa_i \quad (v = 1, \ldots, p). \tag{5.18}$$

This equation can be solved iteratively (for example, by Newton–Raphson or a quasi-Newton procedure) to yield $\kappa_v^{(k+1)}$ $(v = 1, \ldots, p)$ and hence $\boldsymbol{K}^{(k+1)}$; that is, $\boldsymbol{\Psi}_2^{(k+1)}$.

We may wish to consider a multicycle version of this ECM algorithm, where an E-step is introduced between the two CM-steps. The effect of this additional E-step is to replace

$$z_{ij}^{(k)} = \tau_i(\boldsymbol{w}_j;\ \boldsymbol{\Psi}^{(k)})$$

by $\tau_i(\boldsymbol{w}_j;\ \boldsymbol{\Psi}^{(k+1/S)})$ in forming $\boldsymbol{V}_2^{(k+1)}$ from (5.17) for use in (5.18).

5.4.3 Fisher's *Iris* Data

McLachlan (1992, Chapter 6) and McLachlan, Basford, and Green (1993) fitted a mixture of two normals with unrestricted covariance matrices to this data set in order to assess whether the *virginica* species in Fisher's well-known and analyzed *Iris* data should be split into two subspecies. One of their solutions corresponding to a local maximum of the likelihood function produces a cluster containing five observations numbered 6, 18, 19, 23, and 31, with the remaining 45 observations in a second cluster. The observations are labeled 1 to 50 in order of their listing in Table 1.1 in Andrews and Herzberg (1985). In biological applications involving clusters, it is often reasonable to assume that the correlation matrices are the same within each cluster. Therefore in order to avoid potential problems with spurious local maxima as a consequence of clusters having a very small generalized variance, McLachlan and Prado (1995) imposed the restriction of equal correlation matrices of the normal components of the mixture model fitted to this data set. The effect of this restriction on the two clusters above for unrestricted component-covariance matrices was to move the entity numbered 31 to the larger cluster.

5.5 EXAMPLE 5.2: MIXTURE MODELS FOR SURVIVAL DATA

5.5.1 Competing Risks in Survival Analysis

In this example, we illustrate the use of the ECM algorithm for ML fitting of finite mixture models in survival analysis. We suppose that the p.d.f. of the failure time T for a patient has the finite mixture representation

$$f(t;\, a) = \sum_{i=1}^{g} f_i(t;\, a), \tag{5.19}$$

where $f_1(t;\, a), \ldots, f_g(t;\, a)$ denote g component densities occurring in proportions $\pi_1, \ldots, \pi_g$, and $0 \leq \pi_i \leq 1 \quad (i = 1, \ldots, g)$. Here a denotes a covariate associated with the patient, for instance, his or her age at some designated time, for example, at the time of therapy for a medical condition.

One situation where the mixture model is directly applicable for the p.d.f. $f(t;\, a)$ of the failure time T is where the adoption of one parametric family for the distribution of failure time is inadequate. A way of handling this is to adopt a mixture of parametric families. Another situation where the mixture model (5.19) is directly applicable for the p.d.f. of the failure time is where a patient is exposed to g competing risks or types of failure. The p.d.f. of the failure time T has the mixture form (5.19), where $f_i(t;\, a)$ is the p.d.f. of T conditional on the failure being of type i, and π_i is the prior probability of a type i failure $(i = 1, \ldots, g)$. In effect, this mixture approach assumes that a patient will fail from a particular risk, chosen by a stochastic mechanism at the outset. For example, after therapy for lung cancer, an uncured patient might be destined to die from lung cancer, while a cured patient will eventually die from other causes. Farewell (1982, 1986) and Larson and Dinse (1985) were among the first to use finite-mixture models to handle competing risks; see McLachlan and McGiffin (1994) for a recent survey of the use of mixture models in survival analysis.

5.5.2 A Two-Component Mixture Regression Model

In the following, we consider the case of $g = 2$ components in the mixture model (5.19), corresponding to two mutually exclusive groups of patients G_1 and G_2, where G_1 corresponds to those patients who die from a particular disease and G_2 to those who die from other causes.

The effect of the covariate a (say, age) on the mixing proportions is modeled by the logistic model, under which

$$\pi_1(a; \boldsymbol{\beta}) = e^{\beta_0 + \beta_1 a}/(1 + e^{\beta_0 + \beta_1 a}), \tag{5.20}$$

where $\boldsymbol{\beta} = (\beta_0,\, \beta_1)^T$ is the parameter vector.

The survivor function,

$$\bar{F}(t;\, a) = \mathrm{pr}\{T > t \mid a\},$$

for a patient aged a, can therefore be represented by the two-component mixture model

$$\bar{F}(t;\, a) = \pi_1(a;\, \boldsymbol{\beta})\bar{F}_1(t;\, a) + \pi_2(a;\, \boldsymbol{\beta})\bar{F}_2(t;\, a), \tag{5.21}$$

where $\bar{F}_i(t;a)$ is the probability that T is greater than t, given that the patient belongs to G_i $(i = 1, 2)$.

5.5.3 Observed Data

For patient j $(j = 1, \ldots, n)$, we observe

$$\boldsymbol{y}_j = (t_j, a_j, \delta_{1j}, \delta_{2j}, \delta_{3j})^T,$$

where δ_{1j}, δ_{2j}, and δ_{3j} are zero-one indicator variables with $\delta_{1j} = 1$ if patient j died at time t_j (after therapy) for the disease under study and zero otherwise, $\delta_{2j} = 1$ if patient j died from other causes at time t_j and zero otherwise, and $\delta_{3j} = 1$ if patient j was still alive at the termination of the study at time t_j and zero otherwise. Thus

$$\delta_{1j} + \delta_{2j} + \delta_{3j} = 1$$

for $j = 1, \ldots, n$. Also, a_j denotes the age of the jth patient. The observed times $t_1, \ldots, t_n$ are regarded as n independent observations on the random variable T defined to be the time to death either from the disease under study or from other causes. For the patients who are still living at the end of the study $(\delta_{3j} = 1)$, their failure times are therefore censored, where it is assumed that the censoring mechanism is noninformative. That is, the censored observation provides only a lower bound on the failure time, and given that bound, the act of censoring imparts no information about the eventual cause of death.

The two-component mixture model (5.21) can be fitted by maximum likelihood if we specify the component survivor functions $\bar{F}_1(t;\, a)$ and $\bar{F}_2(t;\, a)$ up to a manageable number of unknown parameters. We henceforth write $\bar{F}_1(t;\, a)$ and $\bar{F}_2(t;\, a)$ as $\bar{F}_1(t;\, a, \boldsymbol{\theta}_1)$ and $\bar{F}_2(t;\, a, \boldsymbol{\theta}_2)$, respectively, where $\boldsymbol{\theta}_i$ denotes the vector of unknown parameters in the specification of the ith component survivor function $(i = 1, 2)$. We let

$$\boldsymbol{\Psi} = (\boldsymbol{\beta}^T, \boldsymbol{\theta}_1^T, \boldsymbol{\theta}_2^T)^T$$

be the vector containing all the unknown parameters. Then the log likelihood for $\boldsymbol{\Psi}$ formed on the basis of $\boldsymbol{y}_1, \ldots, \boldsymbol{y}_n$ is given by

$$\log L(\boldsymbol{\Psi}) = \sum_{j=1}^{n} \left[\sum_{i=1}^{2} \delta_{ij} \{\log \pi_i(a_j;\, \boldsymbol{\beta}) + \log f_i(t_j;\, a_j, \boldsymbol{\theta}_i)\} \right.$$
$$\left. + \delta_{3j} \log\{\pi_1(a_j;\, \boldsymbol{\beta}) \bar{F}_1(t_j;\, a_j, \boldsymbol{\theta}_1) + \pi_2(a_j;\, \boldsymbol{\beta}) \bar{F}_2(t_j;\, a_j, \boldsymbol{\theta}_2)\} \right], \tag{5.22}$$

where

$$f_i(t;\, a_j, \boldsymbol{\theta}_i) = -d\bar{F}_i(t;\, a_j, \boldsymbol{\theta}_i)/dt$$

is the density function of T in G_i $(i = 1, 2)$.

5.5.4 Application of EM Algorithm

Rather than working directly with the log likelihood (5.22), the MLE of $\boldsymbol{\Psi}$ can be found by an application of the EM algorithm. If $\delta_{3j} = 1$ for patient j (that is, if the survival time t_j is censored), it is computationally convenient to introduce the zero-one indicator variable z_{ij}, where $\boldsymbol{z}_j = (z_{1j}, z_{2j})^T$ and $z_{ij} = 1$ or 0 according as to whether this patient belongs to group G_i or not $(i = 1, 2;\ j = 1, \ldots, n)$. We can then apply the EM algorithm within the framework where $(\boldsymbol{y}_1^T, \boldsymbol{z}_1^T)^T, \ldots, (\boldsymbol{y}_n^T, \boldsymbol{z}_n^T)^T$ are viewed as the complete data. The actual time to failure for those patients with $\delta_{3j} = 1$ is not introduced as an incomplete variable in the complete-data framework, as it does not simplify the calculations.

The complete-data log likelihood is given then by

$$\begin{aligned} \log L_c(\boldsymbol{\Psi}) = \sum_{j=1}^{n} \Bigg[& \sum_{i=1}^{2} \delta_{ij} \{\log \pi_i(a_j;\, \boldsymbol{\beta}) + \log f_i(t_j;\, a_j, \boldsymbol{\theta}_i)\} \\ & + \delta_{3j} \sum_{i=1}^{2} z_{ij} \{\log \pi_i(a_j;\, \boldsymbol{\beta}) + \log \bar{F}_i(t_j;\, a_j, \boldsymbol{\theta}_i)\} \Bigg]. \end{aligned} \tag{5.23}$$

On the E-step at the $(k+1)$th iteration of the EM algorithm, we have to find the conditional expectation of $\log L_c(\boldsymbol{\Psi})$, given the observed data $\boldsymbol{y}_1, \ldots, \boldsymbol{y}_n$, using the current fit $\boldsymbol{\Psi}^{(k)}$ for $\boldsymbol{\Psi}$. This is effected by replacing z_{ij} in $\log L_c(\boldsymbol{\Psi})$ by $z_{ij}^{(k)}$, which is its current conditional expectation given $\boldsymbol{y}_j$, for each patient j that has $\delta_{3j} = 1$. Now

$$\begin{aligned} z_{ij}^{(k)} &= E_{\boldsymbol{\Psi}^{(k)}}(Z_{ij} \mid t_j, \delta_{3j} = 1, a_j) \\ &= \tau_i(t_j;\, a_j, \boldsymbol{\Psi}^{(k)}), \end{aligned} \tag{5.24}$$

where

$$\tau_1(t_j;\, a_j, \boldsymbol{\Psi}^{(k)}) = \frac{\pi_1(a_j;\, \boldsymbol{\beta}^{(k)}) \bar{F}_1(t_j;\, a_j, \boldsymbol{\theta}_1^{(k)})}{\pi_1(a_j;\, \boldsymbol{\beta}^{(k)}) \bar{F}_1(t_j;\, a_j, \boldsymbol{\theta}_1^{(k)}) + \pi_2(a_j;\, \boldsymbol{\beta}^{(k)}) \bar{F}_2(t_j;\, a_j, \boldsymbol{\theta}_2^{(k)})}$$

and $\tau_2(t_j;\, a_j, \boldsymbol{\Psi}^{(k)}) = 1 - \tau_1(t_j;\, a_j, \boldsymbol{\Psi}^{(k)})$.

From above, $Q(\boldsymbol{\Psi};\, \boldsymbol{\Psi}^{(k)})$ is given by the expression for $\log L_c(\boldsymbol{\Psi})$ with z_{ij} replaced by $z_{ij}^{(k)}$ for those patients j with $\delta_{3j} = 1$. The M-step then involves choosing $\boldsymbol{\Psi}^{(k+1)}$ so as to maximize $Q(\boldsymbol{\Psi}; \boldsymbol{\Psi}^{(k)})$ with respect to $\boldsymbol{\Psi}$ to give $\boldsymbol{\Psi}^{(k+1)}$.

5.5.5 M-Step for Gompertz Components

We now pursue the implementation of the M-step in the case where the component survivor functions are modeled by the Gompertz distribution additively adjusted on the log scale for the age a of the patient. That is, $\bar{F}_i(t; a, \boldsymbol{\theta}_i)$ is modeled as

$$\bar{F}_i(t; a, \boldsymbol{\theta}_i) = \exp\{-e^{\lambda_i+\gamma_i a}(e^{\xi_i t} - 1)/\xi_i\}, \tag{5.25}$$

where $\boldsymbol{\theta}_i = (\lambda_i, \xi_i, \gamma_i)^T \quad (i = 1, 2)$.

The identifiability of mixtures of Gompertz distributions has been established by Gordon (1990a, 1990b) in the case of mixing proportions that do not depend on any covariates. The extension to the case of mixing proportions specified by the logistic model (5.20) is straightforward. It follows that a sufficient condition for identifiability of the Gompertz mixture model (5.25) is that the matrix $(\boldsymbol{a}_1^+, \ldots, \boldsymbol{a}_n^+)$ be of full rank, where

$$\boldsymbol{a}_j^+ = (1, a_j)^T.$$

Unfortunately, for Gompertz component distributions, $\boldsymbol{\Psi}^{(k+1)}$ does not exist in closed form. We consider its iterative computation, commencing with $\boldsymbol{\beta}^{(k+1)}$. On equating the derivative of $Q(\boldsymbol{\Psi}; \boldsymbol{\Psi}^{(k)})$ with respect to $\boldsymbol{\beta}$ to zero, we have that $\boldsymbol{\beta}^{(k+1)}$ can be computed iteratively by the Newton–Raphson method as

$$\boldsymbol{\beta}^{(k+1,m+1)} = \boldsymbol{\beta}^{(k+1,m)} + \boldsymbol{B}^{(k+1,m)^{-1}} \boldsymbol{U}^{(k,m)},$$

where

$$\boldsymbol{B}^{(k+1,m)} = \boldsymbol{A}^T \boldsymbol{V}^{(k+1,m)} \boldsymbol{A}, \tag{5.26}$$

$$\boldsymbol{V}^{(k+1,m)} = \text{diag}(v_1^{(k+1,m)}, \ldots, v_n^{(k+1,m)}), \tag{5.27}$$

$$v_j^{(k+1,m)} = \pi_1(a_j; \boldsymbol{\beta}^{(k+1,m)})\, \pi_2(a_j; \boldsymbol{\beta}^{(k+1,m)}), \tag{5.28}$$

$$\boldsymbol{A} = (\boldsymbol{a}_1^+, \ldots, \boldsymbol{a}_n^+)^T,$$

$$\boldsymbol{U}^{(k,m)} = \boldsymbol{w}^{(k)} - \sum_{j=1}^{n} \pi_1(a_j; \boldsymbol{\beta}^{(k+1,m)})\boldsymbol{a}_j^+, \tag{5.29}$$

and

$$\boldsymbol{w}^{(k)} = \sum_{j=1}^{n} \{\delta_{1j} + \delta_{3j}\tau_1(t_j; a_j, \boldsymbol{\Psi}^{(k)})\}\boldsymbol{a}_j^+. \tag{5.30}$$

Provided the sequence $\{\boldsymbol{\beta}^{(k+1,m+1)}\}$ converges, as $m \to \infty$, $\boldsymbol{\beta}^{(k+1)}$ can be taken equal to $\boldsymbol{\beta}^{(k+1,m+1)}$ for m sufficiently large.

On equating the derivatives of $Q(\boldsymbol{\Psi}; \boldsymbol{\Psi}^{(k)})$ with respect to the elements of $\boldsymbol{\theta}_i$, we have that $\lambda_i^{(k+1)}$, $\xi_i^{(k+1)}$, and $\gamma_i^{(k+1)}$ satisfy the following equations

$$\sum_{j=1}^{n}\{\delta_{ij} - (\delta_{ij} + \delta_{3j} z_{ij}^{(k)})(h_{ij}^{(k+1)}/\xi_i^{(k+1)})\} = 0, \tag{5.31}$$

$$\sum_{j=1}^{n}\{\delta_{ij} t_j + (\delta_{ij} + \delta_{3j} z_{ij}^{(k)})(-t_j c_{ij}^{(k+1)}/\xi_i^{(k+1)} + h_{ij}^{(k+1)}/\xi_i^{(k+1)^2})\} = 0, \tag{5.32}$$

and

$$\sum_{j=1}^{n}\{\delta_{ij} a_j - (\delta_{ij} + \delta_{3j} z_{ij}^{(k)}) a_j h_{ij}^{(k+1)}/\xi_i^{(k+1)}\} = 0 \tag{5.33}$$

for $i = 1$ and 2, where

$$h_{ij}^{(k+1)} = \exp(\lambda_i^{(k+1)} + \gamma_i^{(k+1)} a_j)\{\exp(\xi_i^{(k+1)} t_j) - 1\}$$

and

$$c_{ij}^{(k+1)} = \exp(\lambda_i^{(k+1)} + \gamma_i^{(k+1)} a_j + \xi_i^{(k+1)} t_j).$$

From (5.31), $\lambda_i^{(k+1)}$ can be expressed in terms of $\gamma_i^{(k+1)}$ and $\xi_i^{(k+1)}$ as

$$\lambda_i^{(k+1)} = \log\{q_{1i}^{(k+1)}/q_{2i}^{(k+1)}\} \tag{5.34}$$

for $i = 1$ and 2, where

$$q_{1i}^{(k+1)} = \xi_i^{(k+1)} \sum_{j=1}^{n} \delta_{ij}$$

and

$$q_{2i}^{(k+1)} = \sum_{j=1}^{n} (\delta_{ij} + \delta_{3j} z_{ij}^{(k)}) \exp(\gamma_i^{(k+1)} a_j)\{\exp(\xi_i^{(k+1)} t_j) - 1\}.$$

Thus in order to obtain $\boldsymbol{\theta}_i^{(k+1)}$, the two equations (5.32) and (5.33), where $\lambda_i^{(k+1)}$ is given by (5.34), have to be solved. This can be undertaken iteratively using a quasi-Newton method, as in the FORTRAN program of McLachlan, Adams, Ng, McGiffin, and Galbraith (1994).

5.5.6 Application of a Multicycle ECM Algorithm

We partition $\boldsymbol{\Psi}$ as

$$\boldsymbol{\Psi} = (\boldsymbol{\Psi}_1^T, \boldsymbol{\Psi}_2^T)^T,$$

where $\boldsymbol{\Psi}_1 = \boldsymbol{\beta}$ and $\boldsymbol{\Psi}_2 = (\boldsymbol{\theta}_1^T, \boldsymbol{\theta}_2^T)^T$. From above, it can be seen that $\Psi_1^{(k+1)}$ and $\boldsymbol{\Psi}_2^{(k+1)}$ are computed independently of each other on the M-step of the

EM algorithm. Therefore, the latter is the same as the ECM algorithm with two CM-steps, where on the first CM-step, $\boldsymbol{\Psi}_1^{(k+1)}$ is calculated with $\boldsymbol{\Psi}_2$ fixed at $\boldsymbol{\Psi}_2^{(k)}$, and where on the second CM-step, $\boldsymbol{\Psi}_2^{(k+1)}$ is calculated with $\boldsymbol{\Psi}_1$ fixed at $\boldsymbol{\Psi}_1^{(k+1)}$.

In order to improve convergence, McLachlan et al. (1994) use a multicycle version of this ECM algorithm where an E-step is performed after the computation of $\boldsymbol{\beta}^{(k+1)}$ and before the computation of the other subvectors $\boldsymbol{\theta}_1^{(k+1)}$ and $\boldsymbol{\theta}_2^{(k+1)}$ in $\boldsymbol{\Psi}_2^{(k+1)}$. This multicycle E-step is effected here by updating $\boldsymbol{\beta}^{(k)}$ with $\boldsymbol{\beta}^{(k+1)}$ in $\boldsymbol{\Psi}^{(k)}$ in the right-hand of the expression (5.24) for $z_{ij}^{(k)}$.

To assist further with the convergence of the sequence of iterates $\boldsymbol{\beta}^{(k+1,m+1)}$, we can also perform an additional E-step after the computation of $\boldsymbol{\beta}^{(k+1,m+1)}$ before proceeding with the computation of $\boldsymbol{\beta}^{(k+1,m+2)}$ during the iterative computations on the first CM-step. That is, on the right-hand sides of (5.26) to (5.30), $z_{ij}^{(k)} = \tau_1(t_j;\, a_j,\, \boldsymbol{\Psi}^{(k)})$ is replaced by

$$\tau_1(t_j;\, a_j,\, \boldsymbol{\Psi}^{(k+1,m+1)}),$$

where

$$\boldsymbol{\Psi}^{(k+1,m+1)} = (\boldsymbol{\beta}^{(k+1,m+1)^T}, \boldsymbol{\theta}_1^{(k)^T}, \boldsymbol{\theta}_2^{(k)^T})^T.$$

This is no longer a multicycle ECM algorithm, and so the inequality

$$L(\boldsymbol{\Psi}^{(k+1)}) \geq L(\boldsymbol{\Psi}^{(k)}) \tag{5.35}$$

does not necessarily hold. However, it has been observed to hold for the data sets analyzed by McLachlan et al. (1994).

5.6 EXAMPLE 5.3: CONTINGENCY TABLES WITH INCOMPLETE DATA

In the previous two examples, the ECM algorithm has been implemented with the vector of unknown parameters $\boldsymbol{\Psi}$ partitioned into a number of subvectors, where each CM-step corresponds to maximization over one subvector with the remaining subvectors fixed at their current values. We now give an example where this is not the case. It concerns incomplete-data ML estimation of the cell probabilities of a $2 \times 2 \times 2$ contingency table under a log linear model without the three-way interaction term. It is well known that in this case, there are no simple formulas for computing the expected cell frequencies or finding the MLE's of the cell probabilities, and an iterative method such as the Iterative Proportional Fitting has to be used; see, for instance, Christensen (1990).

Let θ_{hij} be the probability for the (h, i, j)th cell $(h, i, j = 1, 2)$, where the parameter space $\boldsymbol{\Omega}$ is the subspace of $\{\theta_{hij},\, h,\, i,\, j = 1, 2\}$ such that the three-way interaction is zero. Starting from the constant table (that is, $\theta_{hij} = 1/8$), given the fully observed cell events $\boldsymbol{y} = \{y_{hij}\}$ and the current estimates $\theta_{hij}^{(k)}$ of the cell probabilities, the $(k+1)$th iteration of Iterative Proportional Fitting is the final output of the following set of three steps:

$$\theta_{hij}^{(k+1/3)} = \theta_{hi(j)}^{(k)} \frac{y_{hi+}}{n}, \tag{5.36}$$

$$\theta_{hij}^{(k+2/3)} = \theta_{h(i)j}^{(k+1/3)} \frac{y_{h+j}}{n}, \tag{5.37}$$

and

$$\theta_{hij}^{(k+3/3)} = \theta_{(h)ij}^{(k+2/3)} \frac{y_{+ij}}{n}, \tag{5.38}$$

where n is the total count,

$$y_{hi+} = \sum_j y_{hij}$$

define the two-way marginal totals for the first two factors,

$$\theta_{hi(j)} = \theta_{hij} / \sum_j \theta_{hij}$$

define the conditional probabilities of the third factor given the first two, etc. It is easy to see that (5.36) corresponds to maximizing the likelihood function subject to the constraints

$$\theta_{hi(j)} = \theta_{hi(j)}^{(k)}$$

for all h, i, j. Similarly, (5.37) and (5.38) correspond to maximizing the likelihood function subject to the constraints

$$\theta_{h(i)j} = \theta_{h(i)j}^{(k+1/3)}$$

and

$$\theta_{(h)ij} = \theta_{(h)ij}^{(k+2/3)},$$

respectively.

Here the notation

$$\theta_{hij}^{(k+s/3)}$$

corresponds to the value of θ_{hij} on the sth of the three CM-steps on the $(k+1)$th iteration of the ECM algorithm. The simplicity of Iterative Proportional Fitting comes from the fact that the constraint of no three-way iteration only imposes restrictions on the conditional probabilities.

Once each iteration of Iterative Proportional Fitting is identified as a set of conditional maximizations, we can immediately add an E-step at each iteration to develop an algorithm to estimate cell probabilities when data are incomplete. For instance, the only difference between the ECM algorithm and Iterative Proportional Fitting for the above example is to replace y_{ij+} by

$$E_{\boldsymbol{\Psi}^{(k)}}(y_{ij+} \mid \boldsymbol{y}),$$

with analogous replacements for y_{h+j} and y_{+ij} at each iteration. Thus in this case, as noted by Meng and Rubin (1993), the ECM algorithm can be viewed as a natural generalization of Iterative Proportional Fitting in the presence of incomplete data.

5.7 ECME ALGORITHM

Liu and Rubin (1994) present an extension of their ECM algorithm called the ECME (expectation–conditional maximization either) algorithm. Here the "either" refers to the fact that with this extension, some or all of the CM-steps of the ECM algorithm are replaced by steps that conditionally maximize the incomplete-data log likelihood function, $\log L(\boldsymbol{\Psi})$, and not the Q-function. Hence with the ECME algorithm, each CM-step either maximizes the conditional expectation of the complete-data log likelihood

$$Q(\boldsymbol{\Psi}; \boldsymbol{\Psi}^{(k)}) = E_{\boldsymbol{\Psi}^{(k)}}\{\log L_c(\boldsymbol{\Psi}) \mid \boldsymbol{y}\},$$

or the actual (incomplete-data) log likelihood function, $\log L(\boldsymbol{\Psi})$, subject to the same constraints on $\boldsymbol{\Psi}$.

Typically, the ECME algorithm is more tedious to code than the ECM algorithm, but the reward of faster convergence is often worthwhile especially because it allows convergence to be more easily assessed. As noted previously, the ECM algorithm is an extension of the EM algorithm that typically converges more slowly than the EM algorithm in terms of iterations, but can be much faster in total computer time. Liu and Rubin (1994) find that their extension is nearly always faster than both the EM and ECM algorithms in terms of the number of iterations and moreover can be faster in total computer time by orders of magnitude.

Analogous convergence results hold for the ECME algorithm as for the ECM and EM algorithms. In particular, the ECME algorithm shares with both the EM and ECM algorithms their stable monotone convergence; that is,

$$L(\boldsymbol{\Psi}^{(k+1)}) \geq L(\boldsymbol{\Psi}^{(k)}).$$

As Meng and van Dyk (1995) note, the proof of the above result in Liu and Rubin (1994) contains a technical error and is valid only when all the CM-steps that act on the Q-function are performed *before* those that act on the actual log likelihood, $\log L(\boldsymbol{\Psi})$.

Liu and Rubin (1994) establish a result on the global speed of convergence of the ECM algorithm analogous to (5.8), which shows that the ECME algorithm typically has a greater speed of convergence than the ECM algorithm. However, they recognize there are situations where the global speed of convergence of the ECME algorithm is less than that of the ECM algorithm.

5.8 EXAMPLE 5.4: MAXIMUM LIKELIHOOD ESTIMATION OF t-DISTRIBUTION WITH UNKNOWN DEGREES OF FREEDOM

5.8.1 Application of EM Algorithm

In Example 2.6 in Section 2.6, we considered the application of the EM algorithm for finding the MLEs of the parameters $\boldsymbol{\mu}$ and $\boldsymbol{\Sigma}$ in the multivariate t-distribution (2.32) with known degrees of freedom ν. We consider here the general case where ν is also unknown, as in Lange et al. (1989). In this more difficult case, Liu and Rubin (1994, 1995) have shown how the MLEs can be found much more efficiently by using the ECME algorithm.

From (2.36) and (2.37), it can be seen that in the general case the E-step on the $(k+1)$th iteration also requires the calculation of the term

$$E_{\boldsymbol{\Psi}^{(k)}}(\log U_j \mid \boldsymbol{w}_j) \tag{5.39}$$

for $j = 1, \ldots, n$.

To calculate this conditional expectation, we need the result (4.152) that if a random variable R has a gamma (α, β) distribution, then

$$E(\log R) = \psi(\alpha) - \log \beta, \tag{5.40}$$

where

$$\psi(s) = \{\partial \Gamma(s)/\partial s\}/\Gamma(s)$$

is the Digamma function.

Applying the result (5.40) to the conditional density of U_j given w_j, as specified by (2.38), it follows that

$$\begin{aligned} E_{\boldsymbol{\Psi}^{(k)}}(\log U_j \mid \boldsymbol{w}_j) &= \psi(\frac{\nu^{(k)}+p}{2}) - \log[\tfrac{1}{2}\{\nu^{(k)} + \delta(\boldsymbol{w}_j, \boldsymbol{\mu}_j^{(k)}; \boldsymbol{\Sigma}^{(k)}\}] \\ &= \log u_j^{(k)} + \{\psi(\frac{\nu^{(k)}+p}{2}) - \log(\frac{\nu^{(k)}+p}{2})\}, \end{aligned} \tag{5.41}$$

where

$$u_j^{(k)} = E_{\boldsymbol{\Psi}^{(k)}}(U_j \mid \boldsymbol{w}_j) \tag{5.42}$$

$$= \frac{\nu^{(k)} + p}{\nu^{(k)} + \delta(\boldsymbol{w}_j, \boldsymbol{\mu}^{(k)}; \boldsymbol{\Sigma}^{(k)})} \tag{5.43}$$

for $j = 1, \ldots, n$. The last term on the right-hand side of (5.41),

$$\psi\left(\frac{\nu^{(k)} + p}{2}\right) - \log\left(\frac{\nu^{(k)} + p}{2}\right),$$

can be interpreted as the correction for just imputing the mean value $u_j^{(k)}$ for u_j in $\log u_j$.

On using the results (2.41) and (5.41) to calculate the conditional expectation of the complete-data log likelihood from (2.36) and (2.35), we have that $Q(\boldsymbol{\Psi}; \boldsymbol{\Psi}^{(k)})$ is given, on ignoring terms not involving ν, by

$$Q(\boldsymbol{\Psi}; \boldsymbol{\Psi}^{(k)}) = -n \log \Gamma(\tfrac{1}{2}\nu) + \tfrac{1}{2} n\nu \log(\tfrac{1}{2}\nu)$$
$$+\tfrac{1}{2} n\nu \left\{ \frac{1}{n} \sum_{j=1}^{n} (\log u_j^{(k)} - u_j^{(k)}) + \psi\left(\frac{\nu^{(k)} + p}{2}\right) - \log\left(\frac{\nu^{(k)} + p}{2}\right) \right\}. \tag{5.44}$$

5.8.2 M-Step

On the M-step at the $(k + 1)$th iteration of the EM algorithm with unknown ν, the computation of $\boldsymbol{\mu}$ and $\boldsymbol{\Sigma}$ is the same as that with known ν. On calculating the left-hand side of the equation,

$$\partial Q(\boldsymbol{\Psi}; \boldsymbol{\Psi}^{(k)}) / \partial \nu = 0,$$

it follows that $\nu^{(k+1)}$ is a solution of the equation

$$-\psi(\tfrac{1}{2}\nu) + \log(\tfrac{1}{2}\nu) + 1 + \frac{1}{n} \sum_{j=1}^{n} (\log u_j^{(k)} - u_j^{(k)})$$
$$+ \psi\left(\frac{\nu^{(k)} + p}{2}\right) - \log\left(\frac{\nu^{(k)} + p}{2}\right) = 0. \tag{5.45}$$

5.8.3 Application of the ECM Algorithm

Liu and Rubin (1995) note that the convergence of the EM algorithm is slow for unknown ν and the one-dimensional search for the computation of

$\nu^{(k+1)}$ is time consuming as discussed and illustrated in Lange et al. (1989). Consequently, they considered extensions of EM that can be more efficient, which we now present.

We consider the ECM algorithm for this problem, where $\boldsymbol{\Psi}$ is partitioned as $(\boldsymbol{\Psi}_1^T, \Psi_2)^T$, with $\boldsymbol{\Psi}_1$ containing $\boldsymbol{\mu}$ and the distinct elements of $\boldsymbol{\Sigma}$ and with Ψ_2 a scalar equal to ν.

On the $(k+1)$th iteration of the ECM algorithm, the E-step is the same as given above for the EM algorithm, but the M-step of the latter is replaced by two CM-steps, as follows.

CM-Step 1. Calculate $\boldsymbol{\Psi}_1^{(k+1)}$ by maximizing $Q(\boldsymbol{\Psi}; \boldsymbol{\Psi}^{(k)})$ with Ψ_2 fixed at $\Psi_2^{(k)}$; that is, ν fixed at $\nu^{(k)}$.

CM-Step 2. Calculate $\Psi_2^{(k+1)}$ by maximizing $Q(\boldsymbol{\Psi}; \boldsymbol{\Psi}^{(k)})$ with $\boldsymbol{\Psi}_1$ fixed at $\boldsymbol{\Psi}_1^{(k+1)}$.

But as $\boldsymbol{\Psi}_1^{(k+1)}$ and $\Psi_2^{(k+1)}$ are calculated independently of each other on the M-step, these two CM-steps of the ECM algorithm are equivalent to the M-step of the EM algorithm. Hence there is no difference between this ECM and the EM algorithms here. But Liu and Rubin (1995) used the ECM algorithm to give two modifications that are different from the EM algorithm. These two modifications are a multicycle version of the ECM algorithm and an ECME extension. The multicycle version of the ECM algorithm has an additional E-step between the two CM-steps. That is, after the first CM-step, the E-step is taken with

$$\begin{aligned}\boldsymbol{\Psi} &= \boldsymbol{\Psi}^{(k+1/2)} \\ &= (\boldsymbol{\Psi}_1^{(k+1)^T}, \Psi_2^{(k)})^T,\end{aligned}$$

instead of with $\boldsymbol{\Psi} = (\boldsymbol{\Psi}_1^{(k)^T}, \Psi_2^{(k)})^T$ as on the commencement of the $(k+1)$th iteration of the ECM algorithm.

5.8.4 Application of ECME Algorithm

The ECME algorithm as applied by Liu and Rubin (1995) to this problem is the same as the ECM algorithm, apart from the second CM-step where $\Psi_2 = \nu$ is chosen to maximize the actual likelihood function $L(\boldsymbol{\Psi})$, as given by (2.33), with $\boldsymbol{\Psi}_1$ fixed at $\boldsymbol{\Psi}_1^{(k+1)}$.

On fixing $\boldsymbol{\Psi}_1$ at $\boldsymbol{\Psi}_1^{(k+1)}$, we have from (2.33)

$$\begin{aligned}\log L(\boldsymbol{\Psi}^{(k+1)}) = &-\tfrac{1}{2}np\log\pi + n\{\log\Gamma(\frac{\nu+p}{2}) - \log\Gamma(\tfrac{1}{2}\nu)\} - \tfrac{1}{2}n\log|\boldsymbol{\Sigma}^{(k+1)}| \\ &+ \tfrac{1}{2}n\log\nu - \tfrac{1}{2}(\nu+p)\sum_{j=1}^{n}\log\{\nu + \delta(\boldsymbol{w}_j, \boldsymbol{\mu}^{(k+1)}; \boldsymbol{\Sigma}^{(k+1)})\}.\end{aligned} \tag{5.46}$$

Thus the second CM-step of the ECME algorithm chooses $\nu^{(k+1)}$ to maximize (5.46) with $\boldsymbol{\mu} = \boldsymbol{\mu}^{(k+1)}$ and $\boldsymbol{\Sigma} = \boldsymbol{\Sigma}^{(k+1)}$. This implies that $\nu^{(k+1)}$ is a solution of the equation

$$-\psi(\tfrac{1}{2}\nu) + \log(\tfrac{1}{2}\nu) + 1 + \frac{1}{n}\sum_{j=1}^{n}\{\log u_j^{(k+1)}(\nu) - u_j^{(k+1)}(\nu)\}$$
$$+\psi\left(\frac{\nu+p}{2}\right) - \log\left(\frac{\nu+p}{2}\right) = 0, \tag{5.47}$$

where

$$u_j^{(k+1)}(\nu) = \frac{\nu+p}{\nu + \delta(\boldsymbol{w}_j, \boldsymbol{\mu}^{(k+1)}; \boldsymbol{\Sigma}^{(k+1)})}.$$

The solution of this equation involves a one-dimensional search as with the EM algorithm. A comparison of (5.45) with (5.47) demonstrates the difference between the second ECM-step and the second ECME-step in their computation of $\nu^{(k+1)}$.

The multicycle ECM algorithm is obtained by performing an E-step before the second CM-step. The multicycle ECME and ECME algorithms are the same in this application, since the second CM-step of the ECME algorithm is with respect to the actual log likelihood function and not the Q-function.

5.8.5 Some Standard Results

In the next subsection where we are to consider the case of missing data, we need the following results that are well known from multivariate normal theory.

Suppose that a random vector $\boldsymbol{W}$ is partitioned into two subvectors $\boldsymbol{W}_1$ and $\boldsymbol{W}_2$ such that

$$\boldsymbol{W} = (\boldsymbol{W}_1^T, \boldsymbol{W}_2^T)^T,$$

where $\boldsymbol{W}_i$ is of dimension p_i, and $p_1 + p_2 = p$. Further, let $\boldsymbol{\mu} = (\boldsymbol{\mu}_1^T, \boldsymbol{\mu}_2^T)^T$ and

$$\boldsymbol{\Sigma} = \begin{pmatrix} \boldsymbol{\Sigma}_{11} & \boldsymbol{\Sigma}_{12} \\ \boldsymbol{\Sigma}_{21} & \boldsymbol{\Sigma}_{22} \end{pmatrix}$$

be the corresponding partitions of $\boldsymbol{\mu}$ and $\boldsymbol{\Sigma}$.

If $\boldsymbol{W} \sim N(\boldsymbol{\mu}, \boldsymbol{\Sigma})$, then the conditional distribution of $\boldsymbol{W}_1$ given $\boldsymbol{W}_2 = \boldsymbol{w}_2$ is p_1-variate normal with mean

$$\boldsymbol{\mu}_{1\cdot 2} = \boldsymbol{\mu}_1 + \boldsymbol{\Sigma}_{12}\boldsymbol{\Sigma}_{22}^{-1}(\boldsymbol{w}_2 - \boldsymbol{\mu}_2) \tag{5.48}$$

and covariance matrix

$$\boldsymbol{\Sigma}_{11\cdot 2} = \boldsymbol{\Sigma}_{11} - \boldsymbol{\Sigma}_{12}\boldsymbol{\Sigma}_{22}^{-1}\boldsymbol{\Sigma}_{21}. \tag{5.49}$$

5.8.6 Missing Data

We now consider the implementation of the EM algorithm and its extensions when some of the $\boldsymbol{w}_j$ have missing data (assumed to be missing at random). For a vector $\boldsymbol{w}_j$ with missing data, we partition it as

$$\boldsymbol{w}_j = (\boldsymbol{w}_{1j}^T, \boldsymbol{w}_{2j}^T)^T, \tag{5.50}$$

where $\boldsymbol{w}_{2j}$ is the subvector of dimension p_j containing the p_j elements of $\boldsymbol{w}_j$ for which the observations are missing. We let the corresponding partition of $\boldsymbol{\mu}$ and $\boldsymbol{\Sigma}$ be given by

$$\boldsymbol{\mu} = (\boldsymbol{\mu}_{1j}^T, \boldsymbol{\mu}_{2j}^T)^T$$

and

$$\boldsymbol{\Sigma} = \begin{pmatrix} \boldsymbol{\Sigma}_{11;j} & \boldsymbol{\Sigma}_{12;j} \\ \boldsymbol{\Sigma}_{21;j} & \boldsymbol{\Sigma}_{22;j} \end{pmatrix}.$$

In order to implement the E-step, we have to first find the conditional expectations of U_j, $\log U_j$, $U_j \boldsymbol{W}_j$, and $U_j \boldsymbol{W}_j \boldsymbol{W}_j^T$ given the observed data $\boldsymbol{y}$ (effectively, $\boldsymbol{w}_{1j}$), where now $\boldsymbol{y}$ is given by

$$\boldsymbol{y} = (\boldsymbol{w}_{11}^T, \ldots, \boldsymbol{w}_{1n}^T)^T.$$

The calculation of $u_j^{(k)} = E_{\boldsymbol{\Psi}^{(k)}}(U_j \mid \boldsymbol{w}_{1j})$ and $E_{\boldsymbol{\Psi}^{(k)}}(\log U_j \mid \boldsymbol{w}_{1j})$ is straightforward in that we simply replace $\boldsymbol{w}_j$ by its observed subvector $\boldsymbol{w}_{1j}$ in (5.43) and (5.41), respectively.

To calculate $E_{\boldsymbol{\Psi}^{(k)}}(U_j \boldsymbol{W}_j \mid \boldsymbol{w}_{1j})$, we first take the expectation of $U_j \boldsymbol{W}_j$ conditional on U_j as well as $\boldsymbol{w}_{1j}$ and note that the conditional expectation of $\boldsymbol{W}_j$ does not depend on u_j. This gives

$$\begin{aligned} E_{\boldsymbol{\Psi}^{(k)}}(U_j \boldsymbol{W}_j \mid \boldsymbol{w}_{1j}) &= E_{\boldsymbol{\Psi}^{(k)}}(U_j \mid \boldsymbol{w}_{1j}) E_{\boldsymbol{\Psi}^{(k)}}(\boldsymbol{W}_j \mid \boldsymbol{w}_{1j}) \\ &= u_j^{(k)} \boldsymbol{w}_{2j}^{(k)}, \end{aligned} \tag{5.51}$$

where

$$\begin{aligned} \boldsymbol{w}_j^{(k)} &= E_{\boldsymbol{\Psi}^{(k)}}(\boldsymbol{W}_j \mid \boldsymbol{w}_{1j}) \\ &= E_{\boldsymbol{\Psi}^{(k)}}(\boldsymbol{W}_j \mid \boldsymbol{w}_{1j}, u_j) \\ &= (\boldsymbol{w}_{1j}^T, \boldsymbol{w}_{2j}^{(k)^T})^T, \end{aligned} \tag{5.52}$$

and where

$$\begin{aligned} \boldsymbol{w}_{2j}^{(k)} &= E_{\boldsymbol{\Psi}^{(k)}}(\boldsymbol{W}_{2j} \mid \boldsymbol{w}_{1j}, u_j) \\ &= \boldsymbol{\mu}_{1j}^{(k)} + \boldsymbol{\Sigma}_{12;j}^{(k)} \boldsymbol{\Sigma}_{22;j}^{(k)^{-1}} (\boldsymbol{w}_{2j}^{(k)} - \boldsymbol{\mu}_{2j}^{(k)}). \end{aligned} \tag{5.53}$$

This last result is obtained on using (5.48).

The expectation in the remaining term $E_{\boldsymbol{\Psi}^{(k)}}(U_j \boldsymbol{W}_j \boldsymbol{W}_j^T \mid \boldsymbol{w}_{1j})$ is also approached by first conditioning on U_j as well as $\boldsymbol{w}_{1j}$ to give

$$\begin{aligned}
E_{\boldsymbol{\Psi}^{(k)}}(U_j \boldsymbol{W}_j \boldsymbol{W}_j^T \mid \boldsymbol{w}_{1j}) &= E_{\boldsymbol{\Psi}^{(k)}}\{U_j E_{\boldsymbol{\Psi}^{(k)}}(\boldsymbol{W}_j \boldsymbol{W}_j^T \mid \boldsymbol{w}_{1j},\, U_j)\} \\
&= E_{\boldsymbol{\Psi}^{(k)}}[U_j\{\mathrm{cov}_{\boldsymbol{\Psi}^{(k)}}(\boldsymbol{W}_j \mid \boldsymbol{w}_{1j},\, U_j) + \boldsymbol{w}_j(k)\boldsymbol{w}_j^{(k)^T}\}] \\
&= E_{\boldsymbol{\Psi}^{(k)}}\{U_j \mathrm{cov}_{\boldsymbol{\Psi}^{(k)}}(\boldsymbol{W}_j \mid \boldsymbol{w}_{1j},\, U_j)\} + u_j^{(k)} \boldsymbol{w}_j^{(k)} \boldsymbol{w}_j^{(k)^T}\} \\
&= \boldsymbol{c}_j^{(k)} + u_j^{(k)} \boldsymbol{w}_j^{(k)} \boldsymbol{w}_j^{(k)^T}, \qquad (5.54)
\end{aligned}$$

where

$$\boldsymbol{c}_j^{(k)} = E_{\boldsymbol{\Psi}^{(k)}}\{U_j \mathrm{cov}_{\boldsymbol{\Psi}^{(k)}}(\boldsymbol{W}_j \mid \boldsymbol{w}_{1j},\, U_j\}.$$

The (h, i)th element of $\boldsymbol{c}_j^{(k)}$ is zero if either $(\boldsymbol{w}_j)_h$ or $(\boldsymbol{w}_j)_i$ is observed and, if both are missing, it is the corresponding element of

$$\boldsymbol{\Sigma}_{11;j}^{(k)} - \boldsymbol{\Sigma}_{12;j}^{(k)} \boldsymbol{\Sigma}_{22;j}^{(k)^{-1}} \boldsymbol{\Sigma}_{21;j}^{(k)},$$

which is obtained on using (5.49). Alternatively, it can be found numerically by applying the sweep operator to $\boldsymbol{\mu}_j^{(k)}$ and $\boldsymbol{\Sigma}_j^{(k)}$ to predict $\boldsymbol{w}_{2j}$ by its linear regression on $\boldsymbol{w}_{1j}$; see Goodnight (1979) and Little and Rubin (1987, Chapter 6).

With the conditional expectations calculated as above, we now can compute the updated estimates $\boldsymbol{\mu}^{(k+1)}$ and $\boldsymbol{\Sigma}^{(k+1)}$, of $\boldsymbol{\mu}$ and $\boldsymbol{\Sigma}$, respectively. It is not difficult to see that, corresponding to (2.42), $\boldsymbol{\mu}^{(k+1)}$ is given by

$$\boldsymbol{\mu}^{(k+1)} = \sum_{j=1}^{n} u_j^{(k)} \boldsymbol{w}_j^{(k)} \Big/ \sum_{j=1}^{n} u_j^{(k)}.$$

That is, we simply impute current conditional expectations for any missing observations in $\boldsymbol{w}_j$.

The modification to (2.43) for the updated estimate of $\boldsymbol{\Sigma}$ is not as straightforward in that it effectively involves imputing the current conditional expectations of cross-product terms like $\boldsymbol{W}_j \boldsymbol{W}_j^T$. It can be confirmed that

$$\begin{aligned}
\boldsymbol{\Sigma}^{(k+1)} &= n^{-1} \sum_{j=1}^{n} [E_{\boldsymbol{\Psi}^{(k)}}(U_j \boldsymbol{W}_j \boldsymbol{W}_j^T \mid \boldsymbol{w}_{1j}) \\
&\quad - \left(\sum_{j=1}^{n} u_j^{(k)} \boldsymbol{w}_j^{(k)}\right) \left(\sum_{j=1}^{n} u_j^{(k)} \boldsymbol{w}_j^{(k)}\right)^T \Big/ \sum_{j=1}^{n} u_j^{(k)} \\
&= n^{-1} \sum_{j=1}^{n} \{u_j^{(k)} (\boldsymbol{w}_j^{(k)} - \boldsymbol{\mu}_j^{(k+1)})(\boldsymbol{w}_j^{(k)} - \boldsymbol{\mu}_j^{(k+1)})^T + \boldsymbol{c}_j^{(k}\}, \qquad (5.55)
\end{aligned}$$

on using (5.54).

Similarly, as above, the E- and CM-steps of the ECM and ECME algorithms can be implemented for this problem when there are missing data.

5.8.7 Numerical Examples

We report here first the numerical example of Liu and Rubin (1994) who consider the data set in Table 1 of Cohen, Dalal, and Tukey (1993), which contains 79 bivariate observations $\boldsymbol{w}_j (j = 1, \ldots, 79)$. Liu and Rubin (1994) create missing data by deleting the second component of every third observation starting from $j = 1$; that is, $j = 1, 4, 7, \ldots,$ and the first component of every third observation starting from $j = 2$; that is, $j = 2, 5, 8, \ldots,$ and treating the deleted observations as being missing at random. Liu and Rubin (1994) fitted the t-distribution to this missing data set via the EM (=ECM) and ECME algorithms, starting each from $\nu^{(0)} = 1000$ and $\boldsymbol{\mu}^{(0)}$ and $\boldsymbol{\Sigma}^{(0)}$ equal to their MLEs under the bivariate normal model. The values of $\log L(\hat{\boldsymbol{\Psi}})$ and $\log \hat{\nu}$ corresponding to each of these algorithms are displayed in Figure 5.1, where the solid and dashed lines represent the values from the EM and ECME algorithms, respectively. The dramatically faster convergence of the ECME algorithm over the EM algorithm is obvious, and is so dramatic that the fact that the starting values are the same is lost in the plots, which display the results as functions of log number of iterations. The CPU time until convergence for the ECME algorithm was nearly one-tenth for the EM algorithm.

In another example, Liu and Rubin (1995) compare the performances of the ECME, EM, and multicycle EM algorithms in their application to an artificial data set created by appending four extreme observations to the artificial bivariate data set of Murray (1977), given in Section 3.6.1. They find that the speed of convergence of the ECME algorithm is significantly faster and, in terms of the actual computational time, to have a sevenfold advantage over the other methods. In more recent work to be considered shortly in Section 5.11, Meng and van Dyk (1995) report results in which the ECME algorithm offers little advantage over the multicycle EM algorithm for this problem.

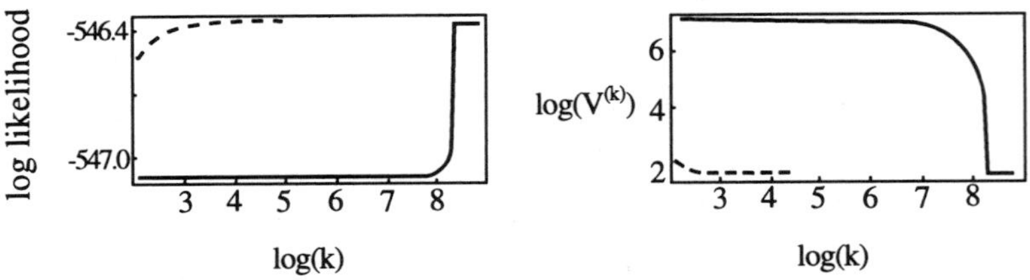

Figure 5.1. Multivariate t: convergence of EM (solid line) and ECME (dashed line) for log likelihood and $\log \nu^{(k)}$. (From Liu and Rubin, 1994.)

5.8.8 Theoretical Results on the Rate of Convergence

Liu and Rubin (1994) provide explicit large-sample results for the rate of convergence (equivalently, the speed of convergence) for ML estimation of the parameters of the univariate t-distribution. They show that the ECME algorithm has a smaller rate of convergence for this problem than the EM algorithm. In the case of known degrees of freedom ν, they note that the large-sample rate of convergence is $3/(3+\nu)$, which was obtained also by Dempster et al. (1980); Meng and Rubin (1994) give similar calculations for a contamination model.

5.9 EXAMPLE 5.5: VARIANCE COMPONENTS

5.9.1 A Variance Components Model

We consider the following variance components, or repeated measures model, as considered previously by Harville (1977), Laird and Ware (1982), Laird, Lange, and Stram (1987), and Liu and Rubin (1994). Specifically, we let $\boldsymbol{y}_j$ denote the $n_j \times 1$ vector of n_j measurements observed on the jth experimental unit, where it is assumed that

$$\boldsymbol{y}_j = \boldsymbol{X}_j\boldsymbol{\beta} + \boldsymbol{Z}_j\boldsymbol{b}_j + \boldsymbol{e}_j \quad (j = 1, \ldots, m). \tag{5.56}$$

Here $\boldsymbol{X}_j$ and $\boldsymbol{Z}_j$ are known $n_j \times p$ and $n_j \times q$ design matrices, respectively, and $\boldsymbol{\beta}$ is a $p \times 1$ vector of fixed effects to be estimated, and $n = \sum_{j=1}^{m} n_j$. The $q \times 1$ random effects vector $\boldsymbol{b}_j$ is distributed $N(\mathbf{0}, \mathbf{D})$, independently of the error vector $\boldsymbol{e}_j$, which is distributed

$$N(\mathbf{0}, \sigma^2\boldsymbol{R}_j),$$

where $\boldsymbol{R}_j$ is a known $n_j \times n_j$ positive-definite symmetric matrix and where $\boldsymbol{D}$ is an unknown $q \times q$ positive-definite symmetric matrix of parameters to be estimated along with the unknown positive scalar σ^2 $(j = 1, \ldots, m)$. It is assumed further that $\boldsymbol{b}_j$ and $\boldsymbol{e}_j$ are distributed independently for $j = 1, \ldots, m$.

For this problem, the observed data vector is $\boldsymbol{y} = (\boldsymbol{y}_1^T, \ldots, \boldsymbol{y}_m^T)^T$, for which the incomplete-data log likelihood, $\log L(\boldsymbol{\Psi})$, is given by

$$\log L(\boldsymbol{\Psi}) = -\tfrac{1}{2}\sum_{j=1}^{m} \log |\boldsymbol{\Sigma}_{11;j}| \\ -\tfrac{1}{2}\sum_{j=1}^{m} (\boldsymbol{y}_j - \boldsymbol{X}_j\boldsymbol{\beta})^T \boldsymbol{\Sigma}_{11;j}^{-1} (\boldsymbol{y}_j - \boldsymbol{X}_j\boldsymbol{\beta}), \tag{5.57}$$

where

$$\boldsymbol{\Sigma}_{11;j} = \boldsymbol{Z}_j \boldsymbol{D} \boldsymbol{Z}_j^T + \sigma^2 \boldsymbol{R}_j \quad (j = 1, \ldots, m).$$

An obvious choice for the complete-data vector is

$$\boldsymbol{x} = (\boldsymbol{x}_1^T, \ldots, \boldsymbol{x}_m^T)^T$$

where

$$\boldsymbol{x}_j = (\boldsymbol{y}_j^T, \boldsymbol{b}_j^T)^T \quad (j = 1, \ldots, m);$$

that is,

$$\boldsymbol{z} = (\boldsymbol{b}_1^T, \ldots, \boldsymbol{b}_m^T)^T$$

is the missing-data vector. Under the assumptions of the model (5.56), it follows that $\boldsymbol{x}_j$ has a multivariate normal distribution with mean

$$\boldsymbol{\mu}_j = \begin{pmatrix} \boldsymbol{X}_j \boldsymbol{\beta} \\ \boldsymbol{0} \end{pmatrix} \tag{5.58}$$

and covariance matrix

$$\boldsymbol{\Sigma}_j = \begin{pmatrix} \boldsymbol{Z}_j \boldsymbol{D} \boldsymbol{Z}_j^T + \sigma^2 \boldsymbol{R}_j & \boldsymbol{Z}_j \boldsymbol{D} \\ \boldsymbol{D} \boldsymbol{Z}_j^T & \boldsymbol{D} \end{pmatrix} \tag{5.59}$$

for $j = 1, \ldots, m$. Note that here $\boldsymbol{X}_j$ is a design matrix and is not the random vector corresponding to $\boldsymbol{x}_j$.

The vector $\boldsymbol{\Psi}$ of unknown parameters is given by the elements of $\boldsymbol{\beta}$, σ^2, and the distinct elements of $\boldsymbol{D}$. From (5.58) and (5.59), the complete-data log likelihood is given, apart from an additive constant, by

$$\begin{aligned} \log L_c(\boldsymbol{\Psi}) &= \sum_{j=1}^{m} \log \phi(\boldsymbol{x}_j; \boldsymbol{\mu}_j, \boldsymbol{\Sigma}_j) \\ &= -\tfrac{1}{2} \sum_{j=1}^{m} \{\log |\boldsymbol{\Sigma}_j| + (\boldsymbol{x}_j - \boldsymbol{\mu}_j)^T \boldsymbol{\Sigma}_j^{-1} (\boldsymbol{x}_j - \boldsymbol{\mu}_j)\}. \end{aligned} \tag{5.60}$$

5.9.2 E-Step

Considering the E-step of the EM algorithm, it can be seen from (5.60), that in order to compute the conditional expectation of (5.60) given the observed data $\boldsymbol{y}$, we require the following conditional moments of the missing random effects vector $\boldsymbol{b}_j$, namely

$$E_{\boldsymbol{\Psi}^{(k)}}(\boldsymbol{b}_j \mid \boldsymbol{y}_j)$$

and

$$E_{\boldsymbol{\Psi}^{(k)}}(\boldsymbol{b}_j \boldsymbol{b}_j^T \mid \boldsymbol{y}_j).$$

These are directly obtainable from the well-known results in multivariate theory given by (5.48) and (5.49).

From these results applied to the joint distribution of $\boldsymbol{y}_j$ and $\boldsymbol{b}_j$, we have that the conditional distribution of $\boldsymbol{b}_j$ given $\boldsymbol{y}_j$ is multivariate normal with mean

$$\begin{aligned}\boldsymbol{\mu}_{2\cdot1;j} &= \boldsymbol{D}\boldsymbol{Z}_j^T(\boldsymbol{Z}_j\boldsymbol{D}\boldsymbol{Z}_j^T + \sigma^2\boldsymbol{R}_j)^{-1}(\boldsymbol{y}_j - \boldsymbol{X}_j\boldsymbol{\beta}) \\ &= \boldsymbol{D} - \boldsymbol{D}\boldsymbol{Z}_j^T(\boldsymbol{R}_j^{-1}\boldsymbol{Z}_j\boldsymbol{D}\boldsymbol{Z}_j^T + \sigma^2\boldsymbol{I}_{n_j})^{-1}\boldsymbol{R}_j^{-1}(\boldsymbol{y}_j - \boldsymbol{X}_j\boldsymbol{\beta}) \end{aligned} \tag{5.61}$$

and covariance matrix

$$\boldsymbol{\Sigma}_{22\cdot1;j} = \boldsymbol{D} - \boldsymbol{D}\boldsymbol{Z}_j^T(\boldsymbol{Z}_j\boldsymbol{D}\boldsymbol{Z}_j^T + \sigma^2\boldsymbol{R}_j)^{-1}\boldsymbol{Z}_j\boldsymbol{D} \tag{5.62}$$

for $j = 1, \ldots, m$.

To show the close connection between least-squares computation and (5.61) and (5.62), we make use of the identity

$$(\boldsymbol{Z}_j^T\boldsymbol{Z}_j + \sigma^2\boldsymbol{D}^{-1})\boldsymbol{D}\boldsymbol{Z}_j^T = \boldsymbol{Z}_j^T(\boldsymbol{Z}_j\boldsymbol{D}\boldsymbol{Z}_j^T + \sigma^2\boldsymbol{I}_{n_j}),$$

from which

$$\boldsymbol{\mu}_{2\cdot1;j} = (\boldsymbol{Z}^T\boldsymbol{R}_j^{-1}\boldsymbol{Z}_j + \sigma^2\boldsymbol{D}^{-1})^{-1}\boldsymbol{Z}_j^T\boldsymbol{R}_j^{-1}(\boldsymbol{y}_j - \boldsymbol{X}_j\boldsymbol{\beta}), \tag{5.63}$$

and

$$\begin{aligned}\boldsymbol{\Sigma}_{22\cdot1;j} &= \boldsymbol{D} - (\boldsymbol{Z}_j^T\boldsymbol{R}_j^{-1}\boldsymbol{Z}_j + \sigma^2\boldsymbol{D}^{-1})^{-1}\boldsymbol{Z}_j^T\boldsymbol{R}_j^{-1}\boldsymbol{Z}_j\boldsymbol{D} \\ &= (\boldsymbol{Z}_j^T\boldsymbol{R}_j^{-1}\boldsymbol{Z}_j + \sigma^2\boldsymbol{D}^{-1})^{-1}\{(\boldsymbol{Z}_j^T\boldsymbol{R}_j^{-1}\boldsymbol{Z}_j + \sigma^2\boldsymbol{D}^{-1})\boldsymbol{D} - \boldsymbol{Z}^T\boldsymbol{R}_j^{-1}\boldsymbol{Z}_j\boldsymbol{D}\} \\ &= (\sigma^{-2}\boldsymbol{Z}_j^T\boldsymbol{R}_j^{-1}\boldsymbol{Z}_j + \boldsymbol{D}^{-1})^{-1}. \end{aligned} \tag{5.64}$$

From (5.63) and (5.64), we have that

$$E_{\boldsymbol{\Psi}^{(k)}}(\boldsymbol{b}_j \mid \boldsymbol{y}_j) = \boldsymbol{b}_j^{(k)},$$

where

$$\boldsymbol{b}_j^{(k)} = (\boldsymbol{Z}_j^T\boldsymbol{R}_j^{-1}\boldsymbol{Z}_j + \sigma^{(k)^2}\boldsymbol{D}^{(k)^{-1}})^{-1}\boldsymbol{Z}_j^T\boldsymbol{R}_j^{-1}(\boldsymbol{y}_j - \boldsymbol{X}_j\boldsymbol{\beta}^{(k)}) \tag{5.65}$$

and

$$\begin{aligned}E_{\boldsymbol{\Psi}^{(k)}}(\boldsymbol{b}_j\boldsymbol{b}_j^T \mid \boldsymbol{y}_j) &= \mathrm{cov}_{\boldsymbol{\Psi}^{(k)}}(\boldsymbol{b}_j \mid \boldsymbol{y}_j) + \boldsymbol{b}_j^{(k)}\boldsymbol{b}_j^{(k)^T} \\ &= \sigma^{(k)^2}(\sigma^{(k)^{-2}}\boldsymbol{Z}_j^T\boldsymbol{R}_j^{-1}\boldsymbol{Z}_j + \boldsymbol{D}^{(k)^{-1}})^{-1} + \boldsymbol{b}_j^{(k)}\boldsymbol{b}_j^{(k)^T}. \end{aligned} \tag{5.66}$$

5.9.3 M-Step

On the M-step at the $(k+1)$th iteration, we have to choose $\boldsymbol{\beta}^{(k+1)}$, $\sigma^{(k+1)^2}$, and $\boldsymbol{D}^{(k+1)}$ to maximize $Q(\boldsymbol{\Psi};\ \boldsymbol{\Psi}^{(k+1)})$. Now

$$\boldsymbol{\beta}^{(k+1)} = \left(\sum_{j=1}^{m} \boldsymbol{X}_j^T \boldsymbol{R}_j^{-1} \boldsymbol{X}_j\right)^{-1} \sum_{j=1}^{m} \boldsymbol{X}_j^T \boldsymbol{R}_j^{-1} (\boldsymbol{y}_j - \boldsymbol{Z}_j \boldsymbol{b}_j^{(k)}).$$

The iterates $\boldsymbol{D}^{(k+1)}$ and $\sigma^{(k+1)^2}$ can be obtained either by direct differentiation of $Q(\boldsymbol{\Psi};\ \boldsymbol{\Psi}^{(k)})$ or by noting that $L_c(\boldsymbol{\Psi})$ belongs to the exponential family and considering the conditional expectations of the complete-data sufficient statistics for $\boldsymbol{\beta}$, $\boldsymbol{D}$, and σ^2.

It follows that

$$\boldsymbol{D}^{(k+1)} = \frac{1}{m} \sum_{j=1}^{m} E_{\boldsymbol{\Psi}^{(k)}}\{\boldsymbol{b}_j \boldsymbol{b}_j^T \mid \boldsymbol{y}_j\} \tag{5.67}$$

and

$$\sigma^{(k+1)^2} = \frac{1}{n} \sum_{j=1}^{m} E_{\boldsymbol{\Psi}^{(k)}}\{\boldsymbol{e}_j^T \boldsymbol{R}_j^{-1} \boldsymbol{e}_j \mid \boldsymbol{y}_j\}.$$

Now

$$\begin{aligned} E_{\boldsymbol{\Psi}^{(k)}}\{\boldsymbol{e}_j^T \boldsymbol{R}_j^{-1} \boldsymbol{e}_j \mid \boldsymbol{y}_j\} &= \operatorname{tr} \boldsymbol{R}_j^{-1} E_{\boldsymbol{\Psi}^{(k)}}\{\boldsymbol{e}_j \boldsymbol{e}_j^T \mid \boldsymbol{y}_j\} \\ &= \operatorname{tr} \boldsymbol{R}_j^{-1} [\operatorname{cov}_{\boldsymbol{\Psi}^{(k)}}\{\boldsymbol{e}_j \mid \boldsymbol{y}_j\} + \boldsymbol{e}_j^{(k)} \boldsymbol{e}_j^{(k)^T}] \\ &= \operatorname{tr}[\boldsymbol{R}_j^{-1} \operatorname{cov}_{\boldsymbol{\Psi}^{(k)}}\{\boldsymbol{e}_j \mid \boldsymbol{y}_j\}] + \boldsymbol{e}_j^{(k)^T} \boldsymbol{R}_j^{-1} \boldsymbol{e}_j^{(k)}, \end{aligned}$$

where

$$\begin{aligned} \boldsymbol{e}_j^{(k)} &= E_{\boldsymbol{\Psi}^{(k)}}(\boldsymbol{e}_j \mid \boldsymbol{y}_j) \\ &= \boldsymbol{y}_j - \boldsymbol{X}_j \boldsymbol{\beta}^{(k)} - \boldsymbol{Z}_j \boldsymbol{b}_j^{(k)}, \end{aligned}$$

and where

$$\begin{aligned} \operatorname{cov}_{\boldsymbol{\Psi}^{(k)}}(\boldsymbol{e}_j \mid \boldsymbol{y}_j) &= \operatorname{cov}_{\boldsymbol{\Psi}^{(k)}}(\boldsymbol{Z}_j \boldsymbol{b}_j \mid \boldsymbol{y}_j) \\ &= \boldsymbol{Z}_j (\sigma^{(k)^{-2}} \boldsymbol{Z}_j^T \boldsymbol{R}_j^{-1} \boldsymbol{Z}_j + \boldsymbol{D}^{(k)^{-1}})^{-1} \boldsymbol{Z}_j^T. \end{aligned}$$

Thus

$$E_{\boldsymbol{\Psi}^{(k)}}(\boldsymbol{e}^T \boldsymbol{R}_j^{-1} \boldsymbol{e}_j \mid \boldsymbol{y}_j) = tr\{\boldsymbol{Z}_j^T \boldsymbol{R}_j^{-1} \boldsymbol{Z}_j (\sigma^{(k)^2} \boldsymbol{Z}_j^T \boldsymbol{R}_j^{-1} \boldsymbol{Z}_j + \boldsymbol{D}^{(k)^{-1}})\} + \boldsymbol{e}_j^{(k)^T} \boldsymbol{R}_j^{-1} \boldsymbol{e}_j^{(k)}.$$

5.9.4 Application of Two Versions of ECME Algorithm

For the variance components model (5.56), Liu and Rubin (1994) consider two versions of the ECME algorithm that can be especially appropriate when the dimension q of $\boldsymbol{D}$ is relatively large. In Version 1 of the ECME algorithm, $\boldsymbol{\Psi}$ is partitioned as $(\boldsymbol{\Psi}_1^T, \boldsymbol{\Psi}_2^T)^T$, where $\boldsymbol{\Psi}_1$ contains the distinct elements of $\boldsymbol{D}$ and σ^2 and $\boldsymbol{\Psi}_2 = \boldsymbol{\beta}$. The two CM-steps are as follows.

CM-Step 1. Calculate $\boldsymbol{D}^{(k+1)}$ and $\sigma^{(k+1)^2}$ as above on the M-step of the EM algorithm.

CM-Step 2. Calculate $\boldsymbol{\beta}^{(k+1)}$ as

$$\boldsymbol{\beta}^{(k+1)} = \left\{ \sum_{j=1}^{m} \boldsymbol{X}_j^T \boldsymbol{\Sigma}_{11;j}^{(k+1)^{-1}} \boldsymbol{X}_j \right\}^{-1} \left\{ \sum_{j=1}^{m} \boldsymbol{X}_j^T \boldsymbol{\Sigma}_{11;j}^{(k+1)^{-1}} \boldsymbol{y}_j \right\},$$

where

$$\boldsymbol{\Sigma}_{11;j}^{(k+1)} = \boldsymbol{Z}_j \boldsymbol{D}^{(k+1)} \boldsymbol{Z}_j^T + \sigma^{(k+1)^2} \boldsymbol{R}_j \quad (j = 1, \ldots, m),$$

which is the value of $\boldsymbol{\beta}$ that maximizes the incomplete-data log likelihood $\log L(\boldsymbol{\Psi})$ over $\boldsymbol{\beta}$ with $\boldsymbol{D} = \boldsymbol{D}^{(k+1)}$ and $\sigma^2 = \sigma^{(k+1)^2}$.

As pointed out by Liu and Rubin (1994), this corresponds to the algorithm given by Laird and Ware (1982) and Laird et al. (1987), who mistakenly called it an EM algorithm. The algorithm was called a hybrid EM algorithm by Jennrich and Schluchter (1986), who also realized it is not an EM algorithm.

In Version 2 of the ECME algorithm proposed in Liu and Rubin (1994), $\boldsymbol{\Psi}$ is partitioned as $(\boldsymbol{\Psi}_1^T, \boldsymbol{\Psi}_2^T, \Psi_3)^T$, where $\boldsymbol{\Psi}_1$ consists of the distinct elements of $\boldsymbol{D}$, $\boldsymbol{\Psi}_2 = \boldsymbol{\beta}$, and $\Psi_3 = \sigma^2$. The three CM-steps are as follows.

CM-Step 1. Calculate $\boldsymbol{D}^{(k+1)}$ as on the M-step of EM.

CM-Step 2. Calculate $\hat{\boldsymbol{\beta}}$ as on the CM-Step 2 above of Version 1 of the ECME algorithm.

CM-Step 3. Calculate $\sigma^{(k+1)^2}$ to maximize $L(\boldsymbol{\Psi})$ over σ^2 with $\boldsymbol{D} = \boldsymbol{D}^{(k+1)}$ and $\boldsymbol{\beta} = \boldsymbol{\beta}^{(k+1)}$. The solution to this step does not exist in closed form and so must be calculated using, say, a quasi-Newton method.

5.9.5 Numerical Example

As a numerical example of the above model, we report the example considered by Liu and Rubin (1994). They analyzed the data of Besag (1991) on the yields of 62 varieties of winter wheat in three physically separated complete replicates, using the first-difference Gaussian noise model for the random fertility effects. Hence in the notation of the model (5.57), $m = 3$, $n_j = 61$, and $n = 183$, where $\boldsymbol{y}_j$ gives the 61 differences of the yields in replicate j, where $\boldsymbol{X}_j$ is the corresponding (61×62) design matrix, $\boldsymbol{Z}_j$ is the (61×61) identity matrix, $\boldsymbol{R}_j = \boldsymbol{R}$, the (61×61) Toeplitz matrix with the first row $(2, -1, 0, \ldots, 0)$, and where $\boldsymbol{\beta}$ is the vector of the fixed variety effects with the constraint

$$\sum_{i=1}^{6} \beta_i = 0.$$

The covariance matrix $\boldsymbol{D}$ of the random effects is restricted to have the form

$$\boldsymbol{D} = \alpha^2 \boldsymbol{I}_q, \tag{5.68}$$

where α is a nonnegative unknown scalar. The vector of unknown parameters is therefore

$$\boldsymbol{\Psi} = (\boldsymbol{\beta}^T, \alpha^2, \sigma^2)^T.$$

Under the restriction (5.68), the equation (5.66) for $\boldsymbol{D}^{(k+1)}$ on the $(k+1)$th iteration of the M-step has to be replaced by the following equation for $\alpha^{(k+1)^2}$,

$$\begin{aligned}\alpha^{(k+1)^2} &= \frac{1}{n} \sum_{j=1}^{m} E_{\boldsymbol{\Psi}^{(k)}}\{\boldsymbol{b}_j^T \boldsymbol{b}_j \mid \boldsymbol{y}\} \\ &= \frac{1}{n} \left[\sum_{j=1}^{m} \boldsymbol{b}_j^{(k)^T} \boldsymbol{b}_j^{(k)} + \operatorname{tr}\{(\boldsymbol{R}\sigma^{(k)^2} + \alpha^{(k)^2} \boldsymbol{I}_q)^{-1} \sigma^{(k)^2} \alpha^{(k)^2} \boldsymbol{R}\} \right].\end{aligned}$$

Liu and Rubin (1994) applied the EM algorithm and Versions 1 and 2 above of the ECME algorithm with the common starting value

$$\boldsymbol{\Psi}^{(0)} = (\boldsymbol{\beta}^{(0)^T}, \sigma^{(0)^2}, \alpha^{(0)^2})^T,$$

where $\boldsymbol{\beta}^{(0)} = \boldsymbol{0}$, $\sigma^{(0)^2} = \alpha^{(0)^2} \times 10^{-4}$, and

$$\alpha^{(0)^2} = \frac{1}{n} \sum_{j=1}^{m} \boldsymbol{y}_j^T \boldsymbol{y}_j.$$

The corresponding values of the log likelihood and α are given in Figure 5.2. In the left-side figure, the solid line corresponds to the EM algorithm and the dashed line to Version 1 of the ECME algorithm. In the right-side figure, the dashed line corresponds to Version 1 of the ECME algorithm and the dotted line to Version 2, using an expanded scale. It can be seen from these figures that in terms of number of iterations, Version 2 is substantially faster than

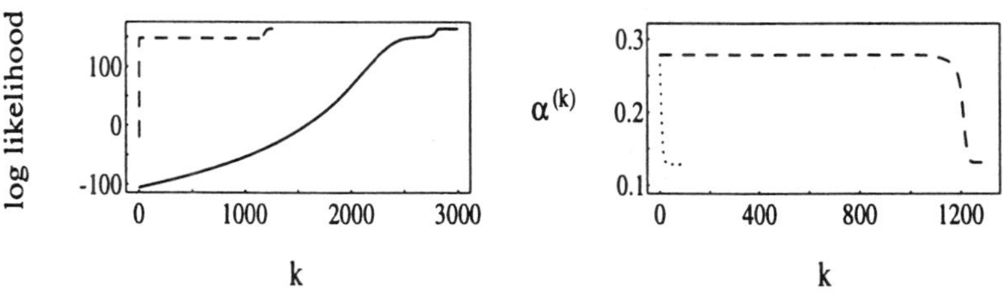

Figure 5.2. Variance components: convergence of EM (solid line) and ECME (dashed line) for log likelihood and ECME-1 (dashed line) and ECME-2 (dotted line) for $\alpha^{(k)}$. (From Liu and Rubin, 1994.)

Version 1 of the ECME algorithm, which is substantially faster than the EM algorithm. In terms of total computer time in this example, Liu and Rubin (1994) find that Version 2 of the ECME algorithm is approximately twice as fast as the EM algorithm, which is approximately an order of magnitude faster than Version 1 of the ECME algorithm.

5.10 EXAMPLE 5.6: FACTOR ANALYSIS

5.10.1 ECM Algorithm for Factor Analysis

The model usually used in factor analysis is

$$\boldsymbol{Y}_j = \boldsymbol{\alpha} + \boldsymbol{B}\boldsymbol{Z}_j + \boldsymbol{e}_j \quad (j = 1, \ldots, n),$$

where $\boldsymbol{Y}_j$ is p-dimensional observation vector, $\boldsymbol{Z}_j$ is a q-dimensional ($q < p$) vector of unobservable variables called factor scores and $\boldsymbol{B}$ is a $p \times q$ matrix of factor loadings (parameters). It is assumed that

$$(\boldsymbol{Y}_1^T, \boldsymbol{Z}_1^T)^T, \ldots, (\boldsymbol{Y}_n^T, \boldsymbol{Z}_n^T)^T$$

are independently and identically distributed. The $\boldsymbol{Z}_j$ are assumed to be independently and identically distributed as $N(\boldsymbol{0}, \boldsymbol{I}_q)$, independently of the errors $\boldsymbol{e}_j$, which are assumed to be independently and identically distributed as $N(\boldsymbol{0}, \boldsymbol{D})$, where $\boldsymbol{D} = \operatorname{diag}(\sigma_1^2, \ldots, \sigma_p^2)$. The σ_i^2 are called the uniquenesses. Thus, conditional on the Z_j, the Y_j are independently distributed as $N(\boldsymbol{\alpha} + \boldsymbol{Z}_j\boldsymbol{B}, \boldsymbol{D})$. Unconditionally, the $\boldsymbol{Y}_j$ are independently and identically distributed according to a normal distribution with mean $\boldsymbol{\alpha}$ and covariance matrix $\boldsymbol{B}\boldsymbol{B}^T + \boldsymbol{D}$. Hence the MLE of $\boldsymbol{\alpha}$ is the sample mean vector $\bar{\boldsymbol{y}}$. Thus, it is customary to replace $(\boldsymbol{y} - \boldsymbol{\alpha})$ by $(\boldsymbol{y} - \bar{\boldsymbol{y}})$ in factor analysis (Rubin and Thayer, 1982) and to take the parameter vector $\boldsymbol{\Psi}$ as consisting of the elements of $\boldsymbol{B}$ and the diagonal elements of $\boldsymbol{D}$. The observed data vector is

$\boldsymbol{y} = (\boldsymbol{y}_1^T, \ldots, \boldsymbol{y}_n^T)^T$, and the (incomplete-data) log likelihood is, apart from an additive constant,

$$\log L(\boldsymbol{\Psi}) = -\tfrac{1}{2}n\{\log |\boldsymbol{B}\boldsymbol{B}^T + \boldsymbol{D}| + \sum_{j=1}^{m}(\boldsymbol{y}_j - \bar{\boldsymbol{y}})^T(\boldsymbol{B}\boldsymbol{B}^T + \boldsymbol{D})^{-1}(\boldsymbol{y}_j - \bar{\boldsymbol{y}})\}.$$

Dempster, Laird, and Rubin (1977) formulate the EM algorithm for this problem with the complete-data vector as

$$\boldsymbol{x} = (\boldsymbol{x}_1^T, \ldots, \boldsymbol{x}_m^T)^T,$$

where

$$\boldsymbol{x}_j = (\boldsymbol{y}_j^T, \boldsymbol{z}_j^T)^T \quad (j = 1, \ldots, m);$$

that is,

$$\boldsymbol{z} = (\boldsymbol{z}_1^T, \ldots, \boldsymbol{z}_m^T)$$

is the missing-data vector. The complete-data log likelihood is, but for an additive constant,

$$\log L_c(\boldsymbol{\Psi}) = -\tfrac{1}{2}n \log |\boldsymbol{D}| - \tfrac{1}{2}\sum_{j=1}^{m}\{(\boldsymbol{y}_j - \boldsymbol{B}\boldsymbol{z}_j)^T\boldsymbol{D}^{-1}(\boldsymbol{y}_j - \boldsymbol{B}\boldsymbol{z}_j) + \boldsymbol{z}_j^T\boldsymbol{z}_j\}.$$

The complete-data density belongs to the exponential family and the complete-data sufficient statistics are $\boldsymbol{C}_{yy}, \boldsymbol{C}_{yz}$, and $\boldsymbol{C}_{zz}$, where

$$\boldsymbol{C}_{yy} = \sum_{j=1}^{n}(\boldsymbol{y}_j - \bar{\boldsymbol{y}})(\boldsymbol{y}_j - \bar{\boldsymbol{y}})^T;\ \boldsymbol{C}_{yz} = \sum_{j=1}^{m}(\boldsymbol{y}_j - \bar{\boldsymbol{y}})\boldsymbol{z}_j^T; \boldsymbol{C}_{zz} = \sum_{j=1}^{m}\boldsymbol{z}_j\boldsymbol{z}_j^T.$$

For this problem, DLR outline the implementation of the EM algorithm and Rubin and Thayer (1982) provide further details. The EM algorithm is implemented as follows.

E-Step. Given the current fit $\boldsymbol{\Psi}^{(k)}$ for $\boldsymbol{\Psi}$, calculate the conditional expectation of these sufficient statistics as follows:

$$E_{\boldsymbol{\Psi}^{(k)}}(\boldsymbol{C}_{yy} \mid \boldsymbol{y}) = \boldsymbol{C}_{yy},$$
$$E_{\boldsymbol{\Psi}^{(k)}}(\boldsymbol{C}_{yz} \mid \boldsymbol{y}) = \boldsymbol{C}_{yy}\boldsymbol{\gamma}^{(k)},$$

and

$$E_{\boldsymbol{\Psi}^{(k)}}(\boldsymbol{C}_{zz} \mid \boldsymbol{y}) = \boldsymbol{\gamma}^{(k)^T}\boldsymbol{C}_{yy}\boldsymbol{\gamma}^{(k)} + n\boldsymbol{\Delta}^{(k)},$$

where

$$\boldsymbol{\gamma}^{(k)} = \{\boldsymbol{D}^{(k)} + \boldsymbol{B}^{(k)}\boldsymbol{B}^{(k)^T}\}^{-1}\boldsymbol{B}^{(k)}$$

and

$$\boldsymbol{\Delta}^{(k)} = \boldsymbol{I}_q - \boldsymbol{B}^{(k)^T}(\boldsymbol{D}^{(k)} + \boldsymbol{B}^{(k)}\boldsymbol{B}^{(k)^T})\boldsymbol{B}^{(k)},$$

which are respectively, the regression coefficients and the residual covariance matrix of $\boldsymbol{z}$ on $\boldsymbol{y}$.

M-Step. With the values of current $\boldsymbol{\gamma}$ and $\boldsymbol{\Delta}$, calculate

$$\boldsymbol{B}^{(k+1)} = (\boldsymbol{\gamma}^T C_{yy}\boldsymbol{\gamma} + \boldsymbol{\Delta})^{-1}\boldsymbol{\gamma}^T C_{yy}$$

and

$$\boldsymbol{D}^{(k+1)} = \text{diag}\,\{C_{yy} - C_{yy}\boldsymbol{\gamma}(\boldsymbol{\gamma}^T C_{yy}\boldsymbol{\gamma} + \boldsymbol{\Delta})^{-1}\boldsymbol{\gamma}^T C_{yy}\}.$$

Although the EM algorithm is numerically stable, unlike the Newton–Raphson type algorithms for factor analysis, it has two unpleasant features. They are, as mentioned in Section 3.9.3, the possibility of multiple solutions (which are not necessarily rotations of each other) and slow convergence rate. The slowness of convergence could be attributed to the typically large fraction of missing data. Rubin and Thayer (1982, 1983) illustrate this EM algorithm with a data set from Lawley and Maxwell (1963), using some computations of Jöreskog (1969). They suggest that with the EM algorithm multiple local maxima can be reached from different starting points and question the utility of second-derivative-based classical methods. Bentler and Tanaka (1983), however, question the multiplicity of solutions obtained and attribute it to the slowness of convergence. Horng (1987), as pointed out in Section 3.9.3, proves sublinear convergence of the EM algorithm in certain situations. Recently, Duan and Simonato (1993) investigate the issue of severity of the multiplicity of solutions with an examination of eight data sets and conclude that multiplicity is indeed a common phenomenon in factor analysis, supportive of the Bentler-Tanaka conclusion. They also show examples of different solutions obtained under different convergence criteria.

Liu and Rubin (1994) develop the ECME algorithm for this problem. Here the partition of $\boldsymbol{\Psi}$ is $(\boldsymbol{\Psi}_1^T, \boldsymbol{\Psi}_2^T)^T$, where $\boldsymbol{\Psi}_1$ contains the elements of $\boldsymbol{B}$ and $\boldsymbol{\Psi}_2 = (\sigma_1^2, \ldots, \sigma_p^2)^T$. This choice is motivated by the simplicity of numerical maximization of the actual log likelihood over the p-dimensional $\boldsymbol{D}$, since it is the log likelihood for a normal distribution with restrictions on the covariance matrix compared to that of maximization over the matrix-valued $\boldsymbol{B}$ or over $\boldsymbol{B}$ and $\boldsymbol{D}$. In this version of the ECM algorithm, the E-step is as in the EM algorithm above. There are two CM-steps.

CM-Step 1. Calculate $\boldsymbol{B}^{(k+1)}$ as above.

CM-Step 2. Maximize the constrained actual log likelihood given $\boldsymbol{B}^{(k+1)}$ by an algorithm such as Newton-Raphson.

5.10.2 Numerical Example

In the example from Lawley and Maxwell (1963), the lower half of the symmetric matrix C_{yy} as given by Rubin and Thayer (1982) is as follows. There are $p = 9$ variables.

$$
Cyy = \begin{bmatrix}
1.0 & & & & & & & & \\
0.554 & 1.0 & & & & & & & \\
0.227 & 0.296 & 1.0 & & & & & & \\
0.189 & 0.219 & 0.769 & 1.0 & & & & & \\
0.461 & 0.479 & 0.237 & 0.212 & 1.0 & & & & \\
0.479 & 0.530 & 0.243 & 0.226 & 0.520 & 1.0 & & & \\
0.243 & 0.425 & 0.304 & 0.291 & 0.514 & 0.473 & 1.0 & & \\
0.280 & 0.311 & 0.718 & 0.681 & 0.313 & 0.348 & 0.374 & 1.0 & 0.672 \\
0.241 & 0.311 & 0.730 & 0.661 & 0.245 & 0.290 & 0.306 & 0.672 & 1.0
\end{bmatrix}
$$

The model used is with $q = 4$. Liu and Rubin (1994) use the same starting values as in Rubin and Thayer (1982, 1983) and apply the above ECME algorithm. Results of the log likelihood and estimates of σ_3 over iterations are given in Figure 5.3 (EM in solid lines and ECME in dashed lines), from which it appears that the ECME algorithm converges much faster than the EM algorithm. In terms of CPU time, the ECME algorithm was only 25 per cent faster than the EM algorithm.

5.11 EFFICIENT DATA AUGMENTATION

5.11.1 Motivation

Meng and van Dyk (1995) consider speeding up the convergence of the EM algorithm through the choice of the complete data, or effectively, the choice of the missing data in the specification of the complete-data problem in the EM framework. Their idea is to search for an efficient way of augmenting the observed data, where by "efficient", they mean less augmentation of the observed data while maintaining the stability and simplicity of the EM algorithm.

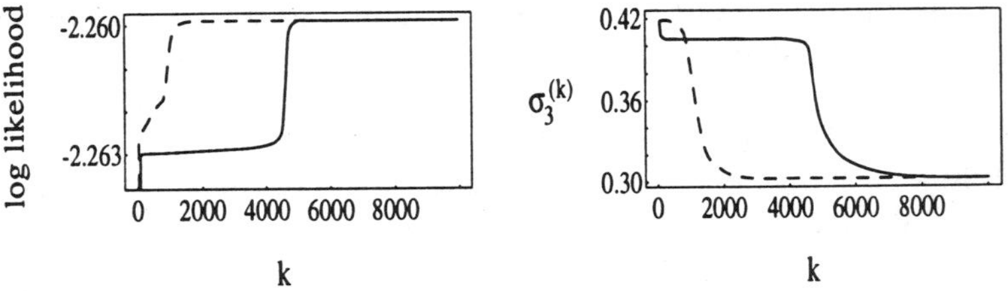

Figure 5.3. Factor analysis: convergence of EM (solid line) and ECME (dashed line) for log likelihood and $\sigma_3^{(k)}$. (From Liu and Rubin, 1994.)

As shown in Section 3.9, the rate of convergence (equivalently, the speed of convergence) of the EM algorithm depends on the proportion of missing information in the prescribed EM framework. The smaller this proportion, the greater the speed of convergence. Hence by augmenting the observed data less, the speed of convergence of the resulting EM algorithm will be greater. However, a disadvantage of less augmentation of the observed data is that the resulting E- or M-steps, or both, may be made appreciably more difficult to implement. But if the E- and M-steps are equally (or only slightly less) simple to implement and the gain in speed is relatively substantial, then there is no reason not to use the faster EM algorithm. To this end, Meng and van Dyk (1995) introduce a working parameter in their specification of the complete data, which thus indexes a class of EM algorithms corresponding to the different values of the working parameter. The aim is to select a value of the working parameter that increases the speed of convergence (that is, provides less data augmentation), without appreciably affecting the stability and simplicity of the resulting EM algorithm.

5.11.2 Maximum Likelihood Estimation of *t*-Distribution

We shall now consider one of the examples that Meng and van Dyk (1995) use to illustrate their approach. It concerns the problem of ML estimation of the parameters of the multivariate t-distribution with known degrees of freedom ν, which was initially introduced as Example 2.6 in Section 2.6 and considered further in the previous sections of this chapter in the case of ν being also unknown.

Suppose as in Example 2.6, that $\boldsymbol{w}_1, \ldots, \boldsymbol{w}_n$ denote an observed random sample from the multivariate t-distribution defined by (2.32). Then from (2.30), we can write $\boldsymbol{W}_j$ as

$$\boldsymbol{W}_j = \boldsymbol{\mu} + \boldsymbol{C}_j / U_j^{1/2} \quad (j = 1, \ldots, n), \tag{5.69}$$

where $\boldsymbol{C}_j$ is distributed $N(\boldsymbol{0}, \boldsymbol{\Sigma})$ independently of U_j, which is distributed as

$$U_j \sim \text{gamma}\left(\tfrac{1}{2}\nu, \tfrac{1}{2}\nu\right). \tag{5.70}$$

From (5.69), $\boldsymbol{W}_j$ can be expressed further as

$$\begin{aligned} \boldsymbol{W}_j &= \boldsymbol{\mu} + |\, \boldsymbol{\Sigma} \,|^{-\frac{a}{2}} \, \boldsymbol{C}_j / \{|\, \boldsymbol{\Sigma} \,|^{-a} \, U_j\}^{1/2} \\ &= \boldsymbol{\mu} + |\, \boldsymbol{\Sigma} \,|^{-\frac{a}{2}} \, \boldsymbol{C}_j / \{U_j(a)^{1/2}\}, \end{aligned} \tag{5.71}$$

where

$$U_j(a) = |\, \boldsymbol{\Sigma} \,|^{-a} \, U_j \tag{5.72}$$

and $U_j(0) = U_j$, as defined by (5.70).

Here, a is the working parameter used by Meng and van Dyk (1995) to scale the missing-data variable U_j ($j = 1, \ldots, n$) in the complete-data formulation of the problem. The complete-data vector is defined as

$$\boldsymbol{x}(a) = (\boldsymbol{y}^T, \boldsymbol{z}^T(a))^T, \tag{5.73}$$

where

$$\boldsymbol{z}(a) = (u_1(a), \ldots, u_n(a))^T$$

is the missing-data vector.

For a given value of a, the EM algorithm can be implemented with the complete-data specified by (5.73). The case $a = 0$ corresponds to the standard specification as considered previously in Example 2.6. Because the distribution of $U_j(a)$ for $a \neq 0$ depends on $\boldsymbol{\Sigma}$, the M-step of the EM algorithm, and hence its speed of convergence, is affected by the choice of a.

From (3.66) and (3.70), the global speed s of convergence of the EM algorithm in a neighborhood of a limit point $\boldsymbol{\Psi}^*$ is given by the smallest eigenvalue of

$$\mathcal{I}_c^{-1}(\boldsymbol{\Psi}^*; \boldsymbol{y})\boldsymbol{I}(\boldsymbol{\Psi}^*; \boldsymbol{y}). \tag{5.74}$$

Here with the specification of the complete-data indexed by the working parameter a, s will be some function of $a, s(a)$. Meng and van Dyk (1995) suggest using the value of a, a_{opt}, that maximizes the speed $s(a)$; that is, a_{opt} is the value of a that maximizes the smallest eigenvalue of (5.74), in which the term $\boldsymbol{I}(\boldsymbol{\Psi}^*; \boldsymbol{y})$ does not depend on a. They showed that a_{opt} is given by

$$a_{\text{opt}} = 1/(\nu + p), \tag{5.75}$$

which remarkably does not depend on the observed data $\boldsymbol{y}$.

We now consider the implementation of the EM algorithm for a given (arbitrary) value of the working parameter a. The implementation of the E-step is straightforward since the complete-data log likelihood is a linear function of $U_j(a)$, which is distributed as a (constant) multiple of U_j. Hence on the E-step at the $(k+1)$th iteration, we have from (5.72) that

$$E_{\boldsymbol{\Psi}^{(k)}}\{U_j(a) \mid \boldsymbol{y}\} = u_j^{(k)}(a),$$

where

$$u_j^{(k)}(a) = \mid \boldsymbol{\Sigma} \mid^{-a} u_j^{(k)}, \tag{5.76}$$

and $u_j^{(k)}$ is given by (5.43).

The M-step is not straightforward for nonzero a, but rather fortuitously, its implementation simplifies in the case of $a = a_{\text{opt}}$. The only modification required to the current estimates of $\boldsymbol{\mu}$ and $\boldsymbol{\Sigma}$ in the standard situation ($a = 0$)

is to replace the divisor n by

$$\sum_{j=1}^{n} u_j^{(k)} \tag{5.77}$$

in the expression (2.43) for $\boldsymbol{\Sigma}^{(k+1)}$.

This replacement does not affect the limit, as

$$\sum_{j=1}^{n} u_j^{(k)} \to n, \quad k \to \infty. \tag{5.78}$$

This was proved by Kent, Tyler, and Vardi (1994), who used it to modify one of their EM algorithms for fitting the t-distribution. They constructed an EM algorithm via a "curious likelihood identity" originally proposed in Kent and Tyler (1991) for transforming a p-dimensional location-scale t-likelihood into a $(p+1)$-dimensional scale-only t-likelihood. They reported that this algorithm with the modification (5.77), converges faster than the EM algorithm for the usual specification of the complete data. Meng and van Dyk (1995) note that the EM algorithm with $a = a_{\text{opt}}$ is identical to the modified EM algorithm of Kent et al. (1994), hence explaining its faster convergence; see also Arslan, Constable, and Kent (1995).

We now outline the details of the M-step as outlined above for a given value of the working parameter a. The complete-data likelihood function $L_c(\boldsymbol{\Psi})$ formed from the complete-data vector $\boldsymbol{x}(a)$ can be factored into the product of the conditional density of $\boldsymbol{Y} = (\boldsymbol{W}_1, \ldots, \boldsymbol{W}_n)^T$ given $\boldsymbol{z}(a)$ and the marginal density of $\boldsymbol{Z}(a)$. Unlike the case of $a = 0$ in Example 2.6, the latter density for nonzero a depends on $\boldsymbol{\Sigma}$, being a function of $|\,\boldsymbol{\Sigma}\,|^a$. As noted above, the conditional expectation of $\log L_c(\boldsymbol{\Psi})$ given the observed data $\boldsymbol{y}$, is effected by replacing $u_j(a)$ by its current conditional expectation, $u_j^{(k)}(a)$. It follows on ignoring terms not depending on $\boldsymbol{\Psi}$ that the Q-function is given by

$$\begin{aligned} Q(\boldsymbol{\Psi}; \boldsymbol{\Psi}^{(k)}) &= \tfrac{1}{2} n \log |\,\boldsymbol{\Sigma}\,| \{a(p + \nu - 1)\} \\ &\quad - \tfrac{1}{2} |\,\boldsymbol{\Sigma}\,|^a \sum_{j=1}^{n} u_j^{(k)}(a)\{\nu + (\boldsymbol{w}_j - \boldsymbol{\mu})^T \boldsymbol{\Sigma}^{-1} (\boldsymbol{w}_j - \boldsymbol{\mu})\}. \end{aligned} \tag{5.79}$$

The term

$$\sum_{j=1}^{n} u_j^{(k)}(a)(\boldsymbol{w}_j - \boldsymbol{\mu})^T \boldsymbol{\Sigma}^{-1} (\boldsymbol{w}_j - \boldsymbol{\mu})$$

on the right-hand side of (5.79) can be written as

$$\sum_{j=1}^{n} u_j^{(k)}(a)\{(\overline{\boldsymbol{w}}_u^{(k)} - \boldsymbol{\mu})^T \boldsymbol{\Sigma}^{-1}(\overline{\boldsymbol{w}}_u^{(k)} - \boldsymbol{\mu}) + \text{tr}(\boldsymbol{\Sigma}^{-1}\boldsymbol{S}_u^{(k)})\}, \tag{5.80}$$

where

$$\begin{aligned}\overline{\boldsymbol{w}}^{(k)} &= \sum_{j=1}^{n} u_j^{(k)}(a)\boldsymbol{w}_j \Big/ \sum_{j=1}^{n} u_j^{(k)}(a), \\ &= \sum_{j=1}^{n} u_j^{(k)}\boldsymbol{w}_j \Big/ \sum_{j=1}^{n} u_j^{(k)},\end{aligned}$$

and

$$\begin{aligned}\boldsymbol{S}_u^{(k)} &= \sum_{j=1}^{n} u_j^{(k)}(a)(\boldsymbol{w}_j - \overline{\boldsymbol{w}}_u^{(k)})(\boldsymbol{w}_j - \overline{\boldsymbol{w}}_u^{(k)})^T \Big/ \sum_{j=1}^{n} u_j^{(k)}(a) \\ &= \sum_{j=1}^{n} u_j^{(k)}(\boldsymbol{w}_j - \overline{\boldsymbol{w}}_u^{(k)})(\boldsymbol{w}_j - \overline{\boldsymbol{w}}_u^{(k)})^T \Big/ \sum_{j=1}^{n} u_j^{(k)}.\end{aligned} \tag{5.81}$$

It follows from (5.79) and (5.80) that

$$\boldsymbol{\mu}^{(k+1)} = \overline{\boldsymbol{w}}_u^{(k)}.$$

Concerning the calculation of $\boldsymbol{\Sigma}^{(k+1)}$, Meng and van Dyk (1995) show that on differentiation of (5.79) with respect to the elements of $\boldsymbol{\Sigma}^{-1}$ and putting $\boldsymbol{\mu} = \boldsymbol{\mu}^{(k+1)} = \overline{\boldsymbol{w}}_u^{(k)}$, it satisfies the equation

$$\frac{\{a(p+\nu)-1\}}{|\,\boldsymbol{\Sigma}\,|^a\,\overline{u}^{(k)}(a)}\boldsymbol{\Sigma} + \boldsymbol{S}_u^{(k)} = a\{\nu + \text{tr}(\boldsymbol{\Sigma}^{-1}\boldsymbol{S}_u^{(k)})\}\boldsymbol{\Sigma}, \tag{5.82}$$

where

$$\overline{u}^{(k)}(a) = \sum_{j=1}^{n} u_j^{(k)}(a)/n.$$

As remarked by Meng and van Dyk (1995), solving (5.82) for arbitrary a is quite difficult, but there are a few values of a that make (5.82) trivial to solve. One is $a = 0$, while another is $a = a_{\text{opt}}$, for which the first term on the left-hand side of (5.82) is zero, which directly shows that $\boldsymbol{\Sigma}^{(k+1)}$ is proportional to $\boldsymbol{S}_u^{(k)}$. It then follows that

$$\boldsymbol{\Sigma}^{(k+1)} = \boldsymbol{S}_u^{(k)}.$$

We are presently only considering the case where the degrees of freedom ν is known. The extension to the general case of unknown ν is straightforward. For since the presence of the working parameter a does not affect the computation of the parameter ν, it is calculated as in the case of $a = 0$ in Section 5.8.

As the theoretical measure $s(a)$ only measures the speed of convergence near convergence, Meng and van Dyk (1995) perform some simulations to further examine the actual gains in computational time by using $a = a_{\text{opt}}$ over $a = 0$ in fitting the t-distribution with known degrees of freedom ν. Random samples of size $n = 100$ were generated from each of three univariate distributions: (1) the standard normal; (2) a t-distribution with $\mu = 0, \sigma^2 = 1$, and $\nu = 1$ (that is, the Cauchy distribution); and (3) a mixture of a standard normal and an exponential with mean 3 in proportions 2/3 and 1/3. The simulations from distributions (1) and (3) were intended to reflect the fact that, in reality, there is no guarantee that the data are from a t-distribution, nor even a symmetric model. For each configuration, there were 1000 simulation trials, on each of which the EM algorithm for $a = 0$ and a_{opt} was started with μ and σ^2 equal to the sample mean and variance, respectively. The number of iterations N_0 and N_{opt} taken by the EM algorithm in achieving

$$\|\boldsymbol{\Psi}^{(k+1)} - \boldsymbol{\Psi}^{(k)}\|^2 / \|\boldsymbol{\Psi}^{(k)}\|^2 \leq 10^{-10}$$

for $a = 0$ and a_{opt}, respectively, was recorded. A comparison of the number of iterations taken by two algorithms can be misleading, but, in this case, the EM algorithm for both values of a require the same amount of computation per iteration. It was found that on all 6000 simulation trials the EM algorithm for $a = a_{\text{opt}}$ was faster than for $a = 0$. Generally, the improvement was quite significant. On 5997 trials, the improvement was greater than 10 percent, and often reached as high as 50 percent when the Cauchy distribution was being fitted. Meng and van Dyk (1995) note that the EM algorithm in the standard situation of $a = 0$ tends to be slower for smaller values of ν in the t-distribution, and thus the observed improvement is greatest where it is most useful.

Meng and van Dyk (1995) also perform a second simulation experiment to assess the gains in higher dimensions. In this experiment, 1000 random samples of size $n = 100$ were generated from a $p = 10$ dimensional Cauchy distribution. The EM algorithm for the standard situation of $a = 0$ was found to be at least 6.5 times slower on every trial and was usually between eight to ten times as slow. As explained by Meng and van Dyk (1995), the difference between the two versions of the EM algorithm corresponding to $a = 0$ and $a = a_{\text{opt}}$, stems from the fact that the Q-function, $Q(\boldsymbol{\Psi}; \boldsymbol{\Psi}^{(k)})$, in the latter case is flatter because of less data augmentation, and so provides a better approximation to the incomplete-data log likelihood, $\log L(\boldsymbol{\Psi})$.

5.11.3 Variance Components Model

Meng and van Dyk (1995) also consider the efficient choice of the complete-data in the variance components model (5.56), to which the ECME was applied in Section 5.9.1. Analogous to the rescaling of the missing data through the use of a working parameter a in the t-model considered above, Meng and van Dyk (1995) consider rescaling the vector $\boldsymbol{b}$ of random effects by $\boldsymbol{D}^{-a}$, where $\boldsymbol{D}$ is the covariance matrix of $\boldsymbol{b}$ and a is an arbitrary constant. The missing data vector is then declared to be

$$z(a) = | \boldsymbol{D} |^{-a} \boldsymbol{b}.$$

However, as the resulting EM algorithm would be very difficult for arbitrary $\boldsymbol{D}$, Meng and van Dyk (1995) first diagonalize $\boldsymbol{D}$. Let $\boldsymbol{C}$ be a lower triangular matrix such that

$$\boldsymbol{C}\boldsymbol{D}\boldsymbol{C}^T = \text{diag}(d_1, \ldots, d_n). \tag{5.83}$$

Then $\boldsymbol{b}^* = \boldsymbol{C}\boldsymbol{b}$ has the diagonal covariance matrix given by (5.83). Meng and van Dyk (1995) then specify the missing data vector as

$$z(\boldsymbol{a}) = (b_1^*/d_1^{a_1/2}, \ldots, b_n^*/d_n^{a_n/2})^T,$$

where the working parameter

$$\boldsymbol{a} = (a_1, \ldots, a_n)^T$$

is now a vector so as to allow each transformed random effect to be scaled by its own standard deviation. Although in principle, the EM algorithm can be implemented for any given value of $\boldsymbol{a}$, Meng and van Dyk (1995) restrict each a_i to being zero or one in order to keep the resulting EM algorithm simple to implement. They report empirical and theoretical results demonstrating the gains where achievable with this approach.

5.12 ALTERNATING ECM ALGORITHM

Meng and van Dyk (1995) propose an extension of the EM algorithm called the Alternating ECM (AECM) algorithm. It is really an extension of the ECM algorithm, where the specification of the complete-data is allowed to be different on each CM-step. That is, the complete-data vector need not be the same on each CM-step. We have seen that the ECME algorithm allows one to consider maximization of the actual log likelihood function on a CM-step rather than the Q-function, representing the current conditional expectation of the complete-data log likelihood function. Hence it can be viewed as a special case of the AECM algorithm, where the complete data can be specified as the observed data $\boldsymbol{y}$ on a CM-step.

There is also the Space-Alternating Generalized EM (SAGE) algorithm proposed by Fessler and Hero (1994). Their algorithm, which was developed without knowledge of the ECM nor ECME algorithms, was motivated from a different angle to that of the ECM algorithm of Meng and Rubin (1993). They started with the CM algorithm (that is, there is no initial augmentation of the observed data) in the special case where each step corresponds to a partition of the parameter vector into subvectors. They then propose the EM algorithm be applied to implement any CM-step for which an explicit solution does not exist. The specification of the complete-data is therefore usually different for each CM-step. With the SAGE algorithm, only one EM iteration is performed on a given CM-step, which is in the same spirit of the ECM algorithm which performs only one iteration of the CM algorithm on each EM iteration.

Meng and van Dyk (1995) propose generalizing the SAGE and ECME algorithms by combining them into the one algorithm, called the AECM algorithm. It allows the augmentation of the observed data to vary over the CM-steps, which is a key ingredient of the SAGE algorithm. It also allows for CM-steps that go beyond a simple partition of the parameter vector into subvectors (as, for example, in Section 5.6), a key feature of the ECME algorithm.

Meng and van Dyk (1995) establish that monotone convergence of the sequence of likelihood values $\{L(\boldsymbol{\Psi}^{(k)}\}$ is retained with the AECM algorithm and that under standard regularity conditions, the sequence of parameter iterates converges to a stationary point of $L(\boldsymbol{\Psi})$. They also provide more complete results for the ECME and SAGE algorithms.

Meng and van Dyk (1995) illustrate the AECM algorithm by applying it to the ML estimation of the parameters of the t-distribution with all parameters unknown. The application of the ECME algorithm to this problem was described in Section 5.8. They consider two versions of this algorithm for this problem. As with the aforementioned application of the ECME algorithm to this problem, both versions had two CM-steps as set out in Section 5.9, corresponding to the partition

$$\boldsymbol{\Psi} = (\boldsymbol{\Psi}_1, \Psi_2)^T,$$

where $\boldsymbol{\Psi}_1$ contains $\boldsymbol{\mu}$ and the distinct elements of $\boldsymbol{\Sigma}$, and with Ψ_2 a scalar equal to ν. In Version 1 of the AECM algorithm, the complete-data vector is specified on both CM-steps by (5.73) with $a = a_{\text{opt}}$ as given by (5.75). In Version 2, the complete-data vector is taken to be the observed data-vector $\boldsymbol{y}$ on the second CM-step for the calculation of the current estimate of ν, as in the application of the ECME algorithm to this problem.

Meng and van Dyk (1995) perform some simulations to compare these two versions of the AECM algorithm with the multicycle ECM and ECME algorithms. Random samples of size $n = 100$ were generated from a $p = 10$ dimensional t-distribution with $\boldsymbol{\mu} = \mathbf{0}$ and $\nu = 10$, and where $\boldsymbol{\Sigma}$ was selected at the outset of the simulation as a positive definite nondiagonal matrix. The same stopping criterion and the same starting values for $\boldsymbol{\mu}$ and $\boldsymbol{\Sigma}$ as in their simulations described in Section 5.11.2 were used for the four algorithms. The starting value for ν was $\nu^{(0)} = 10$. Version 1 of the AECM algorithm was found to be eight to twelve times faster than either the multicycle ECM or the ECME algorithms. The cost per iteration is less for Version 1 and the multicycle ECM algorithm than for the ECME algorithm. Moreover, Version 2 of the AECM algorithm was only slightly more efficient than Version 1 in terms of the number of iterations required, and less efficient in terms of actual computer time.

In these results, the choice of the complete-data vector for the computation of ν on the second CM-step makes little difference in terms of the number of iterations required for convergence. Yet, as Meng and van Dyk (1995) point out, Liu and Rubin (1994, 1995) have shown that the ECME algorithm can be much more efficient than the multicycle ECM algorithm. However, there are two principal differences between their examples and the simulations of Meng and van Dyk (1995). The latter were for ten-dimensional data with no missing values, whereas the examples of Liu and Rubin (1994, 1995) were for bivariate data which has a very high proportion of missing values.

Meng and van Dyk (1995) replicate the analyses in Liu and Rubin (1995) to investigate the relative merit of the four algorithms in the presence of missing values. These analyses were for the clinical-trial example of Liu and Rubin (1995), where the four-dimensional data were taken from Shih and Weisberg (1986), and the artificial bivariate data set of Murray (1977). As above, Version 1 of the AECM algorithm emerges clearly as their recommended choice.

In other illustrative work on the AECM algorithm, Meng and van Dyk (1995) apply it to the analysis of PET/SPECT data. The application of the EM algorithm to this problem has been considered in Section 2.5.

5.13 EMS ALGORITHM

Silverman, Jones, Wilson, and Nychka (1990) modify the EM algorithm by introducing a smoothing step (the S-step) at each EM iteration. They called the resulting procedure the EMS algorithm. When applied to the problem of estimating the vector $\boldsymbol{\lambda}$ of emission intensities for PET/SPECT data, as considered in Section 2.5, the estimate $\lambda_i^{(k+1)}$ is given by

$$\lambda_s^{(k+1)} = \sum_{i=1}^{n} w_{si} \lambda_i^{(k)} q_i^{-1} \sum_{j=1}^{d} \left\{ y_j p_{ij} \Big/ \sum_{h=1}^{n} \lambda_h^{(k)} p_{hj} \right\} \quad (s = 1, \ldots, n), \tag{5.84}$$

where $\boldsymbol{w} = ((w_{si}))$ is a smoothing matrix. Usually, $\boldsymbol{w}$ is taken to be row-stochastic. If it is doubly stochastic, then the algorithm preserves photon energy and no normalization step is required; see Green (1990b) and Kay (1994).

In partial justification of this procedure, Nychka (1990) shows that an approximation to the EMS algorithm can be viewed as a penalized likelihood approach. Kay (1994) reports some unpublished work in which he has established the convergence of the EMS algorithm and explained its faster rate of convergence than the ordinary EM algorithm.

5.14 ONE-STEP-LATE ALGORITHM

The One-Step-Late (OSL) algorithm was first suggested by Green (1990a) in the context of reconstruction of real tomographic data. It was subsequently proposed by Green (1990b) in a general context for finding the MAP estimate or the MPLE.

With the latter solution, the problem in the context of the analysis of PET/SPECT data as considered in Section 2.5, is to maximize the function

$$\log L(\boldsymbol{\lambda}) - \xi K(\boldsymbol{\lambda}), \tag{5.85}$$

where ξ is some smoothing parameter and $K(\boldsymbol{\lambda})$ is a roughness functional.

Alternatively,

$$\exp\{-\xi K(\boldsymbol{\lambda})\}$$

can be regarded as proportional to a prior distribution for $\boldsymbol{\lambda}$ as proposed, for example, by Geman and McClure (1985). In that case, (5.85) represents the log of the posterior density of $\boldsymbol{\lambda}$, ignoring a normalizing constant not involving $\boldsymbol{\lambda}$.

In applying the EM algorithm to the problem of maximizing (2.26), the E-step is the same as in Section 2.5, in that it effectively requires the calculation of (2.28). But the M-step is now more complicated since $\boldsymbol{\lambda}^{(k+1)}$ must be obtained by solving a set of highly nonlinear, nonlocal equations,

$$\sum_{j=1}^{d} z_{ij}^{(k)}/\lambda_i - \sum_{j=1}^{d} p_{ij} - \xi \partial K(\boldsymbol{\lambda})/\partial \lambda_i = 0 \quad (i = 1, \ldots, n), \tag{5.86}$$

where $z_{ij}^{(k)}$ is the conditional expectation of Z_{ij} given the observed data $\boldsymbol{y}$, using the current fit $\boldsymbol{\lambda}^{(k)}$ for $\boldsymbol{\lambda}$. It is given by (2.28).

Green (1990a) proposes to evaluate the partial derivatives in (5.86) at the current estimate $\boldsymbol{\lambda}^{(k)}$, thus yielding

$$\lambda_i^{(k+1)} = \frac{\sum_{j=1}^{d} z_{ij}^{(k)}}{\sum_{j=1}^{d} p_{ij} + \xi \partial K(\boldsymbol{\lambda}^{(k)})/\partial \lambda_i} \quad (i = 1, \ldots, n). \tag{5.87}$$

Green (1990b) proves a local convergence result for this OSL algorithm provided that the value of ξ is not too large. Lange (1990) extends this work and, by introducing a line-search at each iteration, is able to produce a global convergence result.

5.15 VARIANCE ESTIMATION FOR PENALIZED EM AND OSL ALGORITHMS

5.15.1 Penalized EM Algorithm

Segal, Bacchetti, and Jewell (1994) synthesize the Supplemented EM algorithm of Meng and Rubin (1991) and the OSL algorithm of Green (1990b) for MPL estimation to suggest a method for computing the asymptotic covariance matrix of the MPLE through the EM computations. Let us consider the maximization of

$$\log L(\boldsymbol{\Psi}) - \xi K(\boldsymbol{\Psi})$$

as in equation (5.85). In the EM algorithm for MPL estimation, the E-step is the same as for ML estimation, but on the M-step

$$Q(\boldsymbol{\Psi}; \boldsymbol{\Psi}^{(k)}) - \xi K(\boldsymbol{\Psi}) \tag{5.88}$$

is to be maximized.

We let $\boldsymbol{M}_P(\boldsymbol{\Psi})$ be the mapping induced by the sequence of iterates $\{\boldsymbol{\Psi}^{(k)}\}$, where $\boldsymbol{\Psi}^{(k+1)}$ is obtained by maximization of (5.88). If this is accomplished by equating the derivative of (5.88) to zero, then Green (1990b) shows that the Jacobian $\boldsymbol{J}_P(\boldsymbol{\Psi})$ of this mapping at $\boldsymbol{\Psi} = \tilde{\boldsymbol{\Psi}}$ is given by

$$\boldsymbol{J}_P(\tilde{\boldsymbol{\Psi}}) = \{\mathcal{I}_c(\tilde{\boldsymbol{\Psi}}; \boldsymbol{y}) + \xi \partial^2 \boldsymbol{K}(\tilde{\boldsymbol{\Psi}})/\partial \boldsymbol{\Psi} \partial \boldsymbol{\Psi}^T\}^{-1} \mathcal{I}_m(\tilde{\boldsymbol{\Psi}}; \boldsymbol{y}), \tag{5.89}$$

where $\tilde{\boldsymbol{\Psi}}$ is the MPLE of $\boldsymbol{\Psi}$.

The asymptotic covariance matrix of the MPLE $\tilde{\boldsymbol{\Psi}}$ can be estimated by $\boldsymbol{V}$ where, corresponding to the use of the inverse of the observed information matrix in the unpenalized case, it is defined as the inverse of the negative of the Hessian of the right-hand side of (5.88) evaluated at $\boldsymbol{\Psi} = \tilde{\boldsymbol{\Psi}}$. Segal et al. (1994) exploited the representation (4.71) of the observed information matrix in the unpenalized situation to express the asymptotic covariance matrix $\boldsymbol{V}$ of the MPLE $\hat{\boldsymbol{\Psi}}$ as

$$\boldsymbol{V} = \{\boldsymbol{I}_d - \boldsymbol{J}_P(\tilde{\boldsymbol{\Psi}})\}^{-1} \boldsymbol{V}_c \tag{5.90}$$

$$= \boldsymbol{V}_c + \Delta \boldsymbol{V}, \tag{5.91}$$

where

$$V_c = \{\mathcal{I}_c(\tilde{\boldsymbol{\Psi}}; \boldsymbol{y}) + \xi \partial^2 \boldsymbol{K}(\tilde{\boldsymbol{\Psi}})/\partial \boldsymbol{\Psi} \partial \boldsymbol{\Psi}^T\}^{-1}$$

and

$$\Delta \boldsymbol{V} = \{\boldsymbol{I}_d - \boldsymbol{J}_P(\tilde{\boldsymbol{\Psi}})\}^{-1} \boldsymbol{J}_P(\tilde{\boldsymbol{\Psi}}) \boldsymbol{V}_c.$$

The diagonal elements of $\Delta \boldsymbol{V}$ give the increases in the asymptotic variances of the components of the MPLE $\tilde{\boldsymbol{\Psi}}$ due to not having observed $\boldsymbol{x}$ fully.

As with the Supplemented EM algorithm, $\boldsymbol{J}_P(\tilde{\boldsymbol{\Psi}})$ can be computed by numerical differentiation, using the penalized EM code; $\mathcal{I}_c(\tilde{\boldsymbol{\Psi}}; \boldsymbol{y})$ can be computed by complete-data methods. The term $\partial^2 \boldsymbol{K}(\tilde{\boldsymbol{\Psi}})/\partial \boldsymbol{\Psi} \partial \boldsymbol{\Psi}^T$ has to be evaluated analytically except in simple cases like quadratic $K(\boldsymbol{\Psi})$, when it becomes a constant.

5.15.2 OSL Algorithm

As noted in Section 1.6.3, the MPLE of $\boldsymbol{\Psi}$ can be computed using the OSL algorithm. Green (1990b) has shown that the Jacobian $\boldsymbol{J}_{OSL}(\boldsymbol{\Psi})$ of the mapping induced by the OSL algorithm is given at $\boldsymbol{\Psi} = \tilde{\boldsymbol{\Psi}}$ by

$$\boldsymbol{J}_{OSL}(\tilde{\boldsymbol{\Psi}}) = \mathcal{I}_c^{-1}(\tilde{\boldsymbol{\Psi}}; \boldsymbol{y})\{\mathcal{I}_m(\tilde{\boldsymbol{\Psi}}; \boldsymbol{y}) - \xi \partial^2 K(\tilde{\boldsymbol{\Psi}})/\partial \boldsymbol{\Psi} \partial \boldsymbol{\Psi}^T\}. \tag{5.92}$$

Adapting the result (4.72) to the OSL algorithm, Segal et al. (1994) note that the asymptotic covariance matrix of the MPLE $\tilde{\boldsymbol{\Psi}}$ can be expressed as

$$\{\boldsymbol{I}_d - \boldsymbol{J}_{OSL}(\tilde{\boldsymbol{\Psi}})\}^{-1} \mathcal{I}_c^{-1}(\tilde{\boldsymbol{\Psi}}; \boldsymbol{y}).$$

It is interesting to note that this formula does not directly involve the penalty term. Similarly, as for the Jacobians of the EM and penalized EM maps, $\boldsymbol{J}_{OSL}(\tilde{\boldsymbol{\Psi}})$ can be computed by numerical differentiation, using the code for the OSL algorithm.

5.15.3 Example 5.7 (Example 4.1 Continued)

To give a numerical example, we return to Example 4.1, involving the multinomial data (1.12). This illustration is taken from Segal et al. (1994) who used it to demonstrate the use of (5.90) for providing an estimate of the variance of the MPLE obtained via the EM algorithm.

Following Green (1990b), Segal et al. (1994) added the penalty function

$$-10(\Psi - 0.5)^2 \tag{5.93}$$

to the log likelihood function $\log L(\Psi)$ given by (1.14). The use of (5.93) corresponds to a truncated normal prior for $\boldsymbol{\Psi}$. In this case, the MPLE of Ψ, $\tilde{\Psi}$, is a solution of the equation

$$\frac{y_1}{2+\Psi} - \frac{y_2+y_3}{1-\Psi} + \frac{y_4}{\Psi} - 20(\Psi - 0.5) = 0. \tag{5.94}$$

This equation can be solved directly, being a quartic polynomial in Ψ with relevant root $\tilde{\Psi} = 0.60238108$. On differentiation of the left-hand side of (5.94) and taking the negative inverse evaluated at $\Psi = \tilde{\Psi}$, the estimate of the variance of $\tilde{\Psi}$ so obtained is 0.00256.

We can apply the EM algorithm to this problem in the same manner as in the unpenalized situation in Section 1.4.2. The E-step is the same, effectively requiring the calculation of (1.23) on the $(k+1)$th iteration. The M-step now takes $\Psi^{(k+1)}$ to be the root of the cubic equation

$$\frac{y_{12}^{(k)}}{2+\Psi} - \frac{y_2+y_3}{1-\Psi} + \frac{y_4}{\Psi} - 20(\Psi - 0.5) = 0. \tag{5.95}$$

Details on the convergence of the sequence $\{\Psi^{(k)}\}$ to $\tilde{\Psi}$ are displayed in Table 5.1, along with the ratios $r^{(k)}$ of successive deviations. It can be seen that they are essentially constant for $k \geq 3$ and are accurate to five decimal places at $k = 5$, where $J_P(\tilde{\Psi}) = 0.130732$.

We have seen in Section 4.2.5 that $\mathcal{I}_c(\hat{\Psi}; \boldsymbol{y})$ is given by the binomial variance, so that

$$\begin{aligned}\mathcal{I}_c(\tilde{\Psi}; \boldsymbol{y}) &= \frac{y_{12}^{(\infty)} + y_2 + y_3 + y_4}{\tilde{\Psi}(1-\tilde{\Psi})} \\ &= 431.38,\end{aligned} \tag{5.96}$$

where

$$y_{12}^{(\infty)} = \tfrac{1}{4} y_1 \tilde{\Psi} / (\tfrac{1}{2} + \tfrac{1}{4}\tilde{\Psi}).$$

Table 5.1. Results of the EM Algorithm for DLR Multinomial Data.

Iteration (k)	$\Psi^{(k)}$	$d^{(k)} = \Psi^{(k)} - \tilde{\Psi}$	$r^{(k)} = d^{(k+1)}/d^{(k)}$
0	0.50000000	-0.12038109	0.143085
1	0.60315651	-0.01722467	0.132367
2	0.61810110	-0.00227998	0.130946
3	0.62008253	-0.00029856	0.130760
4	0.62034205	-0.00003904	0.130736
5	0.62037598	-0.00000510	0.130732
6	0.62038042	-0.00000067	0.130762
7	0.62038100	-0.00000009	0.130732
8	0.62038108	-0.00000001	0.130732

Source: Adapted from Segal et al. (1994).

Also, $\xi\partial^2 \boldsymbol{K}(\tilde{\boldsymbol{\Psi}})/\partial\boldsymbol{\Psi}\partial\boldsymbol{\Psi}^T = 20$, and so from (5.90), we finally obtain $V = 0.00255$, which is decomposed according to equation (5.91) as $V_c = 0.00222$ and $\Delta V = 0.00033$.

5.16 LINEAR INVERSE PROBLEMS

Vardi and Lee (1993) note that many problems in science and technology can be posed as linear inverse problems with positivity restrictions, which they referred to as LININPOS problems. They show that all such problems can be viewed as statistical estimation problems from incomplete data based on "infinitely large samples", which could be solved by maximum likelihood via the EM algorithm. In the framework of Vardi and Lee (1993), LININPOS problems require solving the equation

$$g(y) = \int_{D_{g_c}} h(x, y) g_c(x)\, dx, \tag{5.97}$$

where D_{g_c} and D_g are the domains of the nonnegative real-valued functions g_c and g, respectively. In image analysis, g_c represents the true distorted image that would have been recorded, had there been no blurring in the image recording process, and g represents the recorded blurred image. The values of g_c and g are grey-level intensities. The function $h(x, y)$, which is assumed to be a bounded nonnegative function on $D_{g_c} \times D_g$, characterizes the blurring mechanism.

As noted by Vardi and Lee (1993), the class of LININPOS problems and the corresponding class of estimation problems based on incomplete data are rich and include examples for which the EM algorithm has been independently derived. Examples include image reconstruction in PET/SPECT, as discussed in Section 2.5, as well as more traditional statistical estimation problems, such as ML estimation from grouped and truncated data, as considered in Section 2.8.

CHAPTER 6

Miscellaneous Topics

6.1 ITERATIVE SIMULATION ALGORITHMS

In the last decade or so a large body of methods has emerged based on iterative simulation techniques useful especially in computing Bayesian solutions to the kind of incomplete-data problems discussed in the previous chapters. Most of these methods are aimed at estimating the entire posterior density and not just finding the maximum *a posteriori* (MAP) estimate of the parameter vector $\boldsymbol{\Psi}$. As emphasized in Gelman and Rubin (1992), it is wise to find the MLE or the MAP estimate of $\boldsymbol{\Psi}$ before using iterative simulations, because of the difficulties in assessing convergence of iterative simulation methods, particularly when the regions of high density are not known *a priori*.

These iterative simulation techniques are conceptually very similar to the EM algorithm, simply replacing the E- and M-steps by draws from the current conditional distribution of the missing data and $\boldsymbol{\Psi}$, respectively. However, in some methods such as the Monte Carlo EM algorithm, which is one of the simpler extensions of the EM algorithm, only the E-step is so implemented. Many of these methods can be interpreted as iterative simulation analogs of the various versions of the EM algorithm and its extensions. In this chapter, we give a brief outline of these methods.

6.2 MONTE CARLO E-STEP

In some applications of the EM algorithm, the E-step is complex and does not admit a closed-form solution to the computation of the conditional expectation of the complete-data log likelihood, that is, the Q-function, $Q(\boldsymbol{\Psi}; \boldsymbol{\Psi}^{(k)})$; for an example, see Whittemore and Grosser (1986). One way to get over this problem is to resort to numerical integration. However, in some situations, especially when the complete-data density does not belong to the exponential family, numerical integration over the missing-data density does not always preserve the function (Meng and Rubin, 1991). Thus executing the E-step by a Monte Carlo process may be a viable and attractive alternative. Such a method was introduced by Wei and Tanner (1990). An EM algorithm where the E-step is executed by Monte Carlo is known as a Monte Carlo EM (MCEM) algorithm. This method applies whether we seek the MLE or the MAP estimate. Guo and Thompson (1992) use the MCEM algorithm to tackle a complex genetic model. Sinha, Tanner, and Hall (1994) use the MCEM algorithm in an example of grouped survival data. Chan and Ledolter (1995) give an example of a time series model with count data, where the E-step is intractable even by numerical integration in view of the high dimensionality of the integral, and use the MCEM algorithm.

The Q-function that has to be calculated on the E-step of the EM algorithm may be expressed as

$$\begin{aligned} Q(\boldsymbol{\Psi}; \boldsymbol{\Psi}^{(k)}) &= E_{\boldsymbol{\Psi}^{(k)}}\{\log p(\boldsymbol{\Psi} \mid \boldsymbol{x}) \mid \boldsymbol{y}\} \\ &= \int_{\mathcal{Z}} \log p(\boldsymbol{\Psi} \mid \boldsymbol{x}) p(\boldsymbol{z} \mid \boldsymbol{y}; \boldsymbol{\Psi}^{(k)})\, d\boldsymbol{z}, \end{aligned} \tag{6.1}$$

where $\boldsymbol{x}$ is the complete data and $\mathcal{Z}$ is the sample space of $\boldsymbol{Z}$. Here $p(\boldsymbol{\Psi} \mid \boldsymbol{x})$ is equal to the complete-data likelihood for $\boldsymbol{\Psi}$ and $p(\boldsymbol{z} \mid \boldsymbol{y}; \boldsymbol{\Psi})$ is equal to the conditional density of the missing data $\boldsymbol{Z}$ given $\boldsymbol{y}$. In a Bayesian context, $p(\boldsymbol{\Psi} \mid \boldsymbol{x})$ is the complete-data posterior density for $\boldsymbol{\Psi}$ and $p(\boldsymbol{z} \mid \boldsymbol{y}; \boldsymbol{\Psi})$ is the conditional predictive density for $\boldsymbol{z}$, where p is used as a generic symbol for the density function or in the discrete case for the probability mass function.

We explain the MCEM algorithm with Example 1.1, where the single unknown parameter is Ψ and the missing data consists of the single variable $z = y_{12}$ which, conditionally on the observed data $\boldsymbol{y}$, has a binomial distribution,

$$\text{binomial}\left(y_1, \frac{\frac{1}{4}\Psi^{(k)}}{\frac{1}{2} + \frac{1}{4}\Psi^{(k)}}\right), \tag{6.2}$$

where $y_1 = 125$.

In this example, the E-step is easily carried out since the the complete-data log likelihood is linear in z, and its current conditional expectation given the observed data $\boldsymbol{y}$ is simply

$$y_1 \frac{\frac{1}{4}\Psi^{(k)}}{\frac{1}{2} + \frac{1}{4}\Psi^{(k)}}.$$

However, a Monte Carlo E-step would run as follows:

Monte Carlo E-Step. On the kth iteration, draw $z^{(1_k)}, \ldots, z^{(M_k)}$ from $p(z \mid \boldsymbol{y}; \Psi^{(k)})$, which in this case is specified by (6.2). Then approximate the Q-function as

$$Q(\Psi; \Psi^{(k)}) = \frac{1}{M} \sum_{m=1}^{M} \log p(\Psi \mid z^{(m_k)}; \boldsymbol{y}). \tag{6.3}$$

M-Step. Maximize $Q(\Psi; \Psi^{(k)})$ over Ψ to obtain $\Psi^{(k+1)}$. In this example, it turns out that the Q-function as approximated by (6.3) is given by

$$Q(\Psi; \Psi^{(k)}) = (\overline{z}^{(k)} + y_4) \log \Psi + (y_2 + y_3) \log(1 - \Psi),$$

yielding

$$\Psi^{(k+1)} = \frac{\overline{z}^{(k)} + y_4}{\overline{z} + y_4 + y_2 + y_3},$$

where $\overline{z}^{(k)}$ is the arithmetic mean of the sampled $z^{(1_k)}, \ldots, z^{(M_k)}$.

For the MCEM algorithm the monotonicity property is lost, but in certain cases, the algorithm gets close to a maximizer with a high probability; see Chan and Ledolter (1995) for details.

In iterative simulation algorithms of this sort, the choice of M and the monitoring of convergence of the algorithm are somewhat difficult problems. Wei and Tanner (1990) recommend that instead of choosing and fixing M for the entire algorithm, small values of M be used in the initial stages of the algorithm and be increased as the algorithm moves closer to convergence. As to monitoring convergence, they recommend that the values of $\Psi^{(k)}$ be tabulated or plotted against k and when convergence is indicated by the stabilization of the process with random fluctuations about a value $\hat{\Psi}$, the process may be terminated or continued with a larger value of M.

When $M = 1$ and the randomly drawn $\boldsymbol{z}^{(1_k)}$ is replaced by some "good" summary of $p(\boldsymbol{z} \mid \boldsymbol{y}; \boldsymbol{\Psi}^{(k)})$, such as its mode or mean, we obtain what is commonly called an EM-type algorithm (Wei and Tanner, 1990). If we use the mean and if the complete-data posterior density for $\boldsymbol{\Psi}$ (that is, the complete-data log likelihood) is linear in $\boldsymbol{z}$, then we have the EM algorithm.

6.3 STOCHASTIC EM ALGORITHM

Prior to the appearance of the MCEM algorithm, Broniatowski, Celeux, and Diebolt (1983) and Celeux and Diebolt (1985, 1986a, 1986b) considered a modified version of the EM algorithm in the context of computing the MLE for finite mixture models. They call it the Stochastic EM algorithm. It is the same as the MCEM algorithm with $M = 1$. As we shall see later, Tanner and Wong (1987) also consider a Monte Carlo E-step with $M = 1$ in a more general method called Data Augmentation, which enables them to connect it to another iterative simulation technique called the Gibbs sampler. We discuss Celeux and Diebolt's method for the mixture problem here.

We use the notation of Sections 1.4.3 and 2.7, where we let $\boldsymbol{\theta}$ denote the vector containing the parameters known *a priori* to be distinct in the g component densities $f_i(\boldsymbol{w}; \boldsymbol{\theta}_i)$ of the mixture to be fitted. Then the totality of parameters for the problem is

$$\boldsymbol{\Psi} = (\boldsymbol{\theta}^T, \boldsymbol{\pi}^T)^T,$$

where $\boldsymbol{\pi} = (\pi_1, \ldots, \pi_{g-1})^T$ and $\pi_g = 1 - \sum_{i=1}^{g-1} \pi_i$. Suppose we have an observed random sample $\boldsymbol{y} = (\boldsymbol{w}_1^T, \ldots, \boldsymbol{w}_n^T)^T$ from the mixture. In the application of the EM algorithm to this problem, the missing-data vector $\boldsymbol{z}$ is taken to be

$$\boldsymbol{z} = (\boldsymbol{z}_1^T, \ldots, \boldsymbol{z}_n^T)^T,$$

where $\boldsymbol{z}_j$ is the vector of zero-one indicator variables that define the component from which the jth observation $\boldsymbol{w}_j$ arises $(j = 1, \ldots, n)$. On the E-step of the EM algorithm, each $\boldsymbol{z}_j$ vector is replaced (because of the linearity of the complete-data log likelihood in $\boldsymbol{z}$) by its current conditional expectation given the observed data $\boldsymbol{y}$, which is

$$\boldsymbol{z}_j^{(k)} = (z_{1j}^{(k)}, \ldots, z_{gj}^{(k)})^T,$$

where

$$\begin{aligned} z_{ij}^{(k)} &= \tau_i(\boldsymbol{w}_j; \boldsymbol{\Psi}^{(k)}) \\ &= \pi_i^{(k)} f_i(\boldsymbol{w}_j; \boldsymbol{\theta}_i^{(k)}) / f(\boldsymbol{w}_j; \boldsymbol{\Psi}^{(k)}) \end{aligned} \tag{6.4}$$

is the (current) posterior probability that the jth observation arises from the ith component of the mixture $(i = 1, \ldots, g;\ j = 1, \ldots, n)$.

However, with the Stochastic EM algorithm, the current posterior probabilities are used in a Stochastic E-step, wherein a single draw is made from the current conditional distribution of $\boldsymbol{z}$ given the observed data $\boldsymbol{y}$. Because of the assumption of independence of the complete-data observations, this is effected by conducting a draw for each j $(j = 1, \ldots, n)$. That is, a draw $\boldsymbol{z}_j^{(1_k)}$ is made from the multinomial distribution with g categories having probabilities specified by (6.4). This effectively assigns each observation $\boldsymbol{w}_j$ outright to one of the g components of the mixture. The M-step then consists of finding the MLE of $\boldsymbol{\Psi}$ as if $\boldsymbol{w}_1, \ldots, \boldsymbol{w}_n$ were deterministically classified according to $\boldsymbol{z}_1^{(1_k)}, \ldots, \boldsymbol{z}_n^{(1_k)}$. This contrasts with the EM algorithm where these computations are weighted with respect to the components of the mixture according to the current posterior probabilities. Note that with this notation, on the kth iteration, $\boldsymbol{z}_j^{(k)}$ denotes the current conditional expectation of $\boldsymbol{Z}_j$, the random variable associated with $\boldsymbol{z}_j$, while $\boldsymbol{z}_j^{(m_k)}$ denotes the mth random draw of $\boldsymbol{Z}_j$ from its current posterior distribution (in the MCEM algorithm). For the Stochastic EM algorithm, there is only one Monte Carlo sample taken, so that $M = 1$ always.

Thus the Stochastic EM algorithm starts with arbitrary multinomial probabilities $\tau_{ij}^{(0)}$ $(i = 1, \ldots, g)$ for each observation j; for convenience, initially these may be taken to be the same for all j. Then with $\boldsymbol{\Psi}^{(k)}$ denoting the value of $\boldsymbol{\Psi}$ obtained on the kth iteration, the following steps are used:

Stochastic E-step. For each j, a draw is made from the multinomial distribution with probabilities $\tau_{ij}^{(k)}$ $(i = 1, \ldots, g)$, which might be arbitrary for $k = 0$ as explained above and are given by the previous M-step for $k > 0$, as explained below. On the kth iteration, let the draw for the jth observation $\boldsymbol{w}_j$ be $\boldsymbol{z}_j^{(1_k)} = (z_{1j}^{(1_k)}, \ldots, z_{gj}^{(1_k)})^T$. Notice that exactly one of these $z_{ij}^{(1_k)}$ is 1 and the others 0 for each j. This results in a partition of the n observations $\boldsymbol{w}_1, \ldots, \boldsymbol{w}_n$ with respect to the g components of the mixture.

M-step. Calculate

$$\pi_i^{(k+1)} = \frac{1}{n} \sum_{j=1}^{n} z_{ij}^{(1_k)} \quad (i = 1, \ldots, g).$$

The calculation of $\boldsymbol{\theta}^{(k+1)}$ depends upon the form of the densities f_i. For instance if f_i are all p-variate normal with a common covariance matrix as in Section 2.7.2, then formulas (2.50) and (2.51) give $\boldsymbol{\theta}^{(k+1)}$. The posterior probabilities required for the Stochastic E-step in the next iteration are updated as in (1.36) as follows:

$$\begin{aligned}\tau_{ij}^{(k+1)} &= \tau_i(\boldsymbol{w}_j;\, \boldsymbol{\Psi}^{(k+1)}) \\ &= \pi_i^{(k+1)} f_i(\boldsymbol{w}_j;\, \boldsymbol{\theta}_i^{(k+1)})/f(\boldsymbol{w}_j;\, \boldsymbol{\Psi}^{(k+1)}) \quad (i = 1, \ldots, g).\end{aligned}$$

Celeux and Diebolt (1985, 1986b) extend this to the case when g is not known and is to be estimated from the data. This algorithm prevents the sequence from staying near an unstable stationary point of the likelihood function. Thus it avoids the cases of slow convergence observed in some uses of the EM algorithm for the mixture problem. The sequence of estimates obtained by this procedure is an ergodic Markov chain and converges weakly to a stationary distribution. More recently, Chauveau (1995) considers the asymptotic behavior of the stochastic EM algorithm in the case of a two-component normal mixture model fitted to censored data; see also Diebolt and Ip (1996).

6.4 DATA AUGMENTATION ALGORITHM

When the object of the Bayesian analysis is to estimate the entire posterior distribution and not just the posterior mode, Tanner and Wong (1987), Wei and Tanner (1990), and Tanner (1991, 1993) propose a method called the

Data Augmentation algorithm. This is suitable when the incomplete-data posterior density is complicated, but the complete-data posterior density is relatively easy to handle and to draw from.

Suppose that in the MCEM algorithm, the sequence $\{\boldsymbol{\Psi}^{(k)}\}$ has converged to $\hat{\boldsymbol{\Psi}}$. Then an estimate of the posterior distribution can be obtained as

$$\frac{1}{M}\sum_{m=1}^{M} p(\boldsymbol{\Psi} \mid \boldsymbol{z}^{(m)}; \boldsymbol{y}),$$

where $\boldsymbol{z}^{(m)}$ denotes the mth draw from $p(\boldsymbol{z} \mid \boldsymbol{y}; \hat{\boldsymbol{\Psi}})$. This is called the Poor Man's Data Augmentation algorithm 1 (PMDA-1). Thus the PMDA-1 is a noniterative algorithm as follows:

a. Draw $\boldsymbol{z}^{(1)}, \ldots, \boldsymbol{z}^{(M)}$ from $p(\boldsymbol{z} \mid \boldsymbol{y}; \hat{\boldsymbol{\Psi}})$.

b. Compute an approximation to the posterior density as

$$\frac{1}{M}\sum_{m=1}^{M} p(\boldsymbol{\Psi} \mid \boldsymbol{z}^{(m)}, \boldsymbol{y}). \tag{6.5}$$

This gives rise to a more general algorithm called the Data Augmentation algorithm. Here starting from a prior density for the parameter $\boldsymbol{\Psi}$, each stage of the algorithm constructs an estimate of the incomplete-data posterior density. Let this be $p^{(k)}(\boldsymbol{\Psi} \mid \boldsymbol{y})$ at the end of the kth iteration. Choose and fix a positive integer M. Then the next iteration is as follows:

Imputation Step (I-step). (a1) Draw a $\boldsymbol{\Psi}^{(m_k)}$ from the current approximation $p^{(k)}(\boldsymbol{\Psi} \mid \boldsymbol{y})$. (a2) Draw a $\boldsymbol{z}^{(m_k)}$ from the current approximation to the missing-data conditional density $p(\boldsymbol{z} \mid \boldsymbol{y}; \boldsymbol{\Psi}^{(m_k)})$. Execute steps (a1) and (a2) for $m = 1, \ldots, M$.

Posterior Step (P-step). Set the next approximation to the posterior density by averaging the posterior densities obtained from the M draws of the missing data and using the supposed simpler form of the complete-data posterior density, as follows:

$$p^{(k+1)}(\boldsymbol{\Psi} \mid \boldsymbol{y}) = \frac{1}{M}\sum_{m=1}^{M} p(\boldsymbol{\Psi} \mid \boldsymbol{z}^{(m_k)}, \boldsymbol{y}).$$

Note that the I-step is equivalent to drawing $\boldsymbol{z}^{(1_k)}, \ldots, \boldsymbol{z}^{(M_k)}$ from the predictive distribution $p(\boldsymbol{z} \mid \boldsymbol{y})$, by the method of composition using the identity

$$p(\boldsymbol{z} \mid \boldsymbol{y}) = \int_{\boldsymbol{\Omega}} p(\boldsymbol{z} \mid \boldsymbol{y}; \boldsymbol{\Psi}) p(\boldsymbol{\Psi} \mid \boldsymbol{y})\, d\boldsymbol{\Psi},$$

where $\boldsymbol{\Omega}$ is the parameter space.

We return now to Example 1.1, which was developed as Example 1.4 in a Bayesian framework in Section 1.6.2. Suppose that the complete-data formulation is as before and that the prior distribution for the parameter Ψ is beta (ν_1, ν_2). Then from (1.73), the posterior distribution for Ψ is

$$\text{beta } (\nu_1 + y_{12} + y_4, \nu_2 + y_2 + y_3). \tag{6.6}$$

Thus $p(\Psi \mid \boldsymbol{z}^{(m_k)}, \boldsymbol{y})$ corresponds to (6.6) with y_{12} replaced by $y_{12}^{(m_k)}$, the mth draw of y_{12} on the kth iteration. The conditional distribution in (a2) above is binomial $(y_1, \frac{\frac{1}{4}\Psi}{\frac{1}{2}+\frac{1}{4}\Psi})$. If no knowledge of Ψ exists initially, one could start with a uniform prior which is a beta distribution with $\nu_1 = 1$ and $\nu_2 = 1$.

It is easily seen that the Data Augmentation algorithm is the iterative simulation version of the EM algorithm where the I-step is the analog of the E-step and the P-step is the analog of the M-step.

Suppose that in the Data Augmentation algorithm, we have $M = 1$. Then the algorithm can be considered to be a repeated application of the following two steps:

a. Given $\boldsymbol{z}^*$, draw $\boldsymbol{\Psi}^*$ from $p(\boldsymbol{\Psi} \mid \boldsymbol{z}^*; \boldsymbol{y})$, in view of the relationship

$$p(\boldsymbol{\Psi} \mid \boldsymbol{y}) = \int_{\mathcal{Z}} p(\boldsymbol{\Psi} \mid \boldsymbol{z}; \boldsymbol{y}) p(\boldsymbol{z} \mid \boldsymbol{y})\, d\boldsymbol{z}.$$

b. Given $\boldsymbol{\Psi}^*$, draw $\boldsymbol{z}^*$ from $p(\boldsymbol{z} \mid \boldsymbol{y}; \boldsymbol{\Psi}^*)$, in view of the relationship

$$p(\boldsymbol{z} \mid \boldsymbol{y}) = \int_{\boldsymbol{\Omega}} p(\boldsymbol{z} \mid \boldsymbol{y}; \boldsymbol{\Psi}) p(\boldsymbol{\Psi} \mid \boldsymbol{y})\, d\boldsymbol{\Psi}.$$

This is called the Chained Data Augmentation algorithm.

6.5 MULTIPLE IMPUTATION

In the EM algorithm, the missing values or the latent values are 'imputed' in the E-step and complete-data methods are applied on the M-step. Thus the EM algorithm besides providing MLEs of parameters, also provides estimates for the missing values. Similarly, in the Data Augmentation algorithm, the imputations $\boldsymbol{z}^{(m)}$ are drawn from appropriate distributions in the I-step. Although these imputed values may be good for the limited purpose of point estimation, using them for other purposes like testing hypotheses may not be suitable.

The method of multiple imputation (MI) is a solution to this problem. Here, for each missing value, a set of M (chosen and fixed) values are simulated from a suitable distribution. The M completed data sets are then analyzed by complete-data methods to obtain M estimates of the $\boldsymbol{\Psi}$ and the covariance matrix of these estimates. These are then used to get a composite estimate of $\boldsymbol{\Psi}$ and its covariance matrix. The multiple imputation method is generally used in a Bayesian framework and the imputed values are drawn from the predictive posterior distribution of the missing values. The imputation can be a step in an iterative scheme like a Data Augmentation algorithm. In fact, the Data Augmentation algorithm can be looked upon as a combination of EM and MI, where the E-step of the EM algorithm is replaced by MI for the missing values and the M-step by MI for $\boldsymbol{\Psi}$. The technique of MI with applications to survey data analysis is discussed in Rubin (1987) and in Little and Rubin (1989), and the relations between MI and other similar techniques in Rubin (1991).

6.6 SAMPLING–IMPORTANCE RESAMPLING

The Sampling–Importance Resampling (SIR) algorithm is a noniterative algorithm, which is useful in many contexts, one of which is drawing samples from an intractable distribution the form of which is known up to a multiplicative constant. This situation often arises when the constant in a posterior distribution needs complicated integration to be carried out. The SIR algorithm is related to the technique of rejection sampling introduced by John von Neumann in 1951 (see Kennedy and Gentle, 1980). Let $\boldsymbol{\Psi}$ represent the parameter vector or parameter vector together with missing-data values as the case may be. In this technique, to draw from a distribution $p(\boldsymbol{\Psi})$, an approximation $h(\boldsymbol{\Psi})$ to it is chosen such that for a constant C, $(p(\boldsymbol{\Psi})/h(\boldsymbol{\Psi})) \leq C$. Then the technique consists of the following steps:

1. Draw $\boldsymbol{\Psi}$ from $h(\boldsymbol{\Psi})$.
2. Draw θ from the uniform distribution over (0,1) [or from any p.d.f. $g(\theta)$ with appropriate modifications in Step 3 below].
3. If $\theta \leq (p(\boldsymbol{\Psi})/Ch(\boldsymbol{\Psi}))$, then accept $\boldsymbol{\Psi}$; otherwise go to Step 1.

Notice that the technique requires the calculation of the normalizing constants in p and h, and C.

The SIR algorithm consists of three steps, after choosing and fixing the numbers m and $M(> m)$ and a distribution $h(\boldsymbol{\Psi})$, which should be easy to draw from and is easy to evaluate up to a multiplicative constant, and which is an approximation to the distribution $p(\boldsymbol{\Psi})$ of $\boldsymbol{\Psi}$:

1. Draw M values $\boldsymbol{\Psi}_1, \ldots \boldsymbol{\Psi}_M$ from $h(\boldsymbol{\Psi})$.
2. Calculate the M ratios, called the importance ratios:

$$r(\boldsymbol{\Psi}_j) = \frac{p(\boldsymbol{\Psi}_j)}{h(\boldsymbol{\Psi}_j)} \quad (j = 1, \ldots, M).$$

3. Draw m values from the M values $\boldsymbol{\Psi}_1, \ldots, \boldsymbol{\Psi}_M$ in the above step with probability proportional to $r(\boldsymbol{\Psi}_j)$.

It can be shown that as $M/m \to \infty$, the probability distribution of the drawn values is the correct distribution. It is appropriate to use the SIR algorithm only if a reasonable approximation to the actual distribution is available. The advantage of the SIR algorithm over the rejection sampling method is that the normalizing constants of p and h and the constant C as in the rejection sampling need not be evaluated. However, the draws in the SIR algorithm are only from the approximate distribution, whereas in rejection sampling they are from the actual distribution p. Rubin (1991) describes the SIR algorithm and illustrates its use and of the EM, MI, Data Augmentation, and other iterative simulation algorithms for the PET problem, which was introduced in Section 2.5.

In the context of the Data Augmentation algorithm, this technique of Importance Sampling could be used to improve the PMDA-1 algorithm if $p(\boldsymbol{z} \mid \boldsymbol{y})$ is easy to evaluate. Note that equation (6.5) is only an approximation since the $\boldsymbol{z}^{(m)}$ are sampled from $p(\boldsymbol{z} \mid \boldsymbol{y};\, \hat{\boldsymbol{\Psi}})$ rather than from $p(\boldsymbol{z} \mid \boldsymbol{y})$. In this context, importance sampling could be used to calculate the observed posterior as follows: Given $\boldsymbol{z}^{(1)}, \ldots, \boldsymbol{z}^{(M)}$ from $p(\boldsymbol{z} \mid \boldsymbol{y};\, \hat{\boldsymbol{\Psi}})$ assign weights

$$r_m = \frac{p(\boldsymbol{z}^{(m)} \mid \boldsymbol{y})}{p(\boldsymbol{z}^{(m)} \mid \boldsymbol{y};\, \hat{\boldsymbol{\Psi}})}$$

and replace equation (6.5) by a weighted average

$$\frac{\sum_{m=1}^{M} r_m p(\boldsymbol{\Psi} \mid \boldsymbol{z}^{(m)};\, \boldsymbol{y})}{\sum_{m=1}^{M} r_m}.$$

Wei and Tanner (1990) call this the Poor Man's Data Augmentation algorithm 2 (PMDA-2). They also discuss approximations to this approach.

6.7 GIBBS SAMPLER

6.7.1 The Algorithm

A multivariate extension of the above Data Augmentation algorithm with parameter vector $\boldsymbol{\Psi} = (\Psi_1, \ldots, \Psi_d)^T$ can now be formulated as follows: Starting from an initial value $\boldsymbol{\Psi}^{(0)} = (\Psi_1^{(0)}, \ldots, \Psi_d^{(0)})^T$, carry out the following d steps on the kth iteration:

1 Draw $\Psi_1^{(k+1)}$ from $p(\Psi_1 \mid \boldsymbol{y}; \Psi_2^{(k)}, \ldots, \Psi_d^{(k)})$.

2 Draw $\Psi_2^{(k+1)}$ from $p(\Psi_2 \mid \boldsymbol{y}; \Psi_1^{(k+1)}, \Psi_3^{(k)}, \ldots, \Psi_d^{(k)})$.

$\vdots \qquad\qquad \vdots \qquad\qquad \vdots$

d Draw $\Psi_d^{(k+1)}$ from $p(\Psi_d \mid \boldsymbol{y}; \Psi_1^{(k+1)}, \ldots, \Psi_{d-1}^{(k+1)})$.

This is a version of what is known as the Gibbs sampler. The vector sequence $\{\boldsymbol{\Psi}^{(k)}\}$ thus generated is known to be a realization of a homogeneous Markov Chain. Many interesting properties of such a Markov sequence have been established, including geometric convergence to a stationary distribution $p(\Psi_1^{(k)}, \ldots, \Psi_d^{(k)} \mid \boldsymbol{y})$ as $k \to \infty$; see Roberts and Polson (1994). Because of this reason, methods of iterative simulation resulting in the generation of Markov Chain sequences have come to be known as Markov Chain Monte Carlo (MCMC) or simply $(\mathrm{MC})^2$ methods.

As discussed in Besag and Green (1993), the Gibbs sampler was founded on the pioneering ideas of Grenander (1983). It was not until the following year that the Gibbs sampler was so termed by Geman and Geman (1984) in their seminal paper on image processing. The roots of MCMC methods can be traced back to an algorithm of Metropolis, Rosenbluth, Rosenbluth, Teller, and Teller (1953), popularly known as the Metropolis algorithm; see also Metropolis and Ulam (1949). Its initial use was in physics and chemistry to investigate the equilibrium properties of large systems of particles, such as molecules in a gas. The first use of the Metropolis algorithm in a statistical context was by Hastings (1970), who developed it further. Besag (1974) studied the associated Markov field structure.

But the Metropolis–Hastings algorithm seems to have had little use in statistics until the advent of the Gibbs sampler by Geman and Geman (1984). Even then, the Gibbs sampler, which is a special case of the Metropolis-Hastings algorithm, seems to have been used mainly in applications in spatial statistics (for example, as in Besag, York, and Mollié, 1991), until the appearance of the paper by Gelfand and Smith (1990). They brought into focus its tremendous potential in a wide variety of statistical problems. In particular, they observed that almost any Bayesian computation could be carried out via the Gibbs sampler. More recently, Geyer and Thompson (1992) make a similar point about ML calculations; almost any ML computation can be done by some MCMC scheme. Further applications of the Gibbs sampler methods have been explored in the papers by Gelfand, Hills, Racine-Poon, and Smith (1990); Zeger and Karim (1991); Gelfand, Smith, and Lee (1992); Gelfand and Carlin (1993); Gilks, Clayton, Spiegelhalter, Best, McNeil, Sharples, and Kirby (1993); Sobel and Lange (1994); and Carlin and Chib (1995), among several others.

The reader is referred to Casella and George (1992) for a tutorial on the Gibbs sampler, to Arnold (1993) for an elementary introduction to it, to Ritter and Tanner (1992) for methodology on facilitating the Gibbs sampler, and to Ripley (1987) for a good general introduction to MCMC methods. An excellent source of work on MCMC methods is the trio of papers by Smith and Roberts (1993), Besag and Green (1993), and Gilks et al. (1993), which were read before the Royal Statistical Society at a meeting on the Gibbs sampler and other MCMC methods. Tierney (1994) focuses on the theory of MCMC methods. In a pair of articles, Gelman and Rubin (1992) and Geyer (1992) summarize many of the important issues in the implementation of MCMC methods. In particular, Gelman and Rubin (1992) note that the problem of creating a simulation mechanism is clearly separate from the problem of using this mechanism to draw inferences. They provide simple methods for obtaining inferences from iterative simulations, using multiple sequences that are generally applicable to the output of any iterative simulation. More recently, Besag, Green, Higdon, and Mengersen (1995) provide an introduction to MCMC methods and its applications, along with several new ideas. Brooks and Roberts (1995) give a review of convergence diagnostics, while a comprehensive account of MCMC methods is provided in the monograph on the topic by Gilks, Richardson, and Spiegelhalter (1996).

6.7.2 Example 6.1: Gibbs Sampler for the Mixture Problem

The Gibbs sampler is extensively used in many Bayesian problems where the joint distribution is complicated and is difficult to handle, but the conditional distributions are often easy enough to draw from. We give an example here.

Consider a Bayesian version of the finite mixture problem such as the one discussed in Section 2.7. If $\boldsymbol{\theta}_i$ denotes the parameter vector of the component density $f_i(\boldsymbol{w}; \boldsymbol{\theta}_i)$ in a mixture of the form

$$f(\boldsymbol{w}; \boldsymbol{\Psi}) = \sum_{i=1}^{g} \pi_i f_i(\boldsymbol{w}; \boldsymbol{\theta}_i),$$

then the totality of parameters for the problem is

$$\boldsymbol{\Psi} = (\boldsymbol{\theta}^T, \boldsymbol{\pi}^T)^T,$$

where $\boldsymbol{\theta} = (\boldsymbol{\theta}_1^T, \ldots, \boldsymbol{\theta}_g^T)^T$ and $\boldsymbol{\pi} = (\pi_1, \ldots, \pi_{g-1})^T$. If the component densities have common parameters, then $\boldsymbol{\theta}$ is the vector of those parameters known *a priori* to be distinct. Given an observed random sample $\boldsymbol{y} = (\boldsymbol{w}_1^T, \ldots, \boldsymbol{w}_n^T)^T$ from the mixture distribution and a prior density $p(\boldsymbol{\Psi})$ for $\boldsymbol{\Psi}$, the posterior density for $\boldsymbol{\Psi}$,

$$p(\boldsymbol{\Psi} \mid \boldsymbol{y}) \propto \prod_{j=1}^{n} \left\{ \sum_{i=1}^{g} \pi_i f_i(\boldsymbol{w}_j; \boldsymbol{\theta}_i) \right\} p(\boldsymbol{\Psi}),$$

is not easy to handle (Titterington et al., 1985). However, suppose we formulate a complete-data problem as in Section 2.7 by introducing the missing-data vector $z = (z_1^T, \ldots, z_n^T)^T$, where z_j is the vector containing the g zero-one indicator variables that define the component membership of each w_j. Then the components of the conditional model in a Bayesian framework are

$$p(w_j \mid z_j, \Psi) = \prod_{i=1}^{g} p_i^{z_{ij}}(w_j; \theta_i) p(z_j \mid \pi)$$

where

$$p(z_j \mid \pi) = \prod_{i=1}^{g} \pi_i^{z_{ij}}.$$

If $p(\Psi)$ is taken to be a product of a Dirichlet prior for π and an independent prior for θ, then the conditional distributions of $(\pi \mid \theta, z)$, $(\theta \mid \pi, z)$ and $(z \mid \theta, \pi)$ are all easily defined and handled so that a Gibbs sampling procedure can be used to obtain estimates of an otherwise intractable posterior distribution; see Gelman and King (1990), Smith and Roberts (1993), Diebolt and Robert (1994), Escobar and West (1995), and Robert (1996) for more details.

6.7.3 Gibbs Sampler Analogs of ECM and ECME Algorithms

The Gibbs sampler can be looked upon as an iterative simulation analog of the (partitioned) ECM algorithm (see Section 5.2.2), where the d constraint functions on Ψ are of the form $g_1(\Psi_2, \ldots, \Psi_d)$, $g_2(\Psi_1, \Psi_3, \ldots, \Psi_d), \ldots,$ $g_d(\Psi_1, \ldots, \Psi_{d-1})$. This is clearly analogous to the choice of conditioning parameters in the Gibbs sampler. In some versions of the Gibbs sampler, the variables (Ψs) are grouped to reduce the number of steps needed at each iteration, when such grouped variables admit a joint conditional distribution simple enough to draw from. This is analogous to integrating out missing data when maximizing a constrained log likelihood function in the ECME algorithm. For instance, suppose we have parameters $\Psi = (\Psi_1^T, \Psi_2^T)^T$. In the ECM algorithm, the distribution of $Z \mid \Psi_1, \Psi_2$ may be used in the E-step and the distributions of $\Psi_1 \mid Z, \Psi_2$ and $\Psi_2 \mid Z, \Psi_1$ may be used in CM-Step 1 and CM-Step 2, respectively. The Gibbs sampler analogs use the three distributions for the draws. However, in the ECME algorithm, CM-Step 2 may use the distribution of $\Psi_2 \mid \Psi_1$. An analogous Gibbs sampler uses the draws $Z, \Psi_2 \mid \Psi_1$ and $\Psi_1 \mid Z, \Psi_2$; Liu, Wong, and Kong (1994) give details and results regarding more rapid convergence of a collapsed Gibbs sampler compared to a one-step Gibbs sampler. Liu (1994) also shows how to use a collapsed Gibbs sampler in the bivariate normal data problem with missing values, for which we discussed the EM algorithm in Example 2.1 of Section 2.2.1.

For the variance components problem discussed in Section 5.9, wherein we presented the EM algorithm and two versions ECME-1 and ECME-2 of the ECME algorithm, we briefly present below the Gibbs sampler analogs. They are, in fact, special versions of the Gibbs sampler and are actually Data Augmentation algorithms. The parameter vector $\boldsymbol{\Psi}$ for this problem consists of the elements of $\boldsymbol{\beta}$, the (distinct) elements of $\boldsymbol{D}$, and σ^2, and the missing-data vector $\boldsymbol{z}$ is $(\boldsymbol{b}_1^T, \ldots, \boldsymbol{b}_m^T)^T$. We start with independent improper prior distributions for $\boldsymbol{\beta}, \boldsymbol{D}$, and σ^2 with $p(\boldsymbol{\beta}) \propto$ constant, $p(\boldsymbol{D}) \propto | \boldsymbol{D} |^{-1/2}$, and $p(\sigma^2) \propto 1/\sigma$.

Gibbs Sampler Analog of EM Algorithm

In the Gibbs sampler, the missing-data vector $\boldsymbol{z}$ as well as the parameter vector $\boldsymbol{\Psi}$ is sampled. For the Gibbs sampler analogue of the EM algorithm, the partition used is $\boldsymbol{z}$ and $\boldsymbol{\Psi}$. Then the steps of the algorithm are:

Step 1. Given the current draw of $\boldsymbol{\Psi}$, draw the components $\boldsymbol{b}_j$ of $\boldsymbol{z}$ from the independent conditional distributions $p(\boldsymbol{b}_j \mid \boldsymbol{y}_j; \boldsymbol{\beta}, \boldsymbol{D}, \sigma^2)$, which are multivariate normal with means and covariance matrices determined from (5.58) and (5.59).

Step 2. Given the draw of $\boldsymbol{z}$, draw $\boldsymbol{D}, \boldsymbol{\beta}$, and σ^2 conditionally independently using a Wishart for $\boldsymbol{D}^{-1}$, a p-variate normal for $\boldsymbol{\beta}$, and a density proportional to χ^2_{n-1} for σ^2.

Gibbs Sampler Analog of ECME-1 Algorithm

In the version of the Gibbs sampler analogous to Version 1 of the ECME algorithm, $\boldsymbol{\Psi}$ is partitioned into $(\boldsymbol{\Psi}_1^T, \boldsymbol{\Psi}_2^T)^T$, where $\boldsymbol{\Psi}_1 = \boldsymbol{\beta}$ and $\boldsymbol{\Psi}_2$ consists of the distinct elements of $\boldsymbol{D}$ and σ^2. The algorithm then is

Step 1. Given the current draw of $\boldsymbol{\Psi}_2$, draw $\boldsymbol{\Psi}_1$ from the conditional p-variate normal distribution $p(\boldsymbol{\beta} \mid \boldsymbol{y}; \boldsymbol{\Psi}_2)$, and then draw components of $\boldsymbol{z}$ from the conditional distribution given $\boldsymbol{y}$.

Step 2. Given $\boldsymbol{\Psi}_1$ and $\boldsymbol{z}$, draw $\boldsymbol{\Psi}_2$ from the conditional distribution using the conditional independence of $\boldsymbol{D}$ and σ^2 given $\boldsymbol{\Psi}_1$ as in the previous algorithm.

Gibbs Sampler Analog of ECME-2 Algorithm

In the version of the Gibbs sampler analogous to Version 2 of the ECME algorithm, the partition $(\boldsymbol{\Psi}_1^T, \boldsymbol{\Psi}_2^T)^T$ of $\boldsymbol{\Psi}$ has $\boldsymbol{\Psi}_1 = (\sigma^2, \boldsymbol{\beta}^T)^T$ and $\boldsymbol{\Psi}_2$ containing the distinct elements of $\boldsymbol{D}$. The algorithm then is

Step 1. Given the current draw of $\boldsymbol{\Psi}_2$, draw $\boldsymbol{\Psi}_1$ and draw components of $\boldsymbol{z}$ using the factorization

$$p(\sigma^2, \boldsymbol{\beta}, \boldsymbol{z} \mid \boldsymbol{y}; \boldsymbol{D}) = p(\sigma^2 \mid \boldsymbol{y}; \boldsymbol{D}) \cdot p(\boldsymbol{\beta} \mid \boldsymbol{y}; \sigma^2, \boldsymbol{D}) \cdot p(\boldsymbol{z} \mid \boldsymbol{y}; \boldsymbol{\beta}, \sigma^2, \boldsymbol{D}).$$

Step 2. Given $\boldsymbol{\Psi}_1$ and $\boldsymbol{z}$, draw $\boldsymbol{D}$ as in the first algorithm.

6.7.4 Some Other Analogs of the Gibbs Sampler

It is expected that the approaches of the SAGE and AECM algorithms will give rise to a more flexible formulation of the Gibbs sampler incorporating the working parameter concept in the form of a parameter-dependent transformation of the random variable before the Gibbs sampler is implemented (see Meng and van Dyk, 1995). Such work has not yet been done.

There are other similarities between the EM-type algorithms and the Gibbs sampler. For instance, there are similarities in the structures of the Jacobian matrices associated with the ECM algorithm and the Gibbs sampler. As we have seen in Chapter 3, these Jacobian matrices determine the rates of convergence of the algorithms and Amit (1991), in fact, computes the rates of convergence of the Gibbs sampler using the Jacobian matrix; see Meng and Rubin (1992); Meng (1994).

The Gibbs sampler and other such iterative simulation techniques being Bayesian in their point of view consider both parameters and missing values as random variables, whereas in the frequentist framework the parameters and missing values are treated from different standpoints. In the iterative algorithms under a frequentist framework, like the EM-type algorithms, parameters are subjected to a maximization operation and missing values are subjected to an averaging operation. In the iterative simulation algorithms under a Bayesian framework both are subjected to random draw operations. Thus the various versions of the Gibbs sampler can be viewed as stochastic analogs of the EM, ECM, and ECME algorithms. Besides these connections, the EM-type algorithms also come in useful as starting points for iterative simulation algorithms where typically regions of high density are not known *a priori*.

6.8 MISSING-DATA MECHANISM AND IGNORABILITY

In analyzing data with missing values and more generally in incomplete-data problems of the kind we consider in this book, it is important to consider the mechanism underlying the "missingness" or incompleteness. For, failure to properly study it and just assuming the missing phenomenon to be purely random and ignorable may vitiate inferences. Rubin (1976) formulates two types of random missingness. If the missingness depends neither on the observed values nor on the missing values, then it is called "Missing Completely at Random" (MCAR). The broader situation where the missingness is allowed to depend on the observed data is called "Missing at Random" (MAR). Thus MCAR is a special case of MAR. Rubin (1976) argues that if the MAR condition holds and the parameter to be estimated and the parameter of the missingness mechanism are distinct, then likelihood inference ignoring the missingness mechanism is valid. Furthermore, under MCAR, sampling inferences without modeling missing-data mechanisms are also valid.

The phenomenon of incompleteness of data is more general than the missing-data phenomenon. Heitjan and Rubin (1991) extend the MAR concept to the general incompleteness situation by defining a notion of "coarsening at random" and show that under this condition if the coarsening parameters and the parameters to be estimated are distinct, then likelihood and Bayesian inferences ignoring this incompleteness mechanism are valid. Heitjan (1993) gives a number of illustrations of this situation and Heitjan (1994) defines the notion of an "observed degree of coarseness" which helps to extend MCAR to "coarsened completely at random" and illustrates it in a number of applications.

Sarkar (1993) develops sufficiency-based conditions for ignorability of missing data and shows that if these conditions are satisfied, the EM algorithm and the Gibbs sampler are valid. However, in general, the objectives of the analysis need to be taken into account while determining whether missing values can be ignored or not. Estimating parameters only need mild conditions to be satisfied, while imputation of missing values need stronger conditions.

Most of the applications of the EM algorithm and its extensions have been made assuming validity of the inference procedures without checking ignorability conditions. In many situations in practice, the response and incompleteness mechanisms are, in fact, related to the study variables and hence the incompleteness mechanism is nonignorable. In such situations, the inference is biased. A correct way to proceed in these situations is to model the incompleteness mechanisms suitably and formulate a more comprehensive statistical problem. Little (1983a, 1983b), Little and Rubin (1987), and Little (1993, 1994) discuss quite a few such problems.

6.9 COMPETING METHODS AND SOME COMPARISONS WITH THE EM ALGORITHM

6.9.1 Introduction

It is generally acknowledged that the EM algorithm and its extensions are the most suitable for handling the kind of incomplete-data problems discussed in this book. However, quite a few other methods are advocated in the literature as possible alternatives to EM-type algorithms. Some of these like Simulated Annealing are general-purpose optimization methods, some like the Delta algorithm are special-purpose ML methods, and others like the Image Space Reconstruction Algorithm (ISRA) are special-purpose algorithms. Some comparisons of the performance of these algorithms with the EM algorithm have been reported for specific problems like the normal mixture resolution problem. We present a brief review of these methods and the comparisons reported. Some other methods and comparisons not reported here can be found in Davenport, Pierce, and Hathaway (1988), Campillo and Le Gland (1989), Celeux and Diebolt (1990, 1992), Fairclough, Pierni, Ridgway, and Schwertman (1992), Yuille, Stolorz, and Utans (1994), and Xu and Jordan (1996).

6.9.2 Simulated Annealing

Simulated Annealing (SA) was introduced by Kirkpatrick, Gelatt, and Vecchi (1983) and Cerný (1985), but in its original form, it dates back to Pincus (1968, 1970); see Ripley (1990) on this point. This is a stochastic-type technique for global optimization using the Metropolis et al. algorithm based on the Boltzmann distribution in statistical mechanics. It is an algorithm in the MCMC family.

To minimize a real-valued function $h(\boldsymbol{\Psi})$ on a compact subset $\boldsymbol{\Omega}$ of $\mathbb{R}^d$, SA generates an inhomogeneous Markov process on $\boldsymbol{\Omega}$ depending on the positive time-parameter T (called the *temperature* in accordance with the physical analogy and also to differentiate it from the statistical parameter $\boldsymbol{\Psi}$). This Markov process has the following Gibbs distribution as its unique stationary distribution,

$$p_T(\boldsymbol{\Psi}) = \frac{\exp(-h(\boldsymbol{\Psi})/T)}{\int_{\boldsymbol{\Omega}} \exp(-h(\boldsymbol{\Psi})/T)\, d\boldsymbol{\Psi}}, \quad \boldsymbol{\Psi} \in \boldsymbol{\Omega}. \tag{6.7}$$

In the limit as T tends to 0 from above, the stationary distribution tends to a distribution concentrated on the points of global minima of h. In a homogeneous version of the algorithm, a sequence of homogeneous Markov chains is generated at decreasing values of T.

The general algorithm is as follows:

Step 1. Choose an initial value for T, say T_0 and initial value $\boldsymbol{\Psi}_0$ of $\boldsymbol{\Psi}$ with $h(\boldsymbol{\Psi}_0) = h_0$.

Step 2. Select a point $\boldsymbol{\Psi}_1$ at random from a neighborhood of $\boldsymbol{\Psi}_0$ and calculate the corresponding h-value.

Step 3. Compute $\Delta_1 = h(\boldsymbol{\Psi}_1) - h(\boldsymbol{\Psi}_0)$. Choose a random variable from a uniform distribution on $[0, 1]$. Let it be u. Check if $u \leq \exp(-\Delta_1/T_0)$ (called the Metropolis criterion). Thus we move to the new point definitely if its function value is less than that at the earlier point; otherwise, we move to the new point with a certain probability.

Step 4. Repeat Steps 2 and 3 after updating the appropriate quantities, until an equilibrium has been reached by the application of a suitable stopping rule.

Step 5. Lower the temperature according to an "annealing schedule" and start at Step 2 with the equilibrium value at the previous T as the starting value. Again a suitable stopping rule between temperatures is used to decide to stop the algorithm giving the solution to the minimization problem.

The stopping rules, the annealing schedule, and the methods of choosing the initial and next values constitute the "cooling schedule". It is well known that the behavior of the SA approach to optimization is crucially dependent on the choice of cooling schedule. The literature gives some standard methods of arriving at these. The reader is referred to Ripley (1988) and Weiss (1989) for an introduction to SA, and to van Laarhoven and Aarts (1987), Aarts and van Laarhoven (1989), and Bertsimas and Tsitsiklis (1993) for general reviews of the literature. Geman and Geman (1984) adopted the SA approach for MAP estimation in image analysis, using the Gibbs sampler to simulate (6.7).

6.9.3 Comparison of SA and EM Algorithm for Normal Mixtures

Ingrassia (1991, 1992) compares SA and the EM algorithm for univariate normal mixtures by simulation. The study of Ingrassia (1992) is an improved version of Ingrassia (1991) as it chooses a constrained formulation of the mixture problem, uses an improved cooling schedule for the SA algorithm, and considers several normal mixtures. The criteria for comparison are distances like the Kullback–Leibler distance measures between distribution functions—between the one used for simulation (the true one) and the one obtained by the algorithm. It is found that though neither algorithm overwhelmingly outperforms the other, SA performs more satisfactorily in many cases. However, SA is much slower than the EM algorithm. Though SA gives estimates closer to true values compared to the EM algorithm, in a number of these cases, the value of the likelihood is greater at the EM solution. Thus the superiority of SA is not so definitive.

Brooks and Morgan (1995) make a comparison between traditional methods such as quasi-Newton and a sequential Quadratic Programming algorithm with SA as well as with hybrids of SA and these traditional methods (where the SA solution is used as a starting point for the traditional method). They find that the traditional methods repeatedly get stuck at points associated with singularities. They conclude that the hybrid algorithms perform well and better than either the traditional methods or SA; for comparisons of the EM algorithm and other methods for the normal mixture problems, see also Everitt (1984); Davenport et al. (1988); Lindsay and Basak (1993); and Aitkin and Aitkin (1996).

6.10 THE DELTA ALGORITHM

Jørgensen (1984) introduces the Delta algorithm. It is a generalization of the scoring method, is a modification of the Newton–Raphson method, and has an interpretation as an iterative weighted least-squares method.

Let the log likelihood be of the form $\log L(\boldsymbol{\mu}(\boldsymbol{\Psi}))$, where $\boldsymbol{\Psi}$ is a $d \times 1$ vector of parameters to be estimated and $\boldsymbol{\mu}$ is a $n \times 1$ vector function of $\boldsymbol{\Psi}$, n being the sample size. For instance, the mean $\boldsymbol{\mu}$ could be of the form $\boldsymbol{X\Psi}$, where $\boldsymbol{X}$ is an $n \times d$ (design) matrix, as in the case of a linear model. In the Delta algorithm, the matrix of second derivatives usually used in the Newton–Raphson algorithm is replaced by an approximation of the form

$$-\,\partial^2 \log L(\boldsymbol{\mu})/\partial\boldsymbol{\Psi}\partial\boldsymbol{\Psi}^T \approx \boldsymbol{X}^T(\boldsymbol{\Psi})\boldsymbol{K}(\boldsymbol{\mu})\boldsymbol{X}(\boldsymbol{\Psi}), \tag{6.8}$$

where $\boldsymbol{K}(\boldsymbol{\mu})$ is a symmetric positive definite matrix, called the weight matrix. Here $\boldsymbol{X}(\boldsymbol{\Psi}) = \partial\boldsymbol{\mu}/\partial\boldsymbol{\Psi}^T$ is called the local design matrix and $\boldsymbol{K}(\boldsymbol{\mu})$ is to be chosen suitably. The name Delta method is due to the similarity of equation (6.8) with the way asymptotic covariance matrices are obtained when parameters are transformed, using the traditional delta method.

If $\boldsymbol{K}(\boldsymbol{\mu})$ is the expected information matrix for $\boldsymbol{\mu}$, then the Delta method is the scoring method.

Böhning and Lindsay (1988) and Böhning (1993) discuss a method called the Lower Bound algorithm, which is a special case of the Delta algorithm. Here if the Hessian is bounded below in the sense of Lowener ordering (defined by $\boldsymbol{A} \geq \boldsymbol{B}$ if $\boldsymbol{A} - \boldsymbol{B}$ is nonnegative definite), that is,

$$\partial^2 \log L(\boldsymbol{\mu})/\partial\boldsymbol{\Psi}\partial\boldsymbol{\Psi}^T \geq B$$

for all $\boldsymbol{\Psi} \in \boldsymbol{\Omega}$, where $\boldsymbol{B}$ is a symmetric, negative definite matrix not depending on $\boldsymbol{\Psi}$, then $-\boldsymbol{B}$ is used as the $\boldsymbol{K}$ matrix. Such a bound can be used to create a montonically convergent modification of the Newton–Raphson algorithm; see our discussion in Section 1.3.2. Böhning and Lindsay (1988) give examples of such likelihoods and call them Type I likelihoods.

We assume that the matrix $\boldsymbol{X}(\boldsymbol{\Psi})$ has full rank for all $\boldsymbol{\Psi}$. Differentiability conditions to get the score vector and information matrices are assumed. Note that

$$\partial \log L(\boldsymbol{\mu})/\partial\boldsymbol{\Psi} = \boldsymbol{X}^T(\boldsymbol{\Psi})S(\boldsymbol{\mu}),$$

where $S(\boldsymbol{\mu}) = \partial \log L(\boldsymbol{\mu})/\partial\boldsymbol{\mu}$. Let $\boldsymbol{X}(\boldsymbol{\Psi}^{(k)})$ be denoted by $\boldsymbol{X}^{(k)}$. Let $\boldsymbol{\Psi}^{(0)}$ be the starting value of the parameter for the Delta algorithm. The algorithm is defined by

$$\boldsymbol{\Psi}^{(k+1)} = \boldsymbol{\Psi}^{(k)} + a^{(k)}\boldsymbol{\Psi}^{(k)},$$

where

$$\boldsymbol{\Psi}^{(k)} = (\boldsymbol{X}^{(k)^T}\boldsymbol{K}\boldsymbol{X}^{(k)})^{-1}\boldsymbol{X}^{(k)^T}S^{(k)}$$

and $a^{(k)}$ is a number called the step-length chosen to make

$$\log L(\boldsymbol{\mu}^{(k+1)}) \geq \log L(\boldsymbol{\mu}^{(k)}).$$

The step-length $a^{(k)}$ can be taken to be 1 (then called the unit-step Delta algorithm), but it is recommended that a search be made to choose $a^{(k)}$ so as to increase the value of $\log L(\boldsymbol{\mu})$.

Jørgensen (1984) discusses various aspects of the algorithm, including choice of starting values, choice of the weight matrix, computation of standard errors of the estimates and implementation in GLIM. He also gives a few illustrative examples.

Jørgensen (1984) suggests that for computing the MLE for an incomplete-data problem via the EM algorithm, on the M-step one iteration of the Delta algorithm be used with $S^{(k)}$ replaced by its current conditional expectation given the observed data. He shows that a step-length chosen to increase the Q-function also increases the incomplete-data log likelihood. He also shows that the weight matrix may be chosen based on Q or as $\mathcal{I}(\boldsymbol{\mu}^{(k)})$ or $\boldsymbol{I}(\boldsymbol{\mu}^{(k)}; \boldsymbol{y})$.

6.11 IMAGE SPACE RECONSTRUCTION ALGORITHM

In Section 2.5, we considered the EM algorithm for the PET problem. Daube–Witherspoon and Muehllehner (1986) modify the EM algorithm for this problem by replacing the equation (2.29) by

$$\lambda_i^{(k+1)} = \lambda_i^{(k)} \left(\sum_{j=1}^{d} y_j p_{ij} \right) \bigg/ \left\{ \sum_{j=1}^{d} p_{ij} \left(\sum_{h=1}^{n} \lambda_h^{(k)} p_{hj} \right) \right\} \quad (i = 1, \ldots, n), \tag{6.9}$$

under the assumption $\sum_{i=1}^{n} \sum_{j=1}^{d} p_{ij} = 1$. They called it the Image Space Reconstruction Algorithm (ISRA), and obtained it heuristically. The operation $\sum_{j=1}^{d} y_j p_{ij}$ represents a back-projection of the data $\{y_j\}$ from the jth detector. The operation $\sum_{j=1}^{d} p_{ij} \{\sum_{h=1}^{n} \lambda_h^{(k)} p_{hj}\}$ on the other hand, represents a corresponding back-projection of the current fit for $\{y_j\}$ or the calculated projection. In the ISRA, the ratio between these quantities is multiplicatively used to update the variables. Formal justification of this algorithm was provided later by Titterington (1987) and De Pierro (1989). The motivation for the algorithm, however, is that it requires fewer calculations than the EM algorithm in each iteration. The algorithm does not converge to the MLE based on the Poisson model of Section 2.5, but to a nonnegative least-squares estimate. Convergence properties of the algorithm have been studied and some details and references may be found in De Pierro (1995).

Archer and Titterington (1995) define a more general form of the ISRA suitable for a general class of linear inverse problems with positivity restrictions, mentioned in Section 5.15, analogous to equation (6.9). They then apply it and the EM algorithm to a variety of problems considered in Vardi and Lee (1993). They show that the speeds of the EM algorithm and the ISRA are comparable. However, since the ISRA needs less computations in each iteration, it is on the whole more efficient than the EM algorithm.

6.12 FURTHER APPLICATIONS OF THE EM ALGORITHM

6.12.1 Introduction

Since the publication of DLR, the number, variety, and range of applications of the EM algorithm and its extensions have been tremendous. The standard incomplete-data statistical problems such as with missing data, grouped data, censored data, and truncated distributions are being increasingly tackled with the EM-type of algorithms. Then there are applications in a variety of incomplete-data problems arising in standard statistical situations such as linear models, contingency tables and log linear models, random effects models and general variance components models, and time series and stochastic processes, etc., Special statistical problems such as mixture resolution, factor analysis, survival analysis, and survey sampling have seen increasing use of EM-type algorithms. These algorithms have also been profitably used in a variety of special applications in engineering, psychometry, econometrics, epidemiology, genetics, astronomy, etc., such as real-time pattern recognition, image processing and reconstruction, autoradiography and medical imaging, especially positron emission tomography (PET), neural networks, and hidden Markov models, and various aspects of signal processing, communication and computing, and AIDS epidemiology and disease prevalence estimation.

Although we have illustrated the methodology of the earlier chapters with many examples, they do not adequately cover the variety and extent of the applications of the EM algorithm and its extensions. Most recent books or review articles on any of these subjects do include applications of the EM algorithm and its extensions where appropriate. For instance, the mixture resolution problem has attracted a great deal of attention and the books by Everitt and Hand (1981), Titterington et al. (1985), and by McLachlan and Basford (1988), and the review papers by Redner and Walker (1984) and Titterington (1990) contain a great deal of discussion on the EM algorithm in the context of mixture resolution. Some applications like statistical analysis with missing data and analysis of incomplete data from surveys are well covered in Little and Rubin (1987) and Madow, Olkin, and Rubin (1983), respectively. The Little–Rubin book contains EM applications in many different contexts such as regression, linear models, time series, contingency tables, etc. We hope that the exposition of the general theory and methodology in the foregoing chapters will help readers grasp these applications and develop their own when required. We conclude the book with a quick summary of some of the more interesting and topical applications of the EM algorithm.

6.12.2 Hidden Markov Models

From the large number of illustrations of various aspects of the EM algorithm by the mixture problem, it is evident that the EM algorithm has played a significant part in the solution of the problem of ML estimation of parameters of a mixture. In the mixture framework with observations $\boldsymbol{w}_1, \ldots, \boldsymbol{w}_n$, the missing-data vector is given by $\boldsymbol{z} = (\boldsymbol{z}_1^T, \ldots, \boldsymbol{z}_n^T)^T$ where, as before, $\boldsymbol{z}_j$ defines the component of the mixture from which the jth data point $\boldsymbol{w}_j$ arises. We can term this unobserved vector $\boldsymbol{z}$ as the "hidden variable". In the mixture problems considered up to now, we have been concerned with the fitting of a mixture model where $\boldsymbol{y}$ denotes an observed random sample. Thus it has sufficed in the complete-data formulation of the problem to take the complete data $\boldsymbol{x}_j = (\boldsymbol{w}_j^T, \boldsymbol{z}_j^T)^T$ $(j = 1, \ldots, n)$ as a collection of n independent observations.

However, in image analysis, say, where the $\boldsymbol{w}_j$ refer to intensities measured on n pixels in a scene, the associated component indicator vectors $\boldsymbol{z}_j$ will not be independently distributed because of the spatial correlation between neighboring pixels. In speech recognition applications, the $\boldsymbol{z}_j$ may be unknown serially dependent prototypical spectra on which the observed speech signals $\boldsymbol{w}_j$ depend. Hence in such cases the sequence or set of hidden values or states $\boldsymbol{z}_j$ cannot be regarded as independent. In these applications, a stationary Markovian model over a finite state space is generally formulated for the distribution of the hidden variable $\boldsymbol{Z}$. In one dimension, this Markovian model is a Markov chain (see, for example, Holst and Lindgren, 1991), and in two and higher dimensions a Markov random field (MRF); see Besag (1986, 1989). The conditional distribution of the observed vector $\boldsymbol{Y}$ is formulated as before to depend only on the value of $\boldsymbol{Z}$, the state of the Markov process. As a consequence of the dependent structure of $\boldsymbol{Z}$, its probability function does not factor into the product of the marginal probability functions of $\boldsymbol{Z}_1, \ldots, \boldsymbol{Z}_n$ in the manner of equation (1.33). However, $\boldsymbol{W}_1, \ldots, \boldsymbol{W}_n$ are assumed conditionally independent given $\boldsymbol{z}_1, \ldots, \boldsymbol{z}_n$; that is

$$p(\boldsymbol{w}_1, \ldots, \boldsymbol{w}_n \mid z_1, \ldots, z_n; \boldsymbol{\theta}) = \prod_{j=1}^{n} p(\boldsymbol{w}_j \mid z_j; \boldsymbol{\theta}),$$

where p is used as a generic symbol for the probability density or the probability mass function and $\boldsymbol{\theta}$ is the parameter vector containing the parameters in these conditional distributions that are known *a priori* to be distinct.

In image-processing applications, the indexing subscript j generally represents pixel-sites and in speech recognition applications, speech frames. In speech recognition applications, $\boldsymbol{W}_j$ is taken to be a finitely many-valued vector, representing random functions of the underlying (hidden) prototypical spectra; see Rabiner (1989); Juang and Rabiner (1991). In some applications, the hidden variable $\boldsymbol{Z}_j$ takes values over a continuum representing the gray levels or a similar characteristic of the true image and $\boldsymbol{W}_j$ is a blurred version of the hidden true image $\boldsymbol{Z}_j$; see Qian and Titterington (1991). In some other applications, the hidden variable $\boldsymbol{Z}_j$ may represent some discretized characteristic of the pixel, about which inference is required to be made and $\boldsymbol{W}_j$ is an observable feature of a pixel statistically related to the hidden characteristic of the pixel.

We first discuss the use of the EM algorithm in a hidden Markov chain. Let the stationary finite Markov chain for Z have state space $\{S_1, \ldots, S_g\}$ with transition probability matrix $\boldsymbol{A} = ((a_{hi})), h, i = 1, \ldots, g$. Thus if $Z_1, \ldots, Z_n$ is the unknown sequence of states,

$$\text{pr}(Z_{j+1} = S_i \mid Z_j = S_h) = a_{hi} \quad (h, i = 1, \ldots, g).$$

Let the probability distribution of the observed $\boldsymbol{W}$ in state S_i over a common sample space $\{\boldsymbol{v}_1, \ldots, \boldsymbol{v}_M\}$ be $b_i(\boldsymbol{v}_m), m = 1, \ldots, M$. Thus if $\boldsymbol{W}_1, \ldots, \boldsymbol{W}_n$ denote the random sequence of observations, then

$$\text{pr}(\boldsymbol{W}_j = \boldsymbol{v}_m \mid Z_j = S_i) = b_i(\boldsymbol{v}_m) \quad (i = 1, \ldots, g;\ m = 1, \ldots, M),$$

not depending on j. Let the initial distribution of the Markov chain be $(\pi_{11}, \ldots, \pi_{g1})^T$, $(\sum_{i=1}^{g} \pi_{i1} = 1)$. Thus the parameter vector $\boldsymbol{\Psi}$ for the hidden Markov chain consists of the elements of $\boldsymbol{A}, \pi_{11}, \ldots, \pi_{g-1,1}$, and the $b_i(\boldsymbol{v}_m)$ for $i = 1, \ldots, g;\ m = 1, \ldots, M - 1$.

We briefly explain the EM algorithm for this problem, known in the hidden Markov model literature as the Baum–Welch algorithm. Baum and his collaborators formulated this algorithm long before DLR and established convergence properties for this algorithm; see Baum and Petrie (1966), Baum and Eagon (1967), and Baum et al. (1970). We use the following notation:

$$\xi_j(h, i) = \text{pr}_{\boldsymbol{\Psi}}(Z_j = S_h, Z_{j+1} = S_i \mid \boldsymbol{Y} = \boldsymbol{y}),$$

$$\alpha_j(h) = \text{pr}_{\boldsymbol{\Psi}}(\boldsymbol{W}_1 = \boldsymbol{w}_1, \ldots, \boldsymbol{W}_j = \boldsymbol{w}_j, Z_j = S_h),$$

$$\beta_j(h) = \text{pr}_{\boldsymbol{\Psi}}(\boldsymbol{W}_{j+1} = \boldsymbol{w}_{j+1}, \boldsymbol{W}_{j+2} = \boldsymbol{w}_{j+2}, \ldots, \boldsymbol{W}_n = \boldsymbol{w}_n \mid Z_j = S_h),$$

and

$$\gamma_j(h) = \sum_{i=1}^{g} \xi_j(h, i)$$

for $j = 1, \ldots, n - 1;\ h, i = 1, \ldots, g$. Computing $\xi_j(h, i)$ at current parameter values is essentially the E-step. It can be seen that $\xi_j(h, i)$ can be written with this notation as

$$\xi_j(h, i) = \frac{\alpha_j(h) a_{hi} b_i(\boldsymbol{w}_{j+1}) \beta_{j+1}(i)}{\sum_{h=1}^{g} \sum_{i=1}^{g} \alpha_j(h) a_{hi} b_i(\boldsymbol{w}_{j+1}) \beta_{j+1}(i)} \quad (j = 1, \ldots, n - 1), \qquad (6.10)$$

since the numerator is $\text{pr}_{\boldsymbol{\Psi}}(Z_j = S_h, Z_{j+1} = S_i, \boldsymbol{Y} = \boldsymbol{y})$ and the denominator is $\text{pr}_{\boldsymbol{\Psi}}(\boldsymbol{Y} = \boldsymbol{y})$. To compute this, the values of $\alpha_j(h)$ and $\beta_j(h)$ are to be computed. This is done by forward and backward recursions as follows at the kth iteration:

For $\boldsymbol{\alpha}_j^{(k)}(\boldsymbol{h})$:

Initialization.

$$\alpha_1^{(k)}(h) = \pi_h^{(k)} b_h^{(k)}(\boldsymbol{w}_1) \quad (h = 1, \ldots, g)$$

Induction.

$$\alpha_{j+1}^{(k)}(h) = \left[\sum_{i=1}^{g} \alpha_j^{(k)}(i) a_{ih}^{(k)}\right] b_h^{(k)}(\boldsymbol{w}_{j+1})$$

$$(j = 1, \ldots, n-1; \quad h = 1, \ldots, g)$$

Termination.

$$\mathrm{pr}_{\boldsymbol{\Psi}^{(k)}}(\boldsymbol{W}_1 = \boldsymbol{w}_1, \ldots, \boldsymbol{W}_n = \boldsymbol{w}_n) = \sum_{h=1}^{g} \alpha_n^{(k)}(h)$$

For $\boldsymbol{\beta}_j^{(k)}(\boldsymbol{h})$:
Initialization.

$$\beta_n^{(k)}(h) = 1 \quad (h = 1, \ldots, g).$$

Induction.

$$\beta_j^{(k)}(h) = \sum_{i=1}^{g} a_{hi}^{(k)} b_i^{(k)}(\boldsymbol{w}_{j+1}) \beta_{j+1}^{(k)}(i)$$

$$\text{for } j = n-1, n-2, \ldots, 1; \quad h = 1, \ldots, g$$

The final computation in the E-step consists in plugging in these values and the current parameter values in (6.10) as follows:

$$\xi_j^{(k)}(h, i) = \frac{\alpha_j^{(k)}(h) a_{hi}^{(k)} b_i^{(k)}(\boldsymbol{w}_{j+1}) \beta_{j+1}^{(k)}(i)}{\sum_{h=1}^{g} \sum_{i=1}^{g} \alpha_j^{(k)}(h) a_{hi}^{(k)} b_i^{(k)}(\boldsymbol{w}_{j+1}) \beta_{j+1}^{(k)}(i)} \quad (j = 1, \ldots, n-1). \tag{6.11}$$

The M-step consists in finding the updated estimates of the parameters using the following formulas which are a combination of the MLEs for the multinomial parameters and Markov chain transition probabilities based on the number of observed transitions from state h to state i in a finite number of steps observed:

$$\pi_{h1}^{(k+1)} = \gamma_1^{(k)}(h),$$

$$a_{hi}^{(k+1)} = \frac{\sum_{j=1}^{n-1} \xi_j^{(k)}(h, i)}{\sum_{j=1}^{n-1} \gamma_j^{(k)}(h)},$$

and

$$b_i^{(k+1)}(\boldsymbol{v}_m) = \frac{\displaystyle\sum_{\substack{j=1 \\ s.t.\ \boldsymbol{w}_j=\boldsymbol{v}_m}}^{n-1} \gamma_j^{(k)}(i)}{\displaystyle\sum_{j=1}^{n-1} \gamma_j^{(k)}(i)}.$$

This algorithm is described in some detail in Rabiner (1989). The reader is referred to Leroux and Puterman (1992) for some recent work related to this problem. Also, Robert, Celeux, and Diebolt (1993) provide a stochastic Bayesian approach to parameter estimation for a hidden Markov chain.

The EM algorithm for the hidden Markov random field is considerably more difficult; see McLachlan (1992, Chapter 13) and the references therein. Even in the exponential family case (see Section 1.5.3) the E- and M-steps are difficult to carry out even by numerical methods, except in some very simple cases like a one-parameter case; in some cases they may be implemented by suitable Gibbs sampler algorithms. A variety of practical procedures has been considered in the literature. They are reviewed by Qian and Titterington (1991, 1992), who also suggest a Monte Carlo restoration-estimation algorithm.

6.12.3 AIDS Epidemiology

Statistical modeling and statistical methods have played an important role in the understanding of HIV (human immunodeficiency virus) infection and the AIDS (acquired immunodeficiency syndrome) epidemic. Statistical ideas and methods enter naturally in these studies in view of the considerable amount of uncertainty in the data generated for such studies. Further, data from AIDS studies have many new features which need the development of new types of statistical models and methods. For a review of the statistical methods in AIDS studies, see Becker (1992).

Data relating to HIV infection and AIDS epidemics are notoriously incomplete with missing information and delayed reporting and are subject to complex censoring and truncation mechanisms. Furthermore, by the very nature of the disease, there is an unknown and random time-interval between infection and diagnosis. All these aspects result in incompleteness of data and unobservable latent variables, giving rise to EM-type methods. We give a brief summary of the applications of the EM-type algorithms in these problems.

Let us consider the problem of predicting the number of AIDS cases from past data. Let the available data be $\{A_t : t = 1, \ldots, T\}$ the number of cases diagnosed as AIDS during a month t (in a certain well-defined population). Let N_t denote the number of individuals infected with HIV in month t. Let f_d be the probability that the duration of infection is d months; f_d gives the probability distribution of incubation period (time between infection and diagnosis) in months. The N_t are unobservable. There are methods to estimate f_d and hence let us assume that the f_d are known. We have that

$$E(A_t \mid N_1, \ldots, N_t) = \sum_{i=1}^{t} N_i f_{t-i+1}$$

under the assumption of independence of incubation period over individuals. Let $\lambda_i = E(N_i)$ and $\mu_i = E(A_i)$. Then

$$\mu_i = \sum_{i=1}^{t} \lambda_i f_{t-i+1}.$$

If we use A_t to estimate λ_i, then prediction is done simply by using

$$\mu_{T+r} = \sum_{i=1}^{T+r} \lambda_i f_{T+r-i+1}$$

for the month $T + r$. This is the method of backprojection or backcalculation. Estimation of λ_t is generally done by assuming $\{N_i\}$ to be an inhomogeneous Poisson process. This is an example of an ill-posed inverse problem, much like the problem of PET discussed in Example 2.5 in Section 2.5. Becker, Watson, and Carlin (1991) in fact use an EMS algorithm like that of Silverman et al. (1990) for the PET problem outlined in our Section 5.13. Brookmeyer (1991) and Bacchetti, Segal, and Jewell (1992, 1993) use a penalized likelihood approach, and Pagano, De Gruttola, MaWhinney, and Tu (1992) use ridge regression. In the approach of Bacchetti et al. (1992, 1993), the assumption of $\{N_i\}$ as an inhomogeneous Poisson process is continued. Further, the quantities $a_{ijt} = \text{pr}\{\text{diagnosed at time } j \text{ and reported at time } t \mid \text{infected at time } i\}$ are assumed known and are used in the likelihood. A penalized likelihood with penalizing roughness on the log scale is used. An EM algorithm is developed where the complete data consist of x_{ijt}, denoting the number infected in month i, diagnosed in month j, and reported in month t. Methods based on the penalized EM algorithm (Green, 1990b) are used. The results of the EM algorithm are used as a starting point for fitting a more complicated model.

The foregoing methods need an estimate of the incubation distribution. Bacchetti (1990) discusses estimation of such a distribution (discretized into months) by using data on prospective studies that periodically retest initially seronegative subjects. For those subjects that convert from negative (last being in month L_i, say for subject i) to seropositive (first positive being in month R_i, say), the time of seroconversion t_i is interval-censored. Bacchetti (1990) uses a penalized likelihood method combined with Turnbull's (1974) EM algorithm for general censored and truncated data.

Tu, Meng, and Pagano (1993) estimate the survival distribution after AIDS diagnosis from surveillance data. They use a discrete proportional hazards model, in the presence of unreported deaths, right-truncated sampling of deaths information up to the time of analysis, reporting delays and non-availability of time of death. They extend Turnbull's (1976) approach for regression analysis of truncated and censored data and develop an EM algorithm. They also use the Supplemented EM algorithm for calculating the asymptotic variance of the MLE.

6.12.4 Neural Networks

A neural network (NN) is a stochastic system that produces an output vector $\boldsymbol{u}$ given an input vector $\boldsymbol{w}$ according to the probability distribution $p(\boldsymbol{u} \mid \boldsymbol{w};\, \boldsymbol{\Psi})$, where $\boldsymbol{\Psi}$ represents the parameters of the system. In principle, the form of p and the values of $\boldsymbol{\Psi}$ can be chosen to make the input–output relation a desired one. The capability of the system can be enhanced by the introduction of what are called hidden variables z, which do not appear explicitly in the input–output relationship but appear as intermediate variables in the system and are stochastically determined by $\boldsymbol{w}$. Thus a NN with hidden variables or units is specified by the conditional distribution $p(\boldsymbol{u}, \boldsymbol{z} \mid \boldsymbol{w};\, \boldsymbol{\Psi})$ and the input–output relationship by the marginal distribution $p(\boldsymbol{u} \mid \boldsymbol{w};\, \boldsymbol{\Psi}) = \int_{\mathcal{Z}} p(\boldsymbol{u}, \boldsymbol{z} \mid \boldsymbol{w};\, \boldsymbol{\Psi})\, dz$.

A commonly used form of the NN is the Boltzmann machine, which behaves according to a discrete-time, stochastic updating rule. The state of the machine at any given time point is described by a binary n-tuple $\boldsymbol{w}^T = (w_1, \ldots, w_n), w_i \in \{0, 1\}$. At each time-point, i is chosen with equal probability from 1 to n and the value of w_i is updated to 1 with probability

$$p_i = \frac{1}{1 + \exp\{-\frac{1}{2T} \sum_{j=1}^{n} a_{ij} w_j\}}$$

and to 0 with probability $1 - p_i$. The matrix $\boldsymbol{A} = ((a_{ij}))$ is generally chosen as a symmetric zero-diagonal matrix. It can be shown that the Boltzmann machine (rather its sequence of states over the time points) is an ergodic Markov chain with state space $\{0, 1\}^n$, with a unique stationary distribution

$$p(\boldsymbol{w}) = b \exp\left[\frac{1}{T} \sum_{i>j} \sum a_{ij} w_i w_j\right], \tag{6.12}$$

where b is a normalizing constant. The quantity T is the "temperature" as in Simulated Annealing (Section 6.9.2) and can be used to increase the randomness in the system; it can be seen that as T increases for fixed $\boldsymbol{A}$, the stationary distribution tends toward the uniform distribution. In the subsequent discussions, we take T to be 1.

The equation (6.12) parameterizes a very flexible family of distributions. Let us denote the family for a given n by $\mathcal{B}_n$. Suppose a given probability distribution P over $\{0,1\}^v$ is to be realized as a stationary distribution of a Boltzmann machine. Suppose that this $P \notin \mathcal{B}_v$. Then it is possible that, by constructing a Boltzmann machine with n ($> v$) variables, a better approximation to the given P could be realized than the best from $\mathcal{B}_v$, as an appropriate v-dimensional marginal distribution of the n-dimensional stationary distribution of the Boltzmann machine. These $n - v$ variables are called hidden variables.

In order to study this approximation, we need a criterion. Let us use the Kullback–Leibler divergence criterion between two discrete distributions $p(\boldsymbol{w})$ and $q(\boldsymbol{w})$ over the space $\mathcal{W}$, defined as

$$D(p;\ q) = \sum_{\boldsymbol{w}\in\mathcal{W}} p(\boldsymbol{w}) \log\{p(\boldsymbol{w})/q(\boldsymbol{w})\}.$$

Let us denote by $\mathcal{D}$ the class of distributions over $\{0,1\}^n$ which give rise to the given distribution P as the marginal distribution of the first v variables. Evidently, we would like a Boltzmann machine on n variables which best approximates the given P by the marginal distribution of the first v variables. Then the problem is one of finding a $B \in \mathcal{B}_n$ such that $D(P; B_{(v)})$ is minimized, where $B_{(v)}$ represents the first v-variable marginal of B. This minimization is facilitated by the fact that

$$\min_{B\in\mathcal{B}_n} D(P; B_{(v)}) = \min_{B\in\mathcal{B}_n, P_n\in\mathcal{D}} D(P_n; B),$$

which is proved in Amari, Kurata, and Nagaoka (1992) and Byrne (1992).

This minimization can be achieved by an EM algorithm as follows: Choose an initial distribution $d^{(0)}$ arbitrarily in $\mathcal{D}$. Then use the following two steps on the (k+1)th iteration:

M-step. Find

$$b^{(k+1)} = \arg\min_{b\in\mathcal{B}_n} D(d^{(k)};\ b).$$

E-step. Find

$$d^{(k+1)} = \arg\min_{d\in\mathcal{D}} D(d;\ b^{(k)}).$$

Since the distributions are over $\{0,1\}^n$, the M-step can be implemented by the Iterative Proportional Fitting (IPF) algorithm. This algorithm is suggested by Csiszár and Tusnády (1984) and is interpreted as an EM algorithm by them as well as by Byrne (1992) and by Neal and Hinton (1993). This algorithm is also called the alternating-minimization algorithm. An interpretation of the alternating-minimization algorithm as an EM algorithm based on the divergence D can be given in more general contexts also. Anderson and

Titterington (1995) extend this to polytomous NNs. This approach is also applicable to the general case of a NN with hidden units described in the first paragraph of this section (see Amari, 1995).

In a problem of supervised learning in a model of hierarchical mixture of experts, Jordan and Jacobs (1994) and Jordan and Xu (1995) consider the following regression type of model for generating an output $\boldsymbol{u}$ from an input $\boldsymbol{w}$:

$$p(\boldsymbol{u} \mid \boldsymbol{w};\, \boldsymbol{\Psi}) = \sum_{h=1}^{g_1} \pi_h(\boldsymbol{w};\, \boldsymbol{\beta}_1) \sum_{i=1}^{g_2} \pi_{i \cdot h}(\boldsymbol{w};\, \boldsymbol{\beta}_2) p_{hi}(\boldsymbol{u} \mid \boldsymbol{w};\, \boldsymbol{\theta}_{hi}), \tag{6.13}$$

where the $\pi_h(\boldsymbol{w};\, \boldsymbol{\beta}_1)$ are multinomial probabilities and the $\pi_{i \cdot h}(\boldsymbol{w};\, \boldsymbol{\beta}_2)$ are conditional multinomial probabilities, and $\boldsymbol{\Psi}$ includes both the $\boldsymbol{\theta}_{hi}$ parameters (expert network parameters) and the $\boldsymbol{\beta}_1, \boldsymbol{\beta}_2$ parameters (gating network parameters). This is a model for a nested sequence of networks that maps $\boldsymbol{w}$ into $\boldsymbol{u}$. There is an observed random sample of n observations given by $(\boldsymbol{u}_j^T,\, \boldsymbol{w}_j^T)^T$ $(j = 1, \ldots, n)$. To tackle the estimation problem, they introduce missing data in terms of the vectors $\boldsymbol{z}_j$ of zero-one indicator variables with components z_{hj} $(h = 1, \ldots, g_1)$ for $j = 1, \ldots, n$, and the vectors $\boldsymbol{z}_{\cdot h;j}$ with components $z_{i \cdot h;j}$ $(i = 1, \ldots, g_2)$ for $h = 1, \ldots, g_1;\ j = 1, \ldots, n$. We let $z_{hij} = z_{hj} z_{i \cdot h;j}$. These indicator variables z_{hj} and $z_{i \cdot h;j}$ depend on an input $\boldsymbol{u}_j$ through their expectations $\pi_h(\boldsymbol{w}_j;\, \boldsymbol{\beta}_1)$ and $\pi_{i \cdot h}(\boldsymbol{w}_j;\, \boldsymbol{\beta}_2)$, according to the model (6.13). Thus we have z_j at the first level with the corresponding random variable Z_j having a multinomial distribution consisting of one draw over g categories with probabilities $\{\pi_{hj} = \pi_h(\boldsymbol{w}_j;\, \boldsymbol{\beta}_1) : h = 1, \ldots, g_1\}$ for each $j = 1, \ldots, n$. At the second level, we have conditional on $\boldsymbol{z}_j$, the realization $z_{\cdot h;j}$ with corresponding random variable $Z_{\cdot h;j}$ having a multinomial distribution consisting of one draw over g categories with probabilities $\{\pi_{i \cdot h;j} = \pi_{i \cdot h}(\boldsymbol{w}_j; \boldsymbol{\beta}_2) : i = 1, \ldots, g_2\}$, for each (h, j) for $h = 1, \ldots, g_1$ and $j = 1, \ldots, n$. We let $\boldsymbol{z}$ be the vector containing all the labels z_{hj} and $z_{i \cdot h;j}$ $(h = 1, \ldots, g_1; i = 1, \ldots, g_2;\ j = 1, \ldots, n)$.

If $\boldsymbol{z}$ were known, then the MLE problem would separate out into regression problems for each expert network and a multiway classification problem for the multinomials. They can be solved independently of each other. A model for the distribution of the jth data point in the complete-data formulation of the problem is

$$p(\boldsymbol{u}_j, z_j \mid \boldsymbol{w}_j;\, \boldsymbol{\Psi}) = \prod_{h=1}^{g_1} \prod_{i=1}^{g_2} \{\pi_{hj} \pi_{i \cdot h;j}\, p_{hi}(\boldsymbol{u}_j \mid \boldsymbol{w}_j;\, \boldsymbol{\theta}_{hi})\}^{z_{hij}}.$$

Taking the logarithm of this and summing over $j = 1, \ldots, n$, we have that the complete-data log likelihood $\log L_c(\boldsymbol{\Psi})$ is given by

$$\sum_{j=1}^{n} \sum_{h=1}^{g_1} \sum_{i=1}^{g_2} z_{hij} \{\log \pi_{hj} + \log \pi_{i \cdot h;j} + \log p_{hi}(\boldsymbol{u}_j \mid \boldsymbol{w}_j;\, \boldsymbol{\theta}_{hi})\}.$$

The EM algorithm then reduces to the following:

E-step. The current conditional expectation $\pi_{hij}^{(k+1)}$ of Z_{hij} is given by

$$\pi_{hij}^{(k+1)} = \frac{\pi_{hj}^{(k)} \pi_{i\cdot h;j}^{(k)} p_{hi}(\boldsymbol{u}_j \mid \boldsymbol{w}_j; \boldsymbol{\theta}_{hi}^{(k)})}{\sum_{h=1}^{g} \sum_{i=1}^{g} \pi_{hj}^{(k)} \pi_{i\cdot h;j}^{(k)} p_{hi}(\boldsymbol{u}_j \mid \boldsymbol{w}_j; \boldsymbol{\theta}_{hi}^{(k)})}.$$

From these joint probabilities $\pi_{hij}^{(k+1)}$, the marginal probabilities $\pi_{hj}^{(k+1)}$ and the conditional probabilities $\pi_{i\cdot h;j}^{(k+1)}$ can be worked out; they correspond to the expected values of Z_{hj} and $Z_{i\cdot h;j}$, respectively.

M-step. The M-step consists of three separate maximization problems:

$$\boldsymbol{\theta}_{hi}^{(k+1)} = \arg\max_{\boldsymbol{\theta}_{hi}} \sum_{j=1}^{n} \pi_{hij}^{(k+1)} \log p_{hi}(\boldsymbol{u}_j \mid \boldsymbol{w}_j; \boldsymbol{\theta}_{hi}),$$

$$\boldsymbol{\beta}_1^{(k+1)} = \arg\max_{\boldsymbol{\beta}_1} \sum_{j=1}^{n} \sum_{h=1}^{g_1} \pi_{hj}^{(k+1)} \log \pi_h(\boldsymbol{w}_j; \boldsymbol{\beta}_1),$$

and

$$\boldsymbol{\beta}_2^{(k+1)} = \arg\max_{\boldsymbol{\beta}_2} \sum_{j=1}^{n} \sum_{h=1}^{g_1} \sum_{i=1}^{g_2} \pi_{hj}^{(k+1)} \pi_{i\cdot h;j}^{(k+1)} \log \pi_{i\cdot h}(\boldsymbol{w}_j; \boldsymbol{\beta}_2).$$

All these equations require iterative methods and can be solved by Iteratively Reweighted Least Squares for a generalized linear model.

For an instance of the use of mixture models in NNs, see Nowlan and Hinton (1992); for the application of NNs in classification, see Ripley (1994); for a treatment of NNs also from a statistical perspective, see Geman, Bienenstock, and Doursat (1992) and Cheng and Titterington (1994). For other applications of the EM algorithm in NNs, see Cheeseman, Kelly, Self, Stutz, Taylor, and Freeman (1988) and Nowlan (1991) (in clustering) and Specht (1991) (in density estimation).

References

Aarts, E.H.L., and van Laarhoven, P.J.M. (1989). Simulated Annealing: An Introduction. *Statistica Neerlandica*, **43**, 31–52.

Achuthan, N.R., and Krishnan, T. (1992). EM algorithm for segregation analysis. *Biometrical Journal*, **34**, 971–988.

Adamidis, K., and Loukas, S. (1993). ML estimation in the Poisson binomial distribution with grouped data via the EM algorithm. *Journal of Statistical Computation and Simulation*, **45**, 33–39.

Aitkin, M., and Aitkin, I. (1994). Efficient computation of maximum likelihood estimates in mixture distributions, with reference to overdispersion and variance components. In *Proceedings XVIIth International Biometric Conference, Hamilton, Ontario*. Alexandria, Virginia: Biometric Society, pp. 123–138.

Aitkin, M., and Aitkin, I. (1996). A hybrid EM/Gauss–Newton algorithm for maximum likelihood in mixture distributions. *Statistics and Computing* [*in press*].

Amari, S. (1995a). The EM algorithm and information geometry in neural network learning. *Neural Computation*, **7**, 13–18.

Amari, S. (1995b). Information geometry of the EM and em algorithms for neural networks. *Neural Networks*, **8**, 1379–1408.

Amari, S., Kurata, K., and Nagaoka, H. (1992). Information geometry of Boltzmann machines. *IEEE Transactions on Neural Networks*, **3**, 260–271.

Amit, Y. (1991). On rates of convergence of stochastic relaxation for Gaussian and non-Gaussian distributions. *Journal of Multivariate Analysis*, **38**, 82–99.

Anderson, N.H., and Titterington, D.M. (1993). Beyond the binary Boltzmann machine. *IEEE Transactions on Neural Networks*, **6**, 1229–1236.

Andrews, D.F. (1974). A robust method for multiple linear regression. *Technometrics*, **16**, 523–531.

Andrews, D.F., and Herzberg, A.M. (1985). *Data: A Collection of Problems from Many Fields for the Student and Research Worker*. New York: Springer-Verlag.

Archer, G.E.B., and Titterington, D.M. (1995). The iterative image space reconstruction (ISRA) as an alternative to the EM algorithm for solving positive linear inverse problems. *Statistica Sinica*, **5**, 77–96.

Arnold, S.F. (1993). Gibbs sampling. In *Handbook of Statistics* Vol. 9, C.R. Rao (Ed.). New York: Elsevier, pp. 599–625.

Arslan, O., Constable, P.D.L., and Kent, J.T. (1993). Domains of convergence for the EM algorithm: a cautionary tale in a location estimation problem. *Statistical Computing*, **3**, 103–108.

Arslan, O., Constable, P.D.L., and Kent, J.T. (1995). Convergence behavior of the EM algorithm for the multivariate *t*-distribution. *Communications in Statistics—Theory and Methods*, **24**, 2981–3000.

Atkinson, S.E. (1992). The performance of standard and hybrid EM algorithms for ML estimates of the normal mixture model with censoring. *Journal of Statistical Computation and Simulation*, **44**, 105–115.

Avni, Y., and Tananbaum, H. (1986). X-ray properties of optically selected QSO's. *Astrophysics Journal*, **305**, 83–99.

Bacchetti, P. (1990). Estimating the incubation period of AIDS by comparing population infection and diagnosis patterns. *Journal of the American Statistical Association*, **85**, 1002–1008.

Bacchetti, P., Segal, M.R., and Jewell, N.P. (1992). Uncertainty about the incubation period of AIDS and its impact on backcalculation. In *AIDS Epidemiology: Methodological Issues*, N.P. Jewell, K. Dietz, and V.T. Farewell (Eds.). Boston: Birkhäuser, pp. 61–80.

Bacchetti, P., Segal, M.R., and Jewell, N.P. (1993). Backcalculation of HIV infection rates (with discussion). *Statistical Science*, **8**, 82–119.

Baum, L.E., and Eagon, J.A. (1967). An inequality with applications to statistical estimation for probabilistic functions of Markov processes and to a model for ecology. *Bulletin of the American Mathematical Society*, **73**, 360–363.

Baum, L.E., and Petrie, T. (1966). Statistical inference for probabilistic functions of finite Markov chains. *Annals of Mathematical Statistics*, **37**, 1554–1563.

Baum, L.E., Petrie, T., Soules, G., and Weiss, N. (1970). A maximization technique occurring in the statistical analysis of probabilistic functions of Markov chains. *Annals of Mathematical Statistics*, **41**, 164–171.

Beale, E.M.L., and Little, R.J.A. (1975). Missing values in multivariate analysis. *Journal of the Royal Statistical Society B*, **37**, 129–145.

Becker, N.G. (1992). Statistical challenges of AIDS. *Australian Journal of Statistics*, **34**, 129–144.

Becker, N.G., Watson, L.F., and Carlin, J.B. (1991). A method of non-parametric backprojection and its application to AIDS data. *Statistics in Medicine*, **10**, 1527–1542.

Behboodian, J. (1970). On a mixture of normal distributions. *Biometrika*, **57**, 215–217.

Bentler, P.M., and Tanaka, J.S. (1983). Problems with EM algorithms for ML factor analysis. *Psychometrika*, **48**, 247–251.

Berndt, E.K., Hall, B.H., Hall, R.E., and Hausman, J.A. (1974). Estimation and inference in nonlinear structural models. *Annals of Economic and Social Measurement*, **3/4**, 653–665.

Bertsimas, D., and Tsitsiklis, J. (1993). Simulated annealing. *Statistical Science*, **8**, 10–15.

Besag, J. (1974). Spatial interaction and the statistical analysis of lattice systems (with discussion). *Journal of the Royal Statistical Society B*, **36**, 192–236.

Besag, J. (1986). On the statistical analysis of dirty pictures (with discussion). *Journal of the Royal Statistical Society B*, **48**, 259–302.

Besag, J. (1989). Towards Bayesian image analysis. *Journal of Applied Statistics*, **16**, 395–407.

Besag, J. (1991). Spatial statistics in the analysis of agricultural field experiments. In *Spatial Statistics and Digital Image Analysis*, Panel on Spatial Statistics and Image Processing (Eds.). Washington, D.C.: National Research Council, National Academy Press, pp. 109–127.

Besag, J., and Green, P.J. (1993). Spatial statistics and Bayesian computation. *Journal of the Royal Statistical Society B*, **55**, 25–37. Discussion: 53–102.

Besag, J., Green, P., Higdon, D., and Mengerson, K. (1995). Bayesian computation and stochastic systems (with discussion). *Statistical Science* **10**, 3–66.

Besag, J., York, J., and Mollié, A. (1991). Bayesian image restoration with two applications in spatial statistics (with discussion). *Annals of the Institute of Statistical Mathematics*, **43**, 1–59.

Bishop, Y.M.M., Fienberg, S.E., and Holland, P.W. (1975). *Discrete Multivariate Analysis: Theory and Practice*. Cambridge, Massachusetts: MIT Press.

Blight, B.J.N. (1970). Estimation from a censored sample for the exponential family. *Biometrika*, **57**, 389–395.

Böhning, D. (1993). Construction of reliable maximum likelihood algorithms with application to logistic and Cox regression. In *Handbook of Statistics* Vol. 9, C.R. Rao (Ed.). Amsterdam: North-Holland, pp. 409–422.

Böhning, D., Dietz, E., Schaub, R., Schlattmann, P., and Lindsay, B. (1994). The distribution of the likelihood ratio for mixtures of densities from the one-parameter exponential family. *Annals of the Institute of Statistical Mathematics*, **46**, 373–388.

Böhning, D., and Lindsay, B. (1988). Monotonicity of quadratic approximation algorithms. *Annals of the Institute of Statistical Mathematics*, **40**, 641–663.

Boyles, R.A. (1983). On the convergence of the EM algorithm. *Journal of the Royal Statistical Society B*, **45**, 47–50.

Broniatowski, M., Celeux, G., and Diebolt, J. (1983). Reconaissance de densités par un algorithme d'apprentissage probabiliste. In *Data Analysis and Informatics* Vol. 3. Amsterdam: North–Holland, pp. 359–374.

Brookmeyer, R. (1991). Reconstruction and future trends of the AIDS epidemic in the United States. *Science*, **253**, 37–42.

Brooks, S.P., and Morgan, B.J.T. (1995). Optimisation using simulated annealing. *The Statistician*, **44**, 241–257.

Brooks, S.P., and Roberts, G.O. (1995). Review of convergence diagnostics. *Technical Report*. Cambridge: University of Cambridge.

Brownlee, K.A. (1965). *Statistical Theory and Methodology in Science and Engineering*. Second Edition. New York: Wiley.

Buck, S.F. (1960). A method of estimation of missing values in multivariate data suitable for use with an electronic computer. *Journal of the Royal Statistical Society B*, **22**, 302–306.

Byrne, C.L. (1993). Iterative image reconstruction algorithms based on cross-entropy minimization. *IEEE Transactions on Image Processing*, **2**, 96–103.

Byrne, W. (1992). Alternating minimization and Boltzmann machine learning. *IEEE Transactions on Neural Networks*, **3**, 612–620.

Campillo, F., and Le Gland, F. (1989). MLE for partially observed diffusions: Direct maximization vs. the EM algorithm. *Stochastic Processes and their Applications*, **33**, 245–274.

Carlin, B.P., and Chib, S. (1995). Bayesian model choice via Markov chain Monte Carlo methods. *Journal of the Royal Statistical Society B*, **57**, 473–484.

Carlin, J.B. (1987). *Seasonal Analysis of Economic Time Series*. Unpublished Ph.D. Dissertation, Department of Statistics, Harvard University, Cambridge, Massachusetts.

Carter, W.H., Jr., and Myers, R.H. (1973). Maximum likelihood estimation from linear combinations of discrete probability functions, *Journal of the American Statistical Association*, **68**, 203–206.

Casella, G., and George, E.I. (1992). Explaining the Gibbs sampler. *The American Statistician*, **46**, 167–174.

Celeux, G., and Diebolt, J. (1985). The SEM algorithm: a probabilistic teacher algorithm derived from the EM algorithm for the mixture problem. *Computational Statistics Quarterly*, **2**, 73–82.

Celeux, G., and Diebolt, J. (1986a). The SEM and EM algorithms for mixtures: numerical and statistical aspects. *Proceedings of the 7th Franco-Belgium Meeting of Statistics*. Bruxelles: Publication des Facultés Universitaries St. Louis.

Celeux, G., and Diebolt, J. (1986b). L'algorithme SEM: un algorithme d'apprentissage probabiliste pour la reconnaissance de mélanges de densités. *Revue de Statistique Appliquée*, 34, 35–52.

Celeux, G., and Diebolt, J. (1990). Une version de type recuit simule de l'algorithme EM. *Comptes Rendus de l'Academie des Sciences. Serie I. Mathematique*, 119–124.

Celeux, G., and Diebolt, J. (1992). A stochastic approximation type EM algorithm for the mixture problem. *Stochastics and Stochastic Reports*, **41**, 119–134.

Ceppellini, R.M., Siniscalco, S., and Smith, C.A.B. (1955). The estimation of gene frequencies in a random-mating population. *Annals of Human Genetics*, **20**, 97–115.

Cerný, V. (1985). Thermodynamical approach to the traveling salesman problem: an efficient simulation algorithm. *Journal of Optimization Theory and Applications*, **45**, 41–51.

Chan, K.S., and Ledolter, J. (1995). Monte Carlo estimation for time series models involving counts. *Journal of the American Statistical Association*, **90**, 242–252.

Chauveau, D. (1995). A stochastic EM algorithm for mixtures with censored data. *Journal of Statistical Planning and Inference*, **46**, 1–25.

Cheeseman, P., Kelly, J., Self, M., Stutz, J., Taylor, W., and Freeman, D. (1988). Auroclass: a Bayesian classification system. In *Proceedings of the Fifth International Conference on Machine Learning*, Ann Arbor, Michigan. San Mateo, California: Morgan Kaufmann.

Chen, T.T. (1972). *Mixed-up Frequencies in Contingency Tables*. Unpublished Ph.D. Dissertation. University of Chicago, Chicago.

Chen, T.T., and Fienberg, S.E. (1974). Two-dimensional contingency tables with both completely and partially cross-classified data. *Biometrics*, **30**, 629–642.

Cheng, B., and Titterington, D.M. (1994). Neural networks: a review from a statistical perspective (with discussion). *Statistical Science*, **9**, 2–54.

Christensen, R. (1990). *Log-linear Models*. New York: Springer-Verlag.

Cochran, W.G., and Cox, G. (1957). *Experimental Designs*. New York: Wiley.

Cohen, M., Dalal, S.R., and Tukey, J.W. (1993). Robust, smoothly heterogeneous variance regression. *Applied Statistics*, **42**, 339–353.

Cox, D.R., and Hinkley, D. (1974). *Theoretical Statistics*. London: Chapman & Hall.

Cramér, H. (1946). *Mathematical Methods of Statistics*. Princeton, New Jersey: Princeton University Press.

Csiszár, I., and Tusnády, G. (1984). Information geometry and alternating minimization procedures. *Statistics and Decisions, Supplementary Issue No. 1*, 205–237.

Daube–Witherspoon, M.E., and Muehllehner, G. (1986). An iterative image space reconstruction algorithm suitable for volume ECT. *IEEE Transactions on Medical Imaging*, **5**, 61–66.

Davenport, J.W., Pierce, M.A., and Hathaway, R.J. (1988). A numerical comparison of EM and quasi-Newton type algorithms for computing MLE's for a mixture of normal distributions. In *Computer Science and Statistics: Proceedings of the 20th Symposium on the Interface*. Alexandria, Virginia: American Statistical Association, pp. 410–415.

Davidon, W.C. (1959). Variable metric methods for minimization. *AEC Research and Development Report ANL-5990*, Argonne National Laboratory, Chicago.

Day, N.E. (1969). Estimating the components of a mixture of normal distributions. *Biometrika*, **56**, 463–474.

Dempster, A.P., Laird, N.M., and Rubin, D.B. (1977). Maximum likelihood from incomplete data via the EM algorithm (with discussion). *Journal of the Royal Statistical Society B*, **39**, 1–38.

Dempster, A.P., Laird, N.M., and Rubin, D.B. (1980). Iteratively reweighted least squares for linear regression when errors are normal/independent distributed. In *Multivariate Analysis* Vol. 5, P.R. Krishnaiah (Ed.). Amsterdam: North-Holland, pp. 35–57.

Dempster, A.P., and Rubin, D.B. (1983). Rounding error in regression: The appropriateness of Sheppard's corrections. *Journal of the Royal Statistical Society B*, **45**, 51–59.

Dennis, J.E., and Schnabel, R.B. (1983). *Numerical Methods for Unconstrained Optimization and Nonlinear Equations*. Englewood Cliffs, New Jersey: Prentice-Hall.

De Pierro, A.R. (1989). On some nonlinear iterative relaxation methods in remote sensing. *Matemática Aplicada e Computacional*, **8**, 153–166.

De Pierro, A.R. (1995). A modified expectation maximization algorithm for penalized likelihood estimation in emission tomography. *IEEE Transactions on Medical Imaging*, **14**, 132–137.

Diebolt, J., and Ip, E.H.S. (1996). Stochastic EM: method and application. In *Markov Chain Monte Carlo in Practice*, W.R. Gilks, S.Richardson, and D.J. Spiegelhalter (Eds.). London: Chapman & Hall, pp. 259–273.

Diebolt, J., and Robert, C.P. (1994). Estimation of finite mixture distributions through Bayesian sampling. *Journal of the Royal Statistical Society B*, **56**, 363–375.

Do, K., and McLachlan, G.J. (1984). Estimation of mixing proportions: a case study. *Applied Statistics*, **33**, 134–140.

Duan, J.-C., and Simonato, J.-G. (1993). Multiplicity of solutions in maximum likelihood factor analysis. *Journal of Statistical Computation and Simulation*, **47**, 37–47.

Duda, R.O., and Hart, P.E. (1973). *Pattern Classification and Scene Analysis*. New York: Wiley.

Efron, B. (1967). The two sample problem with censored data. In *Proceedings of the 5th Berkeley Symposium on Mathematical Statistics and Probability* Vol. 4. Berkeley, California: University of California Press, pp. 831–853.

Efron, B. (1977). Contribution to the discussion of paper by A.P. Dempster, N.M. Laird, and D.B. Rubin. *Journal of the Royal Statistical Society B*, **39**, 29.

Efron, B. (1979). Bootstrap methods: another look at the jackknife. *Annals of Statistics*, **7**, 1–26.

Efron, B. (1982). Maximum likelihood and decision theory. *Annals of Statistics*, **10**, 340–356.

Efron, B. (1994). Missing data, imputation and the bootstrap (with discussion). *Journal of the American Statistical Association*, **89**, 463–479.

Efron, B., and Hinkley, D.V. (1978). Assessing the accuracy of the maximum likelihood estimator: observed versus expected Fisher information (with discussion). *Biometrika*, **65**, 457–487.

Efron, B., and Tibshirani, R. (1993). *An Introduction to the Bootstrap*. London: Chapman & Hall.

Elandt-Johnson, R. (1971). *Probability Models and Statistical Methods in Genetics*. New York: Wiley.

Escobar, M., and West, M. (1995). Bayesian density estimation and inference using mixtures. *Journal of the American Statistical Association*, **90**, 577–588.

Everitt, B.S. (1984). Maximum likelihood estimation of the parameters in the mixture of two univariate normal distributions: a comparison of different algorithms. *The Statistician*, **33**, 205–215.

Everitt, B.S. (1987). *Introduction to Optimization Methods and their Application in Statistics*. London: Chapman & Hall.

Everitt, B.S. (1988). A Monte Carlo investigation of the likelihood-ratio test for number of classes in latent class analysis. *Multivariate Behavioral Research*, **23**, 531–538.

Everitt, B.S., and Hand, D.J. (1981). *Finite Mixture Distributions*. London: Chapman & Hall.

Fairclough, D.L., Pierni, W.C., Ridgway, G.J., and Schwertman, N.C. (1992). A Monte Carlo approximation of the smoothing, scoring and EM algorithms for dispersion matrix estimation with incomplete growth curve data. *Journal of Statistical Computation and Simulation*, **43**, 77–92.

Farewell, V.T. (1982). The use of mixture models for the analysis of survival data with long-term survivors. *Biometrics*, **38**, 1041–1046.

Farewell, V.T. (1986). Mixture models in survival analysis: Are they worth the risk? *Canadian Journal of Statistics*, **14**, 257–262.

Fessler, J.A., and Hero, A.O. (1994). Space-alternating generalized expectation-maximization algorithm. *IEEE Transactions on Signal Processing*, **42**, 2664–2677.

Fienberg, S.E. (1972). The analysis of multi-way contingency tables. *Biometrics*, **28**, 177–202.

Fisher, R.A. (1922). On the mathematical foundations of theoretical statistics. *Philosophical Transactions of the Royal Society of London A*, **222**, 309–368.

Fisher, R.A. (1925). Theory of statistical estimation. *Proceedings of the Cambridge Philosophical Society*, **22**, 700–725.

Fisher, R.A. (1934). Two new properties of maximum likelihood. *Proceedings of the Royal Society of London A*, **144**, 285–307.

Ganesalingam, S., and McLachlan, G.J. (1978). The efficiency of a linear discriminant function based on unclassified initial samples. *Biometrika*, **65**, 658–662.

Ganesalingam, S., and McLachlan, G.J. (1979). Small sample results for a linear discriminant function estimated from a mixture of normal populations. *Journal of Statistical Computation and Simulation*, **9**, 151–158.

Gelfand, A.E., and Carlin, B.P. (1993). Maximum-likelihood estimation for constrained- or missing-data models. *Canadian Journal of Statistics*, **21**, 303–311.

Gelfand, A.E., Hills, S.E., Racine-Poon, A., and Smith, A.F.M. (1990). Illustration of Bayesian inference in normal data models using Gibbs sampling. *Journal of the American Statistical Association*, **85**, 972–985.

Gelfand, A.E., and Smith, A.F.M. (1990). Sampling-based approaches to calculating marginal densities. *Journal of the American Statistical Association*, **85**, 398–409.

Gelfand, A.E., Smith. A.F.M., and Lee, T.-M. (1992). Bayesian analysis of constrained parameter and truncated data problems using Gibbs sampling. *Journal of the American Statistical Association*, **87**, 523–532.

Gelman, A., Carlin, J., Stern, H., and Rubin, D. (1995). *Bayesian Data Analysis*. London: Chapman & Hall.

Gelman, A., and King, G. (1990). Estimating the electoral consequences of legislative redirecting. *Journal of the American Statistical Association*, **85**, 274–282.

Gelman, A., and Rubin, D.B. (1992). Inference from iterative simulation using multiple sequences. *Statistical Science*, **7**, 457–472. Discussion: 483–511.

Geman, S., Bienenstock, E., and Doursat, R. (1992). Neural networks and the bias/variance dilemma. *Neural Computation*, **4**, 1–58.

Geman, S., and Geman, D. (1984). Stochastic relaxation, Gibbs distributions, and the Bayesian restoration of images. *IEEE Transactions on Pattern Analysis and Machine Intelligence*, **6**, 721–741.

Geman, S., and McClure, D.E. (1985). Bayesian image analysis: An application to single photon emission tomography. In *Proceedings of the Statistical Computing Section, American Statistical Association*. Alexandria, Virginia: American Statistical Association, pp. 12–18.

Geyer, C.J. (1992). Practical Markov chain Monte Carlo (with discussion). *Statistical Science*, **7**, 473–511.

Geyer, C.J., and Thompson, E.A. (1992). Constrained Monte-Carlo maximum likelihood for dependent data (with discussion). *Journal of the Royal Statistical Society B*, **54**, 657–699.

Gilks, W.R., Clayton, D.G., Spiegelhalter, D.J., Best, N.G., McNeil, A.J., Sharples, L.D., and Kirby, A.J. (1993). Modelling complexity: applications of Gibbs sampling in medicine (with discussion). *Journal of the Royal Statistical Society B*, **55**, 39–102.

Gilks, W.R., Richardson, S., and Spiegelhalter, D.J. (Eds.). (1996). *Markov Chain Monte Carlo in Practice*. London: Chapman & Hall.

Gill, R.D. (1989). Non- and semi-parametric maximum likelihood estimators and the von Mises method (Part I) (with discussion). *Scandinavian Journal of Statistics: Theory and Applications*, **16**, 97–128.

Golub, G.H., and Nash, S.G. (1982). Non-orthogonal analysis of variance using a generalized conjugate gradient algorithm. *Journal of the American Statistical Association*, **77**, 109–116.

Goodman, L.A. (1974). The analysis of systems of qualitative variables when some of the variables are unobservable. Part I: a modified latent structure approach. *American Journal of Sociology*, **79**, 1179–1259.

Goodnight, J.H. (1979). A tutorial on the sweep operator. *The American Statistician*, **33**, 149–158.

Gordon, N.H. (1990a). Maximum likelihood estimation for mixtures of two Gompertz distributions when censoring occurs. *Communications in Statistics—Simulation and Computation*, **19**, 737–747.

Gordon, N.H. (1990b). Application of the theory of finite mixtures for the estimation of 'cure' rates of treated cancer patients. *Statistics in Medicine*, **9**, 397–407.

Green, P.J. (1990a). Bayesian reconstructions from emission tomography data using a modified EM algorithm. *IEEE Transactions on Medical Imaging*, **9**, 84–93.

Green, P.J. (1990b). On use of the EM algorithm for penalized likelihood estimation. *Journal of the Royal Statistical Society B*, **52**, 443–452.

Grenander, U. (1983). Tutorial in pattern theory. *Technical Report*. Providence, Rhode Island: Division of Applied Mathematics, Brown University.

Guo, S.W., and Thompson, E.A. (1992). A Monte Carlo method for combined segregation and linkage analysis. *American Journal of Human Genetics*, **51**, 1111–1126.

Haberman, S.J. (1974). Log-linear models for frequency tables derived by indirect observation: maximum likelihood equations. *Annals of Statistics*, **2**, 911–924.

Haberman, S.J. (1976). Iterative scaling procedures for log-linear models for frequency tables derived by indirect observation. In *Proceedings of the Statistical Computing Section, American Statistical Association*. Alexandria, Virginia: American Statistical Association, pp. 45–50.

Haberman, S.J. (1977). Product models for frequency tables involving indirect observation. *Annals of Statistics*, **5**, 1124–1147.

Hartley, H.O. (1958). Maximum likelihood estimation from incomplete data. *Biometrics*, **14**, 174–194.

Hartley, H.O. (1978). Contribution to the discussion of paper by R.E. Quandt and J.B. Ramsey. *Journal of the American Statistical Association*, **73**, 738–741.

Hartley, H.O., and Hocking, R.R. (1971). The analysis of incomplete data (with discussion). *Biometrics*, **27**, 783–808.

Harville, D.A. (1977). Maximum likelihood approaches to variance component estimation and to related problems. *Journal of the American Statistical Association*, **72**, 320–340.

Hasselblad, V. (1966). Estimation of parameters for a mixture of normal distributions. *Technometrics*, **8**, 431–444.

Hasselblad, V. (1969). Estimation of finite mixtures of distributions from the exponential family. *Journal of the American Statistical Association*, **64**, 1459–1471.

Hastings, W.K. (1970). Monte Carlo sampling methods using Markov chains and their applications. *Biometrika*, **57**, 97–109.

Hathaway, R.J. (1983). Constrained maximum likelihood estimation for normal mixtures. In *Computer Science and Statistics: The Interface*, J.E. Gentle (Ed.). Amsterdam: North-Holland, pp. 263–267.

Hathaway, R.J. (1985). A constrained formulation of maximum-likelihood estimation for normal mixture distributions. *Annals of Statistics*, **13**, 795–800.

Hathaway, R.J. (1986). A constrained EM algorithm for univariate normal mixtures. *Journal of Statistical Computation and Simulation*, **23**, 211–230.

Healy, M.J.R., and Westmacott, M. (1956). Missing values in experiments analyzed on automatic computers. *Applied Statistics*, **5**, 203–206.

Heckman, J., and Singer, B. (1984). A method for minimizing the impact of distributional assumptions in econometric models for duration data. *Econometrica*, **52**, 271–320.

Heitjan, D.F. (1989). Inference from grouped continuous data: a review (with discussion). *Statistical Science*, **4**, 164–183.

Heitjan, D.F. (1993). Ignorability and coarse data: some biomedical examples. *Biometrics*, **49**, 1099–1109.

Heitjan, D.F. (1994). Ignorability in general incomplete-data models. *Biometrika*, **81**, 701–708.

Heitjan, D.F., and Rubin, D.B. (1991). Ignorability and coarse data. *Annals of Statistics*, **19**, 2244–2253.

Heyde, C.C., and Morton, R. (1996). Quasi-likelihood and generalizing the EM algorithm. *Journal of the Royal Statistical Society B*, **58**, 317–327.

Holst, U., and Lindgren, G. (1991). Recursive estimation in mixture models with Markov regime. *IEEE Transactions on Information Theory*, **37**, 1683–1690.

Horng, S.C. (1986). *Sublinear Convergence of the EM Algorithm.* Unpublished Ph.D. Dissertation. University of California, Los Angeles, California.

Horng, S.C. (1987). Examples of sublinear convergence of the EM algorithm. *In Proceedings of the Statistical Computing Section, American Statistical Association*. Alexandria, Virginia: American Statistical Association, pp. 266–271.

Hosmer, D.W., Jr. (1973a). On MLE of the parameters of a mixture of two normal distributions when the sample size is small. *Communications in Statistics*, **1**, 217–227.

Hosmer, D.W., Jr. (1973b). A comparison of iterative maximum-likelihood estimates of the parameters of a mixture of two normal distributions under three types of sample. *Biometrics*, **29**, 761–770.

Huber, P.J. (1964). Robust estimation of a location parameter. *Annals of Mathematical Statistics*, **35**, 73–101.

Ingrassia, S. (1991). Mixture decomposition via the simulated annealing algorithm. *Applied Stochastic Models and Data Analysis*, **7**, 317–325.

Ingrassia, S. (1992). A comparison between the simulated annealing and the EM algorithms in normal mixture decompositions. *Statistics and Computing*, **2**, 203–211.

Jamshidian, M., and Jennrich, R.I. (1993). Conjugate gradient acceleration of the EM algorithm. *Journal of the American Statistical Association*, **88**, 221–228.

Jennrich, R.I., and Schluchter, M.D. (1986). Unbalanced repeated-measures models with structured covariance matrices. *Biometrics*, **42**, 805–820.

Jones, P.N., and McLachlan, G.J. (1990). Algorithm AS 254. Maximum likelihood estimation from grouped and truncated data with finite normal mixture models. *Applied Statistics*, **39**, 273–282.

Jones, P.N., and McLachlan, G.J. (1992). Improving the convergence rate of the EM algorithm for a mixture model fitted to grouped truncated data. *Journal of Statistical Computation and Simulation*, **43**, 31–44.

Jordan, M.I., and Jacobs, R.A. (1994). Hierarchical mixtures of experts and the EM algorithm. *Neural Computation*, **6**, 181–214.

Jordan, M.I., and Xu, L. (1995). Convergence results for the EM approach to mixtures of experts architectures. *Neural Networks*, **8**, 1409–1431.

Jöreskog, K.G. (1969). A general approach to confirmatory maximum likelihood factor analysis. *Psychometrika*, **34**, 183–202.

Jørgensen, B. (1984). The delta algorithm and GLIM. *International Statistical Review*, **52**, 283–300.

Juang, B.H., and Rabiner, L.R. (1991). Hidden Markov model for speech recognition. *Technometrics*, **33**, 251–272.

Kay, J. (1994). Statistical models for PET and SPECT data. *Statistical Methods in Medical Research*, **3**, 5–21.

Kempthorne, O. (1957). *An Introduction to Genetic Statistics*. New York: Wiley.

Kennedy, W.J., Jr., and Gentle, J.E. (1980). *Statistical Computing*. New York: Marcel Dekker.

Kent, J.T., and Tyler, D.E. (1991). Redescending M-estimates of multivariate location and scatter. *Annals of Statistics*, **19**, 2102–2119.

Kent, J.T., Tyler, D.E., and Vardi, Y. (1994). A curious likelihood identity for the multivariate t-distribution. *Communications in Statistics—Simulation and Computation*, **23**, 441–453.

Kiefer, N.M. (1978). Discrete parameter variation: efficient estimation of a switching regression model. *Econometrica*, **46**, 427–434.

Kingman, J.F.C. (1993). *Poisson Processes*. Oxford: Oxford University Press.

Kirkpatrick, S., Gelatt, C.D., Jr., and Vecchi, M.P. (1983). Optimization by simulated annealing. *Science*, **220**, 671–680.

Krishnan, T. (1995). EM algorithm in tomography: a review and a bibliography. *Bulletin of Informatics and Cybernetics*, **27**, 5–22.

Laird, N.M. (1978). Nonparametric maximum likelihood estimation of a mixing distribution. *Journal of the American Statistical Association*, **73**, 805–811.

Laird, N.M. (1982). The computation of estimates of variance components using the EM algorithm. *Journal of Statistical Computation and Simulation*, **14**, 295–303.

Laird, N.M., Lange, N., and Stram, D. (1987). Maximum likelihood computations with repeated measures: Application of the EM algorithm. *Journal of the American Statistical Association*, **82**, 97–105.

Laird, N.M., and Ware, J.H. (1982). Random-effects models for longitudinal data. *Biometrics*, **38**, 963–974.

Lange, K. (1990). Convergence of EM image reconstruction algorithm with Gibbs smoothing. *IEEE Transactions on Medical Imaging*, **9**, 439–446.

Lange, K. (1995a). A gradient algorithm locally equivalent to the EM algorithm. *Journal of the Royal Statistical Society B*, **57**, 425–437.

Lange, K. (1995b). A quasi-Newton acceleration of the EM algorithm. *Statistica Sinica*, **5**, 1–18.

Lange, K., and Carson, R. (1984). EM reconstruction algorithms for emission and transmission tomography. *Journal of Computer Assisted Tomography*, **8**, 306–316.

Lange, K., Little, R.J.A., and Taylor, J.M.G. (1989). Robust statistical modeling using the *t* distribution. *Journal of the American Statistical Association*, **84**, 881–896.

Lange, K., and Sinsheimer, S.S. (1993). Normal/independent distributions and their applications in robust regression. *Journal of Computational and Graphical Statistics*, **2**, 175–198.

Lansky, D., Casella, G., McCulloch, C., and Lansky, D. (1992). Convergence and invariance properties of the EM algorithm. In *Proceedings of the Statistical Computing Section, American Statistical Association*. Alexandria, Virginia: American Statistical Association, pp. 28–33.

Larson, M.G., and Dinse, G.E. (1985). A mixture model for the regression analysis of competing risks data. *Applied Statistics*, **34**, 201–211.

Lawley, D.N., and Maxwell, A.E. (1963). *Factor Analysis as a Statistical Method*. London: Butterworth.

Lehmann, E.L. (1983). *Theory of Point Estimation*. New York: Wiley.

Leroux, B.G., and Puterman, M.L. (1992). Maximum-penalized-likelihood estimation for independent and Markov-dependent mixture models. *Biometrics*, **48**, 545–558.

Lindsay, B.G. (1995). *Mixture Models: Theory Geometry, and Applications. NSF-CBMS Regional Conference Series in Probability and Statistics* Vol. 5. Hayward, California: Institute of Mathematical Statistics.

Lindsay, B.G., and Basak, P.K. (1993). Multivariate normal mixtures: a fast consistent method of moments. *Journal of the American Statistical Association*, **88**, 468–476.

Lindstrom, M.J., and Bates, D.M. (1988). Newton–Raphson and EM algorithms for linear mixed-effects models for repeated-measures data. *Journal of the American Statistical Association*, **83**, 1014–1022.

Little, R.J.A. (1983a). Superpopulation models for nonresponse: The ignorable case. In *Incomplete Data in Sample Surveys* Vol. 2, W.G. Madow, I. Olkin, and D.B. Rubin (Eds.). New York: Academic Press, pp. 341–382.

Little, R.J.A. (1983b). Superpopulation models for nonresponse: the nonignorable case. In *Incomplete Data in Sample Surveys* Vol. 2, W.G. Madow, I. Olkin, and D.B. Rubin (Eds.). New York: Academic Press, pp. 383–413.

Little, R.J.A. (1988). Robust estimation of the mean and covariance matrix from data with missing values. *Applied Statistics*, **37**, 23–38.

Little, R.J.A. (1993). Pattern-mixture models for multivariate incomplete data. *Journal of the American Statistical Association*, **88**, 125–134.

Little, R.J.A. (1994). A class of pattern-mixture models for normal incomplete data. *Biometrika*, **81**, 471–483.

Little, R.J.A., and Rubin, D.B. (1987). *Statistical Analysis with Missing Data*. New York: Wiley.

Little, R.J.A., and Rubin, D.B. (1989). The analysis of social science data with missing values. *Sociological Methods and Research*, **18**, 292–326.

Little, R.J.A., and Rubin, D.B. (1990). The analysis of social science data with missing values. In *Modern Methods of Data Analysis*, J. Fox and J. Scott Long (Eds.). Newbury Park, California: Sage Publications. pp. 374–409.

Liu, C., and Rubin, D.B. (1994). The ECME algorithm: a simple extension of EM and ECM with faster monotone convergence. *Biometrika*, **81**, 633–648.

Liu, C., and Rubin, D.B. (1995). ML estimation of the t distribution using EM and its extensions, ECM and ECME. *Statistica Sinica*, **5**, 19–39.

Liu, J.S. (1994). The collapsed Gibbs sampler in Bayesian computations with application to a gene regulation problem. *Journal of the American Statistical Association*, **89**, 958–966.

Liu, J.S., Wong, W.H., and Kong, A. (1994). Covariance structure of the Gibbs sampler with applications to the comparisons of estimators and augmentation schemes. *Biometrika*, **81**, 27–40.

Louis, T.A. (1982). Finding the observed information matrix when using the EM algorithm. *Journal of the Royal Statistical Society B*, **44**, 226–233.

Madow, W.G., Olkin, I., and Rubin, D.B. (Eds.) (1983). *Incomplete Data in Sample Surveys* Vol. 2: *Theory and Bibliographies*. New York: Academic Press.

McColl, J.H., Holmes, A.P., and Ford, I. (1994). Statistical methods in neuroimaging with particular application to emission tomography. *Statistical Methods in Medical Research*, **3**, 63–86.

McKendrick, A.G. (1926). Applications of mathematics to medical problems. *Proceedings of the Edinburgh Mathematical Society*, **44**, 98–130.

McLachlan, G.J. (1975). Iterative reclassification procedure for constructing an asymptotically optimal rule of allocation in discriminant analysis. *Journal of the American Statistical Association*, **70**, 365–369.

McLachlan, G.J. (1977). Estimating the linear discriminant function from initial samples containing a small number of unclassified observations. *Journal of the American Statistical Association*, **72**, 403–406.

McLachlan, G.J. (1982). The classification and mixture maximum likelihood approaches to cluster analysis. In *Handbook of Statistics* Vol. 2, P.R. Krishnaiah and L.N. Kanal (Eds.). Amsterdam: North-Holland, pp. 199–208.

McLachlan, G.J. (1987). On bootstrapping the likelihood ratio test statistic for the number of components in a normal mixture. *Applied Statistics*, **36**, 318–324.

McLachlan, G.J. (1992). *Discriminant Analysis and Statistical Pattern Recognition*. New York: Wiley.

McLachlan, G.J. (1997). *Recent Advances in Finite Mixture Models*. New York: Wiley, *to appear.*

McLachlan, G.J., Adams, P., Ng, S.K., McGiffin, D.C., and Galbraith, A.J. (1994). Fitting mixtures of Gompertz distributions to censored survival data. *Research Report No. 28*. Brisbane: Centre for Statistics, The University of Queensland.

McLachlan, G.J., and Basford, K.E. (1988). *Mixture Models: Inference and Applications to Clustering*. New York: Marcel Dekker.

McLachlan, G.J., Basford, K.E., and Green, M. (1993). On inferring the number of components in normal mixture models. *Research Report No. 9*. Brisbane: Centre for Statistics, The University of Queensland.

McLachlan, G.J., and Jones, P.N. (1988). Fitting mixture models to grouped and truncated data via the EM algorithm. *Biometrics*, **44**, 571–578.

McLachlan, G.J., and McGiffin, D.C. (1994). On the role of finite mixture models in survival analysis. *Statistical Methods in Medical Research*, **3**, 211–226.

McLachlan, G.J., and Prado, P. (1995). On the fitting of normal mixture models with equal correlation matrices. *Research Report No. 36*. Brisbane: Centre for Statistics, The University of Queensland.

Meilijson, I. (1989). A fast improvement to the EM algorithm on its own terms. *Journal of the Royal Statistical Society B*, **51**, 127–138.

Meng, X.L. (1994). On the rate of convergence of the ECM algorithm. *Annals of Statistics*, **22**, 326–339.

Meng, X.L., and Pedlow, S. (1992): EM: a bibliographic review with missing articles. In *Proceedings of the Statistical Computing Section, American Statistical Association*. Alexandria, Virginia: American Statistical Association, pp. 24–27.

Meng, X.L., and Rubin, D.B. (1989). Obtaining asymptotic variance-covariance matrices for missing-data problems using EM. In *Proceedings of the Statistical Computing Section, American Statistical Association*. Alexandria, Virginia: American Statistical Association, pp. 140–144.

Meng, X.L., and Rubin, D.B. (1991). Using EM to obtain asymptotic variance-covariance matrices: the SEM algorithm. *Journal of the American Statistical Association*, **86**, 899–909.

Meng, X.L., and Rubin, D.B. (1992). Recent extensions to the EM algorithm (with discussion). In *Bayesian Statistics* Vol. 4, J.M. Bernardo, J.O. Berger, A.P. Dawid, and A.F.M. Smith (Eds.). Oxford: Oxford University Press, pp. 307–320.

Meng, X.L., and Rubin, D.B. (1993). Maximum likelihood estimation via the ECM algorithm: a general framework. *Biometrika*, **80**, 267–278.

Meng, X.L., and Rubin, D.B. (1994). On the global and componentwise rates of convergence of the EM algorithm. *Linear Algebra and its Applications*, **199**, 413–425.

Meng, X.L., and van Dyk, D. (1995). The EM algorithm—An old folk song sung to a fast new tune. *Technical Report No. 408*. Chicago: Department of Statistics, University of Chicago, Chicago.

Metropolis, N., Rosenbluth, A.W., Rosenbluth, M.N., Teller, A.H., and Teller, E. (1953). Equations of state calculations by fast computing machines. *Journal of Chemical Physics*, **21**, 1087–1092.

Metropolis, N., and Ulam, S. (1949). The Monte Carlo method. *Journal of the American Statistical Association*, **44**, 335–341.

Mitchell, T.J., and Turnbull, B.W. (1979). Log-linear models in the analysis of disease prevalence data from survival/sacrifice experiments. *Biometrics*, **35**, 221–234.

Mosimann, J.E. (1962). On the compound multinomial distribution, the multivariate β-distribution, and correlations among proportions. *Biometrika*, **49**, 65–82.

Murray, G.D. (1977). Contribution to the discussion of paper by A.P. Dempster, N.M. Laird, and D.B. Rubin. *Journal of the Royal Statistical Society B*, **39**, 27–28.

Narayanan, A. (1991). Algorithm AS 266: maximum likelihood estimation of parameters of the Dirichlet distribution. *Applied Statistics*, **40**, 365–374.

Neal, R.N., and Hinton, G.E. (1993). A new view of the EM algorithm that justifies incremental and other variants. *Technical Report*. Toronto: Department of Computer Science, University of Toronto.

Nelder, J.A. (1977). Contribution to the discussion of paper by A.P. Dempster, N.M. Laird, and D.B. Rubin. *Journal of the Royal Statistical Society B*, **39**, 23–24.

Newcomb, S. (1886). A generalized theory of the combination of observations so as to obtain the best result. *American Journal of Mathematics*, **8**, 343–366.

Nowlan, S.J. (1991). Soft competitive algorithm: Neural network learning algorithms based on fitting statistical mixtures. *Technical Report No. CMU-CS-91-126*. Pittsburgh: Carnegie-Mellon University.

Nowlan, S.J., and Hinton, G.E. (1992). Simplifying neural networks by soft weight-sharing. *Neural Computation*, **4**, 473–493.

Nychka, D. (1990). Some properties of adding a smoothing step to the EM algorithm. *Statistics & Probability Letters*, **9**, 187–193.

Orchard, T., and Woodbury, M.A. (1972). A missing information principle: theory and applications. In *Proceedings of the 6th Berkeley Symposium on Mathematical Statistics and Probability* Vol. 1. Berkeley, California: University of California Press, pp. 697–715.

Ostrowski, A.M. (1966). *Solution of Equations and Systems of Equations*. Second Edition. New York: Academic Press.

Pagano, M., De Gruttola, V., MaWhinney, S., and Tu, X.M. (1992). The HIV epidemic in New York city: Statistical methods for projecting AIDS incidence and prevalence. In *AIDS Epidemiology: Methodological Issues*, N.P. Jewell, K. Dietz, and V.T. Farewell (Eds.). Boston: Birkhäuser, pp. 123–142.

Peters, B.C., and Coberly, W.A. (1976). The numerical evaluation of the maximum-likelihood estimate of mixture proportions. *Communications in Statistics—Theory and Methods*, **5**, 1127–1135.

Peters, B.C., and Walker, H.F. (1978). An iterative procedure for obtaining maximum-likelihood estimates of the parameters for a mixture of normal distributions. *SIAM Journal of Applied Mathematics*, **35**, 362–378.

Pincus, M. (1968). A closed form solution of certain programming problems. *Operations Research*, **16**, 690–694.

Pincus, M. (1970). A Monte-Carlo method for the approximate solution of certain types of constrained optimization problems. *Operations Research*, **18**, 1125–1228.

Powell, M.J.D. (1978). A fast algorithm for nonlinearly constrained optimization calculations. In *Lecture Notes in Mathematics No. 630*, G.A. Watson (Ed.). New York: Springer-Verlag, pp. 144–157.

Qian, W., and Titterington, D.M. (1991). Estimation of parameters in hidden Markov models. *Philosophical Transactions of the Royal Society of London. Series A: Mathematical and Physical Sciences*, **337**, 407–428.

Qian, W., and Titterington, D.M. (1992). Stochastic relaxations and EM algorithms for Markov random fields. *Journal of Statistical Computation and Simulation*, **40**, 55–69.

Rabiner, L.R. (1989). A tutorial on hidden Markov models and selected applications in speech recognition. In *Proceedings of the IEEE*, **77**, 257–286.

Rai, S.N., and Matthews, D.E. (1993). Improving the EM algorithm. *Biometrics*, **49**, 587–591.

Rao, C.R (1973). *Linear Statistical Inference and its Applications*. Second Edition. New York: Wiley.

Ratschek, H., and Rokne, J. (1988). *New Computer Methods for Global Optimization*. New York: Halsted Press.

Redner, R.A., and Walker, H.F. (1984). Mixture densities, maximum likelihood and the EM algorithm. *SIAM Review*, **26**, 195–239.

Ripley, B.D. (1987). *Stochastic Simulation*. New York: Wiley.

Ripley, B.D. (1988). *Statistical Inference for Spatial Processes*. Cambridge: Cambridge University Press.

Ripley, B.D. (1990). Book review of *Simulated Annealing (SA) & Optimization: Modern Algorithms with VLSI, Optimal Design and Missile Defense Applications*, by M.E. Johnson (Ed.). *Journal of Classification*, **7**, 287–290.

Ripley, B.D. (1994). Neural networks and related methods of classification (with discussion). *Journal of the Royal Statistical Society B*, **56**, 409–456.

Ritter, C., and Tanner, M.A. (1992). Facilitating the Gibbs sampler: the Gibbs stopper and the Griddy-Gibbs sampler. *Journal of the American Statistical Association*, **87**, 861–868.

Robert, C.P. (1996). Mixtures of distributions: inference and estimation. In *Markov Chain Monte Carlo in Practice*, W.R. Gilks, S. Richardson, and D.J. Spiegelhalter (Eds.). London: Chapman & Hall, pp. 441–464.

Robert, C.P., Celeux, G., and Diebolt, J. (1993). Bayesian estimation of hidden Markov chains: A stochastic implementation. *Statistics & Probability Letters*, **16**, 77–83.

Roberts, G.O., and Polson, N.G. (1994). On the geometric convergence of the Gibbs sampler. *Journal of the Royal Statistical Society B*, **56**, 377–384.

Rubin, D.B. (1976). Inference with missing data. *Biometrika*, **63**, 581–592.

Rubin, D.B. (1983). Iteratively reweighted least squares. In *Encyclopedia of Statistical Sciences* Vol. 4, S. Kotz, N.L. Johnson, and C.B. Read (Eds.). New York: Wiley, pp. 272–275.

Rubin, D.B. (1987). *Multiple Imputation for Nonresponse in Surveys*. New York: Wiley.

Rubin, D.B. (1991). EM and beyond. *Psychometrika*, **56**, 241–254.

Rubin, D.B., and Thayer, D.T. (1982). EM algorithms for ML factor analysis. *Psychometrika*, **47**, 69–76.

Rubin, D.B., and Thayer, D.T. (1983). More on EM for factor analysis. *Psychometrika*, **48**, 253–257.

Ruppert, D., and Carroll, R.J. (1980). Trimmed least square estimation in the linear model. *Journal of the American Statistical Association*, **75**, 828–838.

Sarkar, A. (1993). On missing data, ignorability and sufficiency. *Technical Report No. 423*. Stanford, California: Department of Statistics, Stanford University.

Schader, M., and Schmid, F. (1985). Computation of M.L. estimates for the parameters of a negative binomial distribution from grouped data: a comparison of the scoring, Newton-Raphson and E-M algorithms. *Applied Stochastic Models and Data Analysis*, **1**, 11–23.

Schafer, J. (1996). *Analysis of Incomplete Multivariate Data by Simulation*. London: Chapman & Hall.

Schmee, J., and Hahn, F.J. (1979). A simple method of regression analysis with censored data. *Technometrics*, **21**, 417–434.

Segal, M.R., Bacchetti, P., and Jewell, N.P. (1994). Variances for maximum penalized likelihood estimates obtained via the EM algorithm. *Journal of the Royal Statistical Society B*, **56**, 345–352.

Shepp, L.A., and Vardi, Y. (1982). Maximum likelihood reconstruction for emission tomography. *IEEE Transactions on Medical Imaging*, **1**, 113–122.

Shih, W.J., and Weisberg, S. (1986). Assessing influence in multiple linear regression with incomplete data. *Technometrics*, **28**, 231–239.

Silverman, B.W., Jones, M.C., Wilson, J.D., and Nychka, D.W. (1990). A smoothed EM approach to indirect estimation problems, with particular reference to stereology and emission tomography. *Journal of the Royal Statistical Society B*, **52**, 271–324.

Sinha, D., Tanner, M.A., and Hall, W.J. (1994). Maximization of the marginal likelihood of grouped survival data. *Biometrika*, **81**, 53–60.

Smith, A.F.M., and Roberts, G.O. (1993). Bayesian computation via the Gibbs sampler and related Markov chain Monte Carlo methods. *Journal of the Royal Statistical Society B*, **55**, 3–23. Discussion: 53–102.

Smith, C.A.B. (1957). Counting methods in genetical statistics, *Annals of Human Genetics*, **21**, 254–276.

Smith, C.A.B. (1977). Contribution to the discussion of the paper by A.P. Dempster, N.M. Laird, and D.B. Rubin. *Journal of the Royal Statistical Society B*, **39**, 24–25.

Sobel, E., and Lange, K. (1994). Metropolis sampling in pedigree analysis. *Statistical Methods in Medical Research*, **2**, 263–282.

Specht, D.F. (1991). A general regression neural network. *IEEE Transactions on Neural Networks*, **2**, 568–576.

Stuart, A., and Ord, J.K. (1994). *Kendall's Advanced Theory of Statistics* Vol. 2: *Classical Inference and Relationship*. Sixth Edition. London: Charles Griffin.

Sundberg, R. (1974). Maximum likelihood theory for incomplete data from an exponential family. *Scandinavian Journal of Statistics: Theory and Applications*, **1**, 49–58.

Sundberg, R. (1976). An iterative method for solution of the likelihood equations for incomplete data from exponential families. *Communications in Statistics—Simulation and Computation*, **5**, 55–64.

Tan, W.Y., and Chang, W.C. (1972). Convolution approach to genetic analysis of quantitative characters of self-fertilized populations. *Biometrics*, **28**, 1073–1090.

Tanner, M.A. (1991). *Tools for Statistical Inference: Observed Data and Data Augmentation Methods*. Lecture Notes in Statistics Vol. 67. New York: Springer-Verlag.

Tanner, M.A. (1993). *Tools for Statistical Inference: Methods for the Exploration of Posterior Distributions and Likelihood Functions*. Second Edition. New York: Springer-Verlag.

Tanner, M.A., and Wong, W.H. (1987). The calculation of posterior distributions by data augmentation (with discussion). *Journal of the American Statistical Association*, **82**, 528–550.

Thisted, R.A. (1988). *Elements of Statistical Computing: Numerical Computation*. London: Chapman & Hall.

Thompson, E.A. (1975). *Human Evolutionary Trees*. Cambridge: Cambridge University Press.

Thompson, E.A. (1977). Contribution to the discussion of the paper by A.P. Dempster, N.M. Laird, and D.B. Rubin. *Journal of the Royal Statistical Society B*, **39**, 33–34.

Tierney, L. (1994). Markov chains for exploring posterior distributions (with discussion). *Annals of Statistics*, **22**, 1701–1762.

Titterington, D.M. (1984). Recursive parameter estimation using incomplete data. *Journal of the Royal Statistical Society B*, **46**, 257–267.

Titterington, D.M. (1987). On the iterative image space reconstruction algorithm for ECT. *IEEE Transactions on Medical Imaging*, **6**, 52–56.

Titterington, D.M. (1990). Some recent research in the analysis of mixture distributions. *Statistics*, **21**, 619–641.

Titterington, D.M., Smith, A.F.M., and Makov, U.E. (1985). *Statistical Analysis of Finite Mixture Distributions*. New York: Wiley.

Tu, X.M., Meng, X.-L., and Pagano, M. (1993). The AIDS epidemic: estimating survival after AIDS diagnosis from surveillance data. *Journal of the American Statistical Association*, **88**, 26–36.

Turnbull, B.W. (1974). Nonparametric estimation of a survivorship function with doubly censored data. *Journal of the American Statistical Association*, **69**, 169–173.

Turnbull, B.W. (1976). The empirical distribution with arbitrarily grouped, censored and truncated data. *Journal of the Royal Statistical Society B*, **38**, 290–295.

Turnbull, B.W., and Mitchell, T.J. (1978). Exploratory analysis of disease prevalence data from survival/sacrifice experiments. *Biometrics*, **34**, 555–570.

Turnbull, B.W., and Mitchell, T.J. (1984). Nonparametric estimation of the distribution of time to onset for specific diseases in survival/sacrifice experiments. *Biometrics*, **40**, 41–50.

van Laarhoven, P.J.M., and Aarts, E.H.L. (1987). *Simulated Annealing: Theory and Applications*. Dordrecht: D.Reidel Publishing Company.

Vardi, Y., and Lee, D. (1993). From image deblurring to optimal investments: Maximum likelihood solutions to positive linear inverse problems (with discussion). *Journal of the Royal Statistical Society B*, **55**, 569–612.

Vardi, Y., Shepp, L.A., and Kaufman, L. (1985). A statistical model for positron emission tomography (with discussion). *Journal of the American Statistical Association*, **80**, 8–37.

Wei, G.C.G., and Tanner, M.A. (1990). A Monte Carlo implementation of the EM algorithm and the poor man's data augmentation algorithms. *Journal of the American Statistical Association*, **85**, 699–704.

Weiss, G.H. (1989). Simulated Annealing. In *Encyclopedia of Statistical Sciences* Supplementary Volume, S. Kotz, N.L. Johnson, and C.B. Read (Eds.). New York: Wiley, pp. 144–146.

Whittemore, A.S., and Grosser, S. (1986). Regression methods for data with incomplete covariates. In *Modern Statistical Methods for Chronic Disease Epidemiology*, S.H. Moolgavkar and R.L. Prentice (Eds.). New York: Wiley, pp. 19–34.

Wolfe, J.H. (1967). NORMIX: Computational methods for estimating the parameters of multivariate normal mixtures of distributions. *Research Memo. SRM 68–2*. San Diego: U.S. Naval Personnel Research Activity.

Wolfe, J.H. (1970). Pattern clustering by multivariate mixture analysis. *Multivariate Behavioral Research*, **5**, 329–350.

Wolynetz, M.S. (1979a). Algorithm AS 138: maximum likelihood estimation from confined and censored normal data. *Applied Statistics*, **28**, 185–195.

Wolynetz, M.S. (1979b). Algorithm AS 139: maximum likelihood estimation in a linear model from confined and censored normal data. *Applied Statistics*, **28**, 195–206.

Wolynetz, M.S. (1980). A remark on Algorithm 138: maximum likelihood estimation from confined and censored normal data. *Applied Statistics*, **29**, 228.

Wu, C.F.J. (1983). On the convergence properties of the EM algorithm. *Annals of Statistics*, **11**, 95–103.

Xu, L., and Jordan, M.I. (1996). On convergence properties of the EM algorithm for Gaussian mixtures. *Neural Computation*, **8**, 129–151.

Yates, F.Y. (1933). The analysis of replicated experiments when the field results are incomplete. *Empire Journal of Experimental Agriculture*, **1**, 129–142.

Yuille, A.L., Stolorz, P., and Utans, J. (1994). Statistical physics, mixtures of distributions and the EM algorithm. *Neural Computation* **6**, 334–340.

Zangwill, W.I. (1969). *Nonlinear Programming: A Unified Approach*. Englewood Cliffs, New Jersey: Prentice-Hall.

Zeger, S.L., and Karim, M.R. (1991). Generalized linear models with random effects: a Gibbs sampling approach. *Journal of the American Statistical Association*, **86**, 79–86.

Author Index

Subject Index

WILEY SERIES IN PROBABILITY AND STATISTICS

ESTABLISHED BY WALTER A. SHEWHART AND SAMUEL S. WILKS

Editors
Vic Barnett, Ralph A. Bradley, Nicholas I. Fisher, J. Stuart Hunter, J. B. Kadane, David G. Kendall, David W. Scott, Adrian F. M. Smith, Jozef L. Teugels, Geoffrey S. Watson

Probability and Statistics

ANDERSON · An Introduction to Multivariate Statistical Analysis, *Second Edition*
*ANDERSON · The Statistical Analysis of Time Series
ARNOLD, BALAKRISHNAN, and NAGARAJA · A First Course in Order Statistics
BACCELLI, COHEN, OLSDER, and QUADRAT · Synchronization and Linearity: An Algebra for Discrete Event Systems
BARTOSZYNSKI and NIEWIADOMSKA-BUGAJ · Probability and Statistical Inference
BERNARDO and SMITH · Bayesian Statistical Concepts and Theory
BHATTACHARYYA and JOHNSON · Statistical Concepts and Methods
BILLINGSLEY · Convergence of Probability Measures
BILLINGSLEY · Probability and Measure, *Second Edition*
BOROVKOV · Asymptotic Methods in Queuing Theory
BRANDT, FRANKEN, and LISEK · Stationary Stochastic Models
CAINES · Linear Stochastic Systems
CAIROLI and DALANG · Sequential Stochastic Optimization
CHEN · Recursive Estimation and Control for Stochastic Systems
CONSTANTINE · Combinatorial Theory and Statistical Design
COOK and WEISBERG · An Introduction to Regression Graphics
COVER and THOMAS · Elements of Information Theory
CSÖRGŐ and HORVÁTH · Weighted Approximations in Probability Statistics
*DOOB · Stochastic Processes
DUDEWICZ and MISHRA · Modern Mathematical Statistics
DUPUIS · A Weak Convergence Approach to the Theory of Large Deviations
ETHIER and KURTZ · Markov Processes: Characterization and Convergence
FELLER · An Introduction to Probability Theory and Its Applications, Volume 1, *Third Edition,* Revised; Volume II, *Second Edition*
FREEMAN and SMITH · Aspects of Uncertainty: A Tribute to D. V. Lindley
FULLER · Introduction to Statistical Time Series, *Second Edition*
FULLER · Measurement Error Models
GHOSH · Sequential Estimation
GIFI · Nonlinear Multivariate Analysis
GUTTORP · Statistical Inference for Branching Processes
HALD · A History of Probability and Statistics and Their Applications before 1750
HALL · Introduction to the Theory of Coverage Processes
HANNAN and DEISTLER · The Statistical Theory of Linear Systems
HEDAYAT and SINHA · Design and Inference in Finite Population Sampling
HOEL · Introduction to Mathematical Statistics, *Fifth Edition*
HUBER · Robust Statistics
IMAN and CONOVER · A Modern Approach to Statistics
JUREK and MASON · Operator-Limit Distributions in Probability Theory
KASS and VOS · Geometrical Foundations of Asymptotic Inference: Curved Exponential Families and Beyond
KAUFMAN and ROUSSEEUW · Finding Groups in Data: An Introduction to Cluster Analysis

*Now available in a lower priced paperback edition in the Wiley Classics Library.

Probability and Statistics (Continued)

KOTZ · Leading Personalities in Statistical Sciences from the Seventeenth Century to the Present
LAMPERTI · Probability: A Survey of the Mathematical Theory, *Second Edition*
LARSON · Introduction to Probability Theory and Statistical Inference, *Third Edition*
LESSLER and KALSBEEK · Nonsampling Error in Surveys
LINDVALL · Lectures on the Coupling Method
MANTON, WOODBURY, and TOLLEY · Statistical Applications Using Fuzzy Sets
MARDIA · The Art of Statistical Science: A Tribute to G. S. Watson
MORGENTHALER and TUKEY · Configural Polysampling: A Route to Practical Robustness
MUIRHEAD · Aspects of Multivariate Statistical Theory
OLIVER and SMITH · Influence Diagrams, Belief Nets and Decision Analysis
*PARZEN · Modern Probability Theory and Its Applications
PRESS · Bayesian Statistics: Principles, Models, and Applications
PUKELSHEIM · Optimal Experimental Design
PURI and SEN · Nonparametric Methods in General Linear Models
PURI, VILAPLANA, and WERTZ · New Perspectives in Theoretical and Applied Statistics
RAO · Asymptotic Theory of Statistical Inference
RAO · Linear Statistical Inference and Its Applications, *Second Edition*
*RAO and SHANBHAG · Choquet-Deny Type Functional Equations with Applications to Stochastic Models
RENCHER · Methods of Multivariate Analysis
ROBERTSON, WRIGHT, and DYKSTRA · Order Restricted Statistical Inference
ROGERS and WILLIAMS · Diffusions, Markov Processes, and Martingales, Volume I: Foundations, *Second Edition;* Volume II: Îto Calculus
ROHATGI · An Introduction to Probability Theory and Mathematical Statistics
ROSS · Stochastic Processes
RUBINSTEIN · Simulation and the Monte Carlo Method
RUBINSTEIN and SHAPIRO · Discrete Event Systems: Sensitivity Analysis and Stochastic Optimization by the Score Function Method
RUZSA and SZEKELY · Algebraic Probability Theory
SCHEFFE · The Analysis of Variance
SEBER · Linear Regression Analysis
SEBER · Multivariate Observations
SEBER and WILD · Nonlinear Regression
SERFLING · Approximation Theorems of Mathematical Statistics
SHORACK and WELLNER · Empirical Processes with Applications to Statistics
SMALL and McLEISH · Hilbert Space Methods in Probability and Statistical Inference
STAPLETON · Linear Statistical Models
STAUDTE and SHEATHER · Robust Estimation and Testing
STOYANOV · Counterexamples in Probability
STYAN · The Collected Papers of T. W. Anderson: 1943–1985
TANAKA · Time Series Analysis: Nonstationary and Noninvertible Distribution Theory
THOMPSON and SEBER · Adaptive Sampling
WELSH · Aspects of Statistical Inference
WHITTAKER · Graphical Models in Applied Multivariate Statistics
YANG · The Construction Theory of Denumerable Markov Processes

Applied Probability and Statistics

ABRAHAM and LEDOLTER · Statistical Methods for Forecasting
AGRESTI · Analysis of Ordinal Categorical Data

*Now available in a lower priced paperback edition in the Wiley Classics Library.

Applied Probability and Statistics (Continued)

AGRESTI · Categorical Data Analysis
AGRESTI · An Introduction to Categorical Data Analysis
ANDERSON and LOYNES · The Teaching of Practical Statistics
ANDERSON, AUQUIER, HAUCK, OAKES, VANDAELE, and WEISBERG · Statistical Methods for Comparative Studies
ARMITAGE and DAVID (editors) · Advances in Biometry
*ARTHANARI and DODGE · Mathematical Programming in Statistics
ASMUSSEN · Applied Probability and Queues
*BAILEY · The Elements of Stochastic Processes with Applications to the Natural Sciences
BARNETT and LEWIS · Outliers in Statistical Data, *Second Edition*
BARTHOLOMEW, FORBES, and McLEAN · Statistical Techniques for Manpower Planning, *Second Edition*
BATES and WATTS · Nonlinear Regression Analysis and Its Applications
BECHHOFER, SANTNER, and GOLDSMAN · Design and Analysis of Experiments for Statistical Selection, Screening, and Multiple Comparisons
BELSLEY · Conditioning Diagnostics: Collinearity and Weak Data in Regression
BELSLEY, KUH, and WELSCH · Regression Diagnostics: Identifying Influential Data and Sources of Collinearity
BERRY · Bayesian Analysis in Statistics and Econometrics: Essays in Honor of Arnold Zellner
BERRY, CHALONER, and GEWEKE · Bayesian Analysis in Statistics and Econometrics: Essays in Honor of Arnold Zellner
BHAT · Elements of Applied Stochastic Processes, *Second Edition*
BHATTACHARYA and WAYMIRE · Stochastic Processes with Applications
BIEMER, GROVES, LYBERG, MATHIOWETZ, and SUDMAN · Measurement Errors in Surveys
BIRKES and DODGE · Alternative Methods of Regression
BLOOMFIELD · Fourier Analysis of Time Series: An Introduction
BOLLEN · Structural Equations with Latent Variables
BOULEAU · Numerical Methods for Stochastic Processes
BOX · R. A. Fisher, the Life of a Scientist
BOX and DRAPER · Empirical Model-Building and Response Surfaces
BOX and DRAPER · Evolutionary Operation: A Statistical Method for Process Improvement
BOX, HUNTER, and HUNTER · Statistics for Experimenters: An Introduction to Design, Data Analysis, and Model Building
BROWN and HOLLANDER · Statistics: A Biomedical Introduction
BUCKLEW · Large Deviation Techniques in Decision, Simulation, and Estimation
BUNKE and BUNKE · Nonlinear Regression, Functional Relations and Robust Methods: Statistical Methods of Model Building
CHATTERJEE and HADI · Sensitivity Analysis in Linear Regression
CHATTERJEE and PRICE · Regression Analysis by Example, *Second Edition*
CLARKE and DISNEY · Probability and Random Processes: A First Course with Applications, *Second Edition*
COCHRAN · Sampling Techniques, *Third Edition*
*COCHRAN and COX · Experimental Designs, *Second Edition*
CONOVER · Practical Nonparametric Statistics, *Second Edition*
CONOVER and IMAN · Introduction to Modern Business Statistics
CORNELL · Experiments with Mixtures, Designs, Models, and the Analysis of Mixture Data, *Second Edition*
COX · A Handbook of Introductory Statistical Methods
*COX · Planning of Experiments

*Now available in a lower priced paperback edition in the Wiley Classics Library.

Applied Probability and Statistics (Continued)

COX, BINDER, CHINNAPPA, CHRISTIANSON, COLLEDGE, and KOTT · Business Survey Methods
CRESSIE · Statistics for Spatial Data, *Revised Edition*
DANIEL · Applications of Statistics to Industrial Experimentation
DANIEL · Biostatistics: A Foundation for Analysis in the Health Sciences, *Sixth Edition*
DAVID · Order Statistics, *Second Edition*
*DEGROOT, FIENBERG, and KADANE · Statistics and the Law
*DEMING · Sample Design in Business Research
DILLON and GOLDSTEIN · Multivariate Analysis: Methods and Applications
DODGE and ROMIG · Sampling Inspection Tables, *Second Edition*
DOWDY and WEARDEN · Statistics for Research, *Second Edition*
DRAPER and SMITH · Applied Regression Analysis, *Second Edition*
DUNN · Basic Statistics: A Primer for the Biomedical Sciences, *Second Edition*
DUNN and CLARK · Applied Statistics: Analysis of Variance and Regression, *Second Edition*
ELANDT-JOHNSON and JOHNSON · Survival Models and Data Analysis
EVANS, PEACOCK, and HASTINGS · Statistical Distributions, *Second Edition*
FISHER and VAN BELLE · Biostatistics: A Methodology for the Health Sciences
FLEISS · The Design and Analysis of Clinical Experiments
FLEISS · Statistical Methods for Rates and Proportions, *Second Edition*
FLEMING and HARRINGTON · Counting Processes and Survival Analysis
FLURY · Common Principal Components and Related Multivariate Models
GALLANT · Nonlinear Statistical Models
GLASSERMAN and YAO · Monotone Structure in Discrete-Event Systems
GNANADESIKAN · Methods for Statistical Data Analysis of Multivariate Observations, *Second Edition*
GREENWOOD and NIKULIN · A Guide to Chi-Squared Testing
GROSS and HARRIS · Fundamentals of Queueing Theory, *Second Edition*
GROVES · Survey Errors and Survey Costs
GROVES, BIEMER, LYBERG, MASSEY, NICHOLLS, and WAKSBERG · Telephone Survey Methodology
HAHN and MEEKER · Statistical Intervals: A Guide for Practitioners
HAND · Discrimination and Classification
*HANSEN, HURWITZ, and MADOW · Sample Survey Methods and Theory, Volume 1: Methods and Applications
*HANSEN, HURWITZ, and MADOW · Sample Survey Methods and Theory, Volume II: Theory
HEIBERGER · Computation for the Analysis of Designed Experiments
HELLER · MACSYMA for Statisticians
HINKELMAN and KEMPTHORNE: · Design and Analysis of Experiments, Volume 1: Introduction to Experimental Design
HOAGLIN, MOSTELLER, and TUKEY · Exploratory Approach to Analysis of Variance
HOAGLIN, MOSTELLER, and TUKEY · Exploring Data Tables, Trends and Shapes
HOAGLIN, MOSTELLER, and TUKEY · Understanding Robust and Exploratory Data Analysis
HOCHBERG and TAMHANE · Multiple Comparison Procedures
HOCKING · Methods and Applications of Linear Models: Regression and the Analysis of Variables
HOEL · Elementary Statistics, *Fifth Edition*
HOGG and KLUGMAN · Loss Distributions
HOLLANDER and WOLFE · Nonparametric Statistical Methods
HOSMER and LEMESHOW · Applied Logistic Regression
HØYLAND and RAUSAND · System Reliability Theory: Models and Statistical Methods
HUBERTY · Applied Discriminant Analysis

*Now available in a lower priced paperback edition in the Wiley Classics Library.

IMAN and CONOVER · Modern Business Statistics
JACKSON · A User's Guide to Principle Components
JOHN · Statistical Methods in Engineering and Quality Assurance
JOHNSON · Multivariate Statistical Simulation
JOHNSON and KOTZ · Distributions in Statistics
Continuous Univariate Distributions—2
Continuous Multivariate Distributions
JOHNSON, KOTZ, and BALAKRISHNAN · Continuous Univariate Distributions, Volume 1, *Second Edition*
JOHNSON, KOTZ, and BALAKRISHNAN · Discrete Multivariate Distributions
JOHNSON, KOTZ, and KEMP · Univariate Discrete Distributions, *Second Edition*
JUDGE, GRIFFITHS, HILL, LÜTKEPOHL, and LEE · The Theory and Practice of Econometrics, *Second Edition*
JUDGE, HILL, GRIFFITHS, LÜTKEPOHL, and LEE · Introduction to the Theory and Practice of Econometrics, *Second Edition*
JUREČKOVÁ and SEN · Robust Statistical Procedures: Aymptotics and Interrelations
KADANE · Bayesian Methods and Ethics in a Clinical Trial Design
KADANE AND SCHUM · A Probabilistic Analysis of the Sacco and Vanzetti Evidence
KALBFLEISCH and PRENTICE · The Statistical Analysis of Failure Time Data
KASPRZYK, DUNCAN, KALTON, and SINGH · Panel Surveys
KISH · Statistical Design for Research
*KISH · Survey Sampling
LAD · Operational Subjective Statistical Methods: A Mathematical, Philosophical, and Historical Introduction
LANGE, RYAN, BILLARD, BRILLINGER, CONQUEST, and GREENHOUSE · Case Studies in Biometry
LAWLESS · Statistical Models and Methods for Lifetime Data
LEBART, MORINEAU., and WARWICK · Multivariate Descriptive Statistical Analysis: Correspondence Analysis and Related Techniques for Large Matrices
LEE · Statistical Methods for Survival Data Analysis, *Second Edition*
LePAGE and BILLARD · Exploring the Limits of Bootstrap
LEVY and LEMESHOW · Sampling of Populations: Methods and Applications
LINHART and ZUCCHINI · Model Selection
LITTLE and RUBIN · Statistical Analysis with Missing Data
LYBERG · Survey Measurement
MAGNUS and NEUDECKER · Matrix Differential Calculus with Applications in Statistics and Econometrics
MAINDONALD · Statistical Computation
MALLOWS · Design, Data, and Analysis by Some Friends of Cuthbert Daniel
MANN, SCHAFER, and SINGPURWALLA · Methods for Statistical Analysis of Reliability and Life Data
MASON, GUNST, and HESS · Statistical Design and Analysis of Experiments with Applications to Engineering and Science
McLACHLAN and KRISHNAN · The EM Algorithm and Extensions
McLACHLAN · Discriminant Analysis and Statistical Pattern Recognition
McNEIL · Epidemiological Research Methods
MILLER · Survival Analysis
MONTGOMERY and MYERS · Response Surface Methodology: Process and Product in Optimization Using Designed Experiments
MONTGOMERY and PECK · Introduction to Linear Regression Analysis, *Second Edition*
NELSON · Accelerated Testing, Statistical Models, Test Plans, and Data Analyses
NELSON · Applied Life Data Analysis
OCHI · Applied Probability and Stochastic Processes in Engineering and Physical Sciences

*Now available in a lower priced paperback edition in the Wiley Classics Library.

Applied Probability and Statistics (Continued)

OKABE, BOOTS, and SUGIHARA · Spatial Tesselations: Concepts and Applications of Voronoi Diagrams
OSBORNE · Finite Algorithms in Optimization and Data Analysis
PANKRATZ · Forecasting with Dynamic Regression Models
PANKRATZ · Forecasting with Univariate Box-Jenkins Models: Concepts and Cases
PORT · Theoretical Probability for Applications
PUTERMAN · Markov Decision Processes: Discrete Stochastic Dynamic Programming
RACHEV · Probability Metrics and the Stability of Stochastic Models
RÉNYI · A Diary on Information Theory
RIPLEY · Spatial Statistics
RIPLEY · Stochastic Simulation
ROSS · Introduction to Probability and Statistics for Engineers and Scientists
ROUSSEEUW and LEROY · Robust Regression and Outlier Detection
RUBIN · Multiple Imputation for Nonresponse in Surveys
RYAN · Modern Regression Methods
RYAN · Statistical Methods for Quality Improvement
SCHOTT · Matrix Analysis for Statistics
SCHUSS · Theory and Applications of Stochastic Differential Equations
SCOTT · Multivariate Density Estimation: Theory, Practice, and Visualization
SEARLE · Linear Models
SEARLE · Linear Models for Unbalanced Data
SEARLE · Matrix Algebra Useful for Statistics
SEARLE, CASELLA, and McCULLOCH · Variance Components
SKINNER, HOLT, and SMITH · Analysis of Complex Surveys
STOYAN, KENDALL, and MECKE · Stochastic Geometry and Its Applications, *Second Edition*
STOYAN and STOYAN · Fractals, Random Shapes and Point Fields: Methods of Geometrical Statistics
THOMPSON · Empirical Model Building
THOMPSON · Sampling
TIERNEY · LISP-STAT: An Object-Oriented Environment for Statistical Computing and Dynamic Graphics
TIJMS · Stochastic Modeling and Analysis: A Computational Approach
TITTERINGTON, SMITH, and MAKOV · Statistical Analysis of Finite Mixture Distributions
UPTON and FINGLETON · Spatial Data Analysis by Example, Volume 1: Point Pattern and Quantitative Data
UPTON and FINGLETON · Spatial Data Analysis by Example, Volume II: Categorical and Directional Data
VAN RIJCKEVORSEL and DE LEEUW · Component and Correspondence Analysis
WEISBERG · Applied Linear Regression, *Second Edition*
WESTFALL and YOUNG · Resampling-Based Multiple Testing: Examples and Methods for p-Value Adjustment
WHITTLE · Optimization Over Time: Dynamic Programming and Stochastic Control, Volume I and Volume II
WHITTLE · Systems in Stochastic Equilibrium
WONNACOTT and WONNACOTT · Econometrics, *Second Edition*
WONNACOTT and WONNACOTT · Introductory Statistics, *Fifth Edition*
WONNACOTT and WONNACOTT · Introductory Statistics for Business and Economics, *Fourth Edition*
WOODING · Planning Pharmaceutical Clinical Trials: Basic Statistical Principles
WOOLSON · Statistical Methods for the Analysis of Biomedical Data
*ZELLNER · An Introduction to Bayesian Inference in Econometrics

*Now available in a lower priced paperback edition in the Wiley Classics Library.

Tracts on Probability and Statistics
BILLINGSLEY · Convergence of Probability Measures
KELLY · Reversability and Stochastic Networks
TOUTENBURG · Prior Information in Linear Models